KW-104-326

FORESTS AT THE WILDLAND–URBAN INTERFACE

Conservation *and* Management

FORESTS AT THE WILDLAND–URBAN INTERFACE

Conservation *and* Management

Edited by
Susan W. Vince, Mary L. Duryea,
Edward A. Macie, and L. Annie Hermansen

CRC PRESS

Boca Raton London New York Washington, D.C.

UNIVERSITY OF PLYMOUTH

9008202404

Library of Congress Cataloging-in-Publication Data

Forests at the wildland-urban interface : conservation and management / edited by Susan W. Vince, Mary L. Duryea, Edward A. Macie, L. Annie Hermansen
p. cm.
Includes bibliographical references and index (p.).
ISBN 1-56670-602-5 (alk. paper)
1. Forest conservation. 2. Forest management. 3. Forest policy. I. Vince, Susan, W. II. Duryea, Mary L. III. Macie, Edward A. IV. Hermansen, L. Annie
SD411.F56 2004
333.75'16'091732-dc21

2004048634

This book contains information obtained from authentic and highly regarded sources. Reprinted material is quoted with permission, and sources are indicated. A wide variety of references are listed. Reasonable efforts have been made to publish reliable data and information, but the author and the publisher cannot assume responsibility for the validity of all materials or for the consequences of their use.

Neither this book nor any part may be reproduced or transmitted in any form or by any means, electronic or mechanical, including photocopying, microfilming, and recording, or by any information storage or retrieval system, without prior permission in writing from the publisher.

All rights reserved. Authorization to photocopy items for internal or personal use, or the personal or internal use of specific clients, may be granted by CRC Press LLC, provided that $1.50 per page photocopied is paid directly to Copyright Clearance Center, 222 Rosewood Drive, Danvers, MA 01923 USA. The fee code for users of the Transactional Reporting Service is ISBN 1-56670-602-5/04/$0.00+$1.50. The fee is subject to change without notice. For organizations that have been granted a photocopy license by the CCC, a separate system of payment has been arranged.

The consent of CRC Press LLC does not extend to copying for general distribution, for promotion, for creating new works, or for resale. Specific permission must be obtained in writing from CRC Press LLC for such copying.

Direct all inquiries to CRC Press LLC, 2000 N.W. Corporate Blvd., Boca Raton, Florida 33431.

Trademark Notice: Product or corporate names may be trademarks or registered trademarks, and are used only for identification and explanation, without intent to infringe.

Visit the CRC Press Web site at www.crcpress.com

© 2005 by CRC Press LLC

No claim to original U.S. Government works
International Standard Book Number 1-56670-602-5
Library of Congress Card Number 2004048634
Printed in the United States of America 1 2 3 4 5 6 7 8 9 0
Printed on acid-free paper

Series statement: Integrative studies in water management and land development

Ecological issues and environmental problems have become exceedingly complex. Today, it is hubris to suppose that any single discipline can provide all the solutions for protecting and restoring ecological integrity. We have entered an age where professional humility is the only operational means for approaching environmental understanding and prediction. As a result, socially acceptable and sustainable solutions must be both imaginative and integrative in scope; in other words, garnered through combining insights gleaned from various specialized disciplines, expressed and examined together.

The purpose of the CRC Press series Integrative Studies in Water Management and Land Development is to produce a set of books that transcends the disciplines of science and engineering alone. Instead, these efforts will be truly integrative in their incorporation of additional elements from landscape architecture, land-use planning, economics, education, environment management, history, and art. The emphasis of the series will be on the breadth of study approach coupled with depth of intellectual vigor required for the investigations undertaken.

Robert L. France
Series Editor
Integrative Studies in Water Management and Land Development
Associate Professor of Landscape Ecology
Science Director of the Center for Technology and Environment
Harvard University
Principal, W.D.N.R.G. Limnetics
Founder, Green Frigate Books

Foreword by the series editor: Blurring the border between nature and culture

As I began reading through the proofs of this book, I was sitting at a table in my sister's house situated a few doors down from the home where we grew up in Winnipeg, Manitoba. Even though most of what we as children had referred to as "the back woods" had since been replaced by suburban sprawl, small parcels remained of the forest, parcels of sufficient size to still house play-forts of new generations of children as well as several families of deer that every evening would emerge to graze upon the green lawns and gardens of the abutting human residents.

Humans love to delineate their world by sharp borders and boundaries, that which ecologists refer to as "ecotones." Nowhere is this more evident than in our collective mindset of that which we regard as being either "nature" or "culture." This book, edited by Susan W. Vince, Mary L. Duryea, Edward A. Macie, and L. Annie Hermansen, ably demonstrates that such a simple dichotomy is false. Here, what the authors refer to as the "wildland–urban interface," or what others might call "urban wilds," are true cultural landscapes worthy of study and management in their own right just as more distant, more imagined (often wrongly so) "natural" areas are. For, as the noted landscape architect Herbert Dreiseitl poetically writes in the Foreword to my recent book, *Deep Immersion: The Experience of Water*, "Why do we have a civilization where our dreams are so far away? Why can't we work in a way that our living space has a quality that this is our new nature — our urban nature?" The present book offers insight into how to bring about the construction of such a new nature.

As the third book in the Integrative Studies in Water Management and Land Development series by CRC Press, the present volume provides a roadmap for managing the growth of "sprawl-scapes" through proactive planning. Contributing authors, drawn from academia, industry, government, and environmental associations, illustrate how to negotiate the delicate balance necessitated through managing forest landscapes adjacent to and even deep within urban centers. It will come as no surprise, as demonstrated in this collection of chapters, that achieving such a complex mandate calls upon all the disciplines of environmental policy, land-use planning, landscape ecology, environmental education, recreation planning, hydrology and forest management. The present book therefore attests to the overall goal of this series as shown in the previous volumes, *Handbook of Water Sensitive Planning and Design* (2002, edited by myself) and *Boreal Shield Watersheds: Lake Trout Ecosystems in a Changing Environment* (2003, edited by Gunn, Steedman, and Ryder), in the need for multidisciplinary (or even better *transdisciplinary*) approaches to complex environmental problem solving.

Returning to my neighborhood in Cambridge, Massachusetts, after reading the bulk of this book on the flight home, I immediately went for a jog through the nearby Alewife Reservation, the largest urban forest in the greater Boston area. There I saw a group of schoolchildren on a nature outing, deeply involved in observing birds and identifying plants. And I was reminded that such wildland–urban interfaces are important not just in their own right, but also in terms of their influence on the sociology of fostering nature awareness and appreciation (also see my edited book, *Facilitating Watershed Management: Fostering Awareness and Stewardship*). By allowing urban dwellers opportunities to experience and steward nature in their everyday lives, such urban wilds help to motivate people to protect more pristine natural areas elsewhere. As Michael Houck writes in the Preface to *Wild in the City: A Guide to Portland's Natural Areas*, playfully turning Thoreau's famous maxim on its head, "in livable cities is the preservation of the wild."

Robert L. France
Harvard University

Preface

The wildland–urban interface in its most general sense is the area where urban lands meet and interact with rural lands. Yet, definitions also vary according to one's perspective. For example, geographers view the interface as isolated patches of development among the forest, as an intermix of forests, rural lands, and development, or as a boundary between urban and rural lands. Resource managers may consider the interface as the area where humans increasingly influence the provision and management of natural resource goods and services. Fire managers view the interface as an area where homes and communities are adjacent to or surrounded by flammable wildland fuels. The wildland–urban interface may be considered a sociopolitical phenomenon as well — the interaction between people with different perceptions about natural resource values.

One thing that everyone agrees upon is that forests at the interface are changing. These forests have more owners and their parcel size is smaller. The impacts of urbanization, including fragmentation of forest tracts, increased exposure to invasive species, greater inputs of pollutants, and the spread of impervious surfaces, are changing the structure and function of these ecosystems. Management for ecological goods and services, such as wildlife, forest products, and clean water, is affected by people adjacent to and occupying the forests. As interface forestlands are converted to nonforest uses, many of their benefits to society are diminished or eliminated.

These changes to the landscape affect just about everyone, including new landowners, people seeking recreational opportunities, traditional rural residents, business leaders, environmentalists, urban dwellers, and developers. Both the subdivisions and the wildlands have new neighbors, and these neighbors' values, lifestyles, and land ethics may differ considerably. Wild and rural lands are part of the American image and support our well-being, and they invoke strong feelings when they are threatened. Likewise, home ownership and private property rights are an integral part of our heritage. Consequently, conflicts are inevitable. When we also consider the broad mix of decision makers in the wildland–urban interface, we see that developing ecologically sound goals and plans for a community's growth can be daunting. However, planning growth while retaining the viability and benefits of wildlands is in everyone's interest.

Natural resource professionals can be critical players in the conservation, management, and even development of forestlands in the interface. Communication skills and involvement in policy and decision making are not ready attributes, but natural resource professionals need to cultivate these characteristics to work effectively in the interface. Attending city council meetings, testifying in front of county commissions, actively engaging the press to report the facts about issues, and educating the public are demanding, and perhaps new, activities for this group.

Against this backdrop, the Urban Forestry Institute at the University of Florida brought together specialists from diverse areas to educate natural resource professionals about the issues, the challenges, and potential solutions to sustaining and managing forests at the interface. After the short course, we began planning with our sponsors, the USDA Forest Service and the Southern Group of State Foresters, to produce a book that would expand upon the course contents and serve as a guide, not only to natural resource professionals but also to planners, policy makers, landscape architects, and others concerned with the conservation of our natural resources. Authors from fields ranging from growth management to fire ecology were invited to contribute. They were charged with the task of presenting innovative approaches and tools that can be used by landowners and communities to conserve and manage forestlands in the face of urbanization. In the fall of 2001, we held a symposium in Gainesville, Florida, "The Wildland–Urban Interface: Sustaining Forests in a Changing Landscape," where the authors offered their ideas and discussed them among scientists and practitioners.

This book has evolved from these educational programs. It explores the issues involved in sustaining and managing forests at the interface, as well as presenting some of the creative solutions being attempted by landowners, communities, and various governmental and nongovernmental organizations. We hope this book will provide models and inspiration for its readers, encouraging them to become more active in efforts to lessen the impacts of urbanization and to sustain the ecological and social benefits of our forests.

Acknowledgments

We thank the USDA Forest Service Southern Research Station and Southern Region, and the Southern Group of State Foresters for their guidance and support. Members of the Southern Wildland–Urban Interface Council helped with planning the symposium and book, and we appreciate their assistance. We are grateful to the School of Forest Resources and Conservation, Institute of Food and Agricultural Sciences, University of Florida, for support.

Additionally, we thank Brian Kenet, David Fausel, and Julie Spadaro at CRC Press for encouraging and watching over this project. Copy editing by Gillian Hillis greatly improved the book's readability, and Ludovica Weaver's assistance with illustrations enhanced the presentation of the text. We also thank Larry Korhnak for providing numerous photographs, which have increased the book's visual appeal.

Each chapter was reviewed by experts in the appropriate topic, and we thank the reviewers for their time and efforts: Janet Ady, Melvin Baughman, John Baust, Dave Bengston, Bill Bentley, Gordon Bradley, Doug Carter, Carol Couch, Brian Day, John DeGrove, Alison Fox, Susan Frankel, Jim Harrell, Wink Hastings, Bruce Hull, Lou Iverson, Susan Jacobson, Jeff Kline, Mark Lapping, Eric Livingston, Paul Mistretta, Bob Neville, David Newman, Ron Nickerson, Doug Porter, Cotton Randall, David Salvesen, Neil Sampson, Jim Smalley, Peter Smallidge, Frank Svoboda, Sarah Thorne, Jack Walstad, Tim White, and Kathy Wolf.

We are especially grateful to the contributing authors of this book, who have been creative and dedicated in their search for solutions to the many problems associated with sustaining forests in the wildland–urban interface.

Susan W. Vince
Mary L. Duryea
L. Annie Hermansen
Gainesville, Florida

Edward A. Macie
Atlanta, Georgia

Contributors

Janaki R.R. Alavalapati
School of Forest Resources and Conservation
University of Florida
Gainesville, Florida

Edward L. Barnard
Florida Department of Agriculture and Consumer Services
Division of Forestry
Gainesville, Florida

Frank C. Beall
University of California Forest Products Laboratory
Richmond, California

Joyce K. Berry
College of Natural Resources
Colorado State University
Fort Collins, Colorado

Deborah J. Chavez
USDA Forest Service
Pacific Southwest Research Station
Riverside, California

Mary L. Duryea
Institute of Food and Agricultural Sciences
University of Florida
Gainesville, Florida

John F. Dwyer
USDA Forest Service
North Central Research Station
Evanston, Illinois

John C. Gordon
Yale School of Forestry and Environmental Studies
New Haven, Connecticut

Sharon G. Haines
International Paper Company
Savannah, Georgia

L. Annie Hermansen
Southern Center for Wildland-Urban Interface Research & Information
USDA Forest Service Southern Research Station
Gainesville, Florida

David A. Hoge
Cooperative Forestry Unit
USDA Forest Service Southern Region
Atlanta, Georgia

William G. Hubbard
Cooperative Extension Service Southern Region
University of Georgia
Athens, Georgia

Larry V. Korhnak
School of Forest Resources and Conservation
University of Florida
Gainesville, Florida

Elizabeth Kramer
Institute of Ecology
University of Georgia
Athens, Georgia

James E. Kundell
Carl Vinson Institute of Government
University of Georgia
Athens, Georgia

Alan J. Long
School of Forest Resources and Conservation
University of Florida
Gainesville, Florida

Edward A. Macie
State and Private Forestry
USDA Forest Service Southern Region
Atlanta, Georgia

David W. Marcouiller
Department of Urban and Regional Planning
University of Wisconsin-Madison
Madison, Wisconsin

Lindell L. Marsh
Siemon, Larsen & Marsh
Irvine, California

Martha C. Monroe
School of Forest Resources and Conservation
University of Florida
Gainesville, Florida

Kenneth R. Munson
International Paper Company
Savannah, Georgia

Margaret Myszewski
Carl Vinson Institute of Government
University of Georgia
Athens, Georgia

Douglas R. Porter
The Growth Management Institute
Chevy Chase, Maryland

Sarah Reichard
Center for Urban Horticulture
University of Washington
Seattle, Washington

R. Neil Sampson
The Sampson Group, Inc.
Alexandria, Virginia

Taylor V. Stein
School of Forest Resources and Conservation
University of Florida
Gainesville, Florida

Susan W. Vince
School of Forest Resources and Conservation
University of Florida
Gainesville, Florida

Dale D. Wade
USDA Forest Service (retired)
Hayesville, North Carolina

Wayne C. Zipperer
Southern Center for Wildland–Urban Interface Research & Information
USDA Forest Service Southern Research Station
Gainesville, Florida

Contents

Part I

The Wildland–Urban Interface

1 Introduction: The City Is Moving to Our Frontier's Doorstep

Mary L. Duryea
Institute of Food and Agricultural Sciences, University of Florida

Susan W. Vince
School of Forest Resources and Conservation, University of Florida

CONTENTS

Today, as we fly across the U.S., we can plainly see what is happening — and what is likely to happen — to our country's landscape. More than 80 percent of the population lives in cities, and these cities are rapidly expanding. Over the outskirts of Atlanta, Georgia, for example, we are struck by the extensive clearing for new development, which creates a mosaic of forest, bare soil, pavement, and houses. When we land at Atlanta's airport, we notice that it is no longer at the edge of the city, but in the middle of a sprawling metropolis.

The causes of urban sprawl are often debated, but commonly cited factors include population growth, land-use policies, transportation infrastructure, and the socioeconomics of the community (e.g., Zhang 2001). What is not disputed is the accelerated rate at which sprawl is occurring. A recent inventory of the nation's land base indicated that 2.2 million acres of rural and open-space land were lost to development each year during the period 1997 to 2001, up from 1.1 million acres annually between 1982 and 1991 (U.S. Department of Agriculture, Natural Resources Conservation Service 2003). Much of this newly developed land had been forested: 46 percent of the land converted by sprawl during 1997 to 2001 and 40 percent of the land developed in the 1980s. Cities and their metropolitan areas are increasing in land size, often at a rate far surpassing population growth. Most of the cities and towns surrounding Seattle, Washington, for example, greatly expanded in land area between 1990 and 2000. One of these, Snoqualmie — a town in the forested Cascade Mountains — expanded from 1.6 to 5.1 square mile (more than a 200 percent increase), while population grew by 5.5 percent (Puget Sound Regional Council 2001). In the Northeast, cities that were once distinct, separated by farms and forestland, have now merged. The city is moving to our frontier's doorstep.

Urban expansion into the countryside has not only displaced farm and forest, it has also mixed with these rural lands. To describe this mosaic, we use the term "wildland–urban interface." Although this expression evokes an image of rural lands in juxtaposition with houses or factories, the interface may take on a variety of aspects (Figure 1.1). In some places, a forest abuts a new subdivision, but in others the forest is interspersed with 5-acre parcels and houses in low-density developments. Sometimes, fragments of the forest remain encircled by development, and large tracts of wild or rural lands are only found many miles from the city.

The wildland–urban interface, in very general terms, is where urban lands meet and interact with wild or

1-56670-602-5/05/$0.00+$1.50
© 2005 by CRC Press LLC

FIGURE 1.1 Three aspects of the wildland–urban interface: (a) a subdivision intersects a large tract of forestland; (b) houses are scattered within the forest; and (c) a remnant patch of forest is surrounded by development. (Photos by Larry Korhnak.)

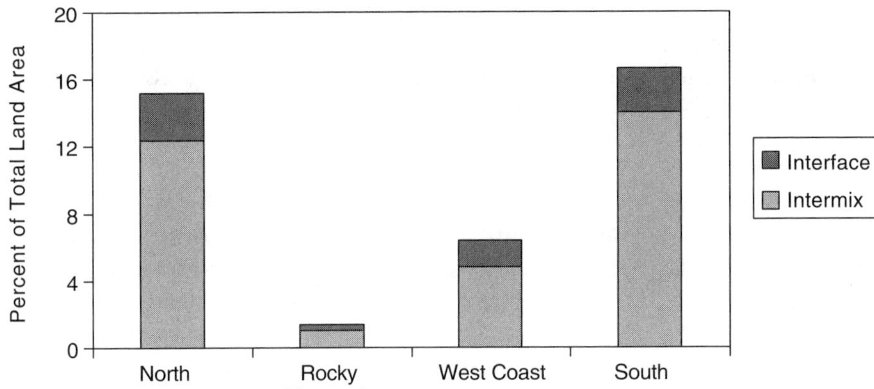

FIGURE 1.2 Percentage of land area that is classified as wildland–urban interface in each of four regions of the continental U.S. Intermix areas are those where housing intermingles with wildland vegetation, and interface areas occur where housing is near a large contiguous block of wildland vegetation. (Adapted from Dwyer et al. 2003.)

rural lands. During the past few decades, the term has taken on more precise meanings depending on our objectives or issues of concern. For those challenged with fire prevention responsibilities, the wildland–urban interface occurs where human-built structures are adjacent to or surrounded by areas prone to wildfire. If the concern is conflict resolution, the interface may be the arena in which people's perceptions, values, and opinions about natural resources interact and compete (Vaux 1982). If forest management is the focus, the interface may be "areas where increased human influence and land use conversion are changing natural resource goods, services, and management" (Macie and Hermansen 2002).

The varying definitions of the wildland–urban interface and the rapid change in our country's landscape make quantification of the total land area in the interface difficult. A recent project supported by the USDA Forest Service and the University of Wisconsin-Madison used 2000 U.S. census data, land-cover maps, and a definition of the wildland–urban interface originating from fire risk assessment to identify and map the interface in the contiguous 48 states (Dwyer et al. 2003). The project defined two types of wildland–urban interface: (1) intermix, areas where housing (more than one per 40 acres) intermingles with wildland (nonagricultural) vegetation and (2) interface, areas with housing and a low density of vegetation that are within fire's reach (1.5 mil) of a large contiguous block of wildland vegetation. Overall, 9.3 percent of the land area of the continental U.S., more than 175 million acres, was classified as wildland–urban interface (intermix and interface combined), and intermix areas predominated (Figure 1.2). Regional differences were considerable (Figure 1.2 and Figure 1.3). The Rocky Mountain states had the least extent of wildland–urban interface and the northeastern and southeastern states the most. In three states in New England — Massachusetts, Rhode Island, and Connsecticut — the wildland–urban interface accounted for more than 60 percent of the land area. On the West Coast, the wildland–urban interface generally comprised a small percentage of the land base,

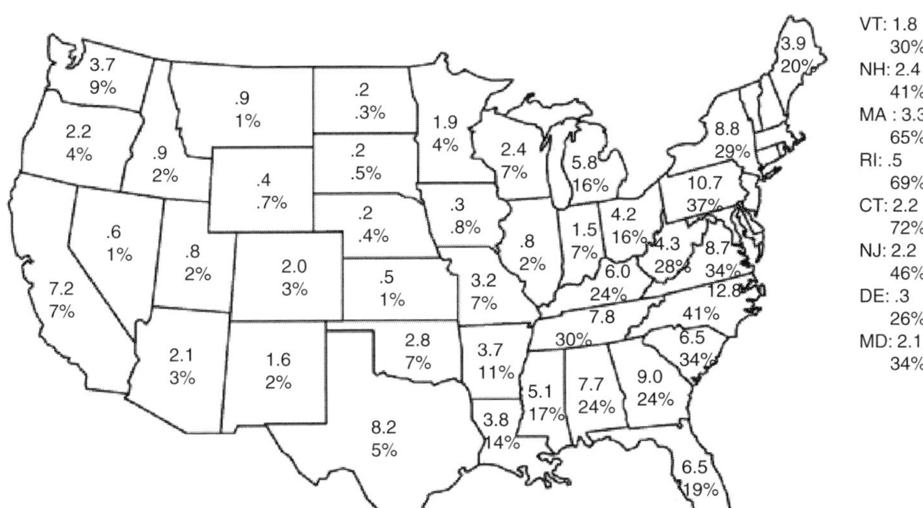

FIGURE 1.3 Area of wildland–urban interface (million acres) and percentage of total land area, by state, in 2000. (Adapted from Dwyer et al. 2003.)

but in some states the acreage involved was high — more than 7 million acres in California. The total forestland in the conterminous U.S. is about 618 million acres (Smith et al. 2001). If we assume that forestlands are geographically distributed in the same manner as total wildlands, then 9.3 percent, or 57 million acres, of forestlands were located in the wildland–urban interface at the beginning of this century.

The distribution of forestlands along an urban–rural gradient, from counties with densely populated metropolitan areas to rural counties with no sizable towns, provides another illustration of the proximity of our nation's forests to urban influences (Figure 1.4). In the East, only 21 percent of forestlands were located in rural counties in 1997 (counties with towns of less than 2500 persons), and 34 percent were in counties with large towns to major metropolitan areas (populations greater than 20,000) (Smith et al. 2001). More forestlands in the West (including Alaska and Hawaii) were found in rural counties (56 percent of the total); yet, more than 80 million acres of forestland existed in counties with large urban centers. It is evident from both analyses that many of our forests are no longer remote from cities; they are surrounded and penetrated by people and development, and are indirectly affected by urbanization as well.

1.1 THE ISSUES AND CHALLENGES

As a forest becomes an interface forest, the most obvious change is its size. Fragmentation or the breaking up of large acreages into smaller pieces may be having the greatest effect on the viability of our forests, next to clearing for development. Sampson and DeCoster (2000) predicted that by 2010, 150 million acres of privately owned forests in the U.S. will be in parcels smaller than 100 acres, and the average size will be 17 acres. More people will own these smaller parcels — an estimated 150,000 new forest landowners each year. Roads and land conversion will physically fragment many forests, resulting in an increasing number of small, isolated forest patches.

Although interface forestlands may continue to supply many goods and services to society, such as recreation, wildlife, and timber products, they are also at risk due to the pressures and impacts of closer human contact. The problems confronting the conservation and management of interface forests are many, including:

- Practicing traditional forest management on lands that are now close to urban people
- Managing for fires on forestlands interspersed with people
- Offering recreational opportunities for a growing and diverse user population
- Conserving biodiversity and managing wildlife populations on disjunct and small parcels of land
- Protecting water supply and water quality for people and natural ecosystems
- Maintaining forest health in the face of increased disturbance and increased exposure to pests such as exotic invasive plants
- Managing for ecosystem services at a landscape scale, across ownership and jurisdictional boundaries
- Identifying and prioritizing critical forestland areas for conservation.

The greatest challenge of all is achieving the diverse objectives of the local community, from more housing and

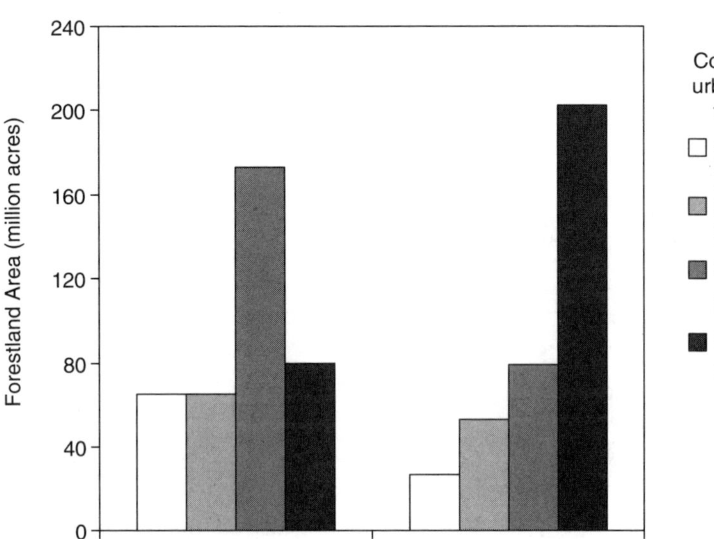

FIGURE 1.4 Forestland area along a gradient from urban to rural counties in the eastern and western U.S. (Adapted from Smith et al. 2001.)

economic development to conservation of wildlands and the ecological and social benefits they provide.

1.2 FORESTS AT THE WILDLAND–URBAN INTERFACE: CONSERVATION AND MANAGEMENT

When we are faced with the numerous challenges for sustaining forests in a dramatically changing landscape, we can be overwhelmed and immobilized. The intent of this book is to offer tools and approaches that will spur planners, managers, and citizens to play an active role in crafting solutions. Today, many sources provide information about new planning approaches to limit urban sprawl, and others suggest innovative ways to conserve natural areas and biodiversity. To successfully conserve and manage forests at the interface, we must pursue both paths. This book seeks to integrate the two, offering strategies to manage growth and at the same time maintain viable forests within a changing landscape. Four sections present the issues and some solutions: (1) The Wildland–Urban Interface; (2) Planning and Managing Growth; (3) Conserving and Managing Forests for Ecological Services and Benefits; and (4) Conserving and Managing Forests under Different Ownerships.

1.2.1 The Wildland–Urban Interface

The first section of the book sets the stage for further discussions by describing the changes at the wildland–urban interface and, in broad terms, explaining why these changes are occurring and what actions are needed. In Chapter 2, Gordon, Sampson, and Berry address the issues of the working landscape — that part of the land in the wildland–urban interface that is currently producing goods and services but is also threatened by urbanization in the next decade. They compare the working forests in this landscape to the Japanese *satoyama* woodlands — areas of forest adjacent to rice farms that produced firewood and other nontimber products. Because the value of the *satoyama* was unrecognized and considered obsolete, these woodlands were destroyed during the expansion of cities. The lesson for the interface is that "only forests that are recognized as valuable and integral parts of the urban infrastructure survive for long within the reach of urban populations." The key is better education of the public. Gordon et al. advocate "local problem analysis, with broad public involvement," noting that solutions must be tailored to the specific region. Some promising actions are easements, collaboration between farm and forest managers, and better communication between resource managers and urban citizens.

The USDA Forest Service initiated an assessment of the southern wildland–urban interface after Florida's damaging fires of 1998 (Macie and Hermansen 2002). In Chapter 3, Hermansen and Macie summarize the assessment. Factors driving landscape change at the interface include: (1) population growth, distribution, age, and ethnicity; (2) changes in the southern economy and taxation policies; and (3) land-use planning and policies. Due to these factors, forests at the interface are reduced in area and fragmented, and their health is jeopardized. Challenges to forest conservation and management include water and watershed management, production of traditional forest products, providing and managing for recreation, fire management including public education, and ecosystem management to conserve wildlife populations. Social, economic, and political issues also arise in the interface. For example, the value and perceived benefits of forests change, and land-use decisions bring more attention and are often controversial. One of the main findings of the assessment, applicable nationwide, is that natural resource managers need to learn new skills (such as managing small parcels), methods of communication to new audiences, and how to develop partnerships to better manage natural resources in the interface. The southern assessment raised many questions and stressed the need to package and disseminate interface information. To answer these concerns, the USDA Forest Service Southern Research Station established the Southern Center for Wildland–Urban Interface Research and Information in Gainesville, Florida, in 2002.

1.2.2 Planning and Managing Growth

Planning and managing growth is at the heart of all the solutions to sustain forests in the interface. This section introduces economic, policy, land-use planning, and landscape assessment tools that can be used to curb urban sprawl and the accompanying forestland loss and degradation. Scenarios for development that retain environmental values, such as sustainable development and smart growth, are described. When incorporated into community and regional planning, these and other growth management approaches show promise for conserving our forests.

Marcouiller's goal in Chapter 4 is to "provide an economic explanation for understanding this rapid land-use change at the wildland–urban interface." Historically as well as today, land-use patterns and changes can be explained by the productive uses of the land and transportation costs of the products. Marcouiller offers economic reasons why land uses such as forestry and housing will be placed at specific distances from market (or city) centers. Land uses will change as people realize that more profit can be gained from a different type of land use — for example, a housing development compared to a silvicultural operation. Economic growth in cities brings new employment, new residents, and new demands for commercial and residential growth; hence, the "highest" and "best" use for land is for residential and commercial development, and urban sprawl occurs. Nonmarket goods and services of land such as recreation, wildlife, and amenities are important, but their values are difficult and elusive to estimate. From an economic perspective, Marcouiller discusses the potential for zoning, urban growth boundaries, publicly provided incentives, and other tools to address land management, urban sprawl, and land-use decisions. He concludes by saying that the "essence of growth management is to balance the need for economic growth with other less market-oriented objectives of society." Land-use planning is the key, and several economic tools are available to help in land-use decision making.

In Chapter 5, Alavalapati explores some of the policies influencing urban sprawl, land use, and forest management at the interface. He emphasizes that sprawl not only results in the loss and restructuring of forestland but also restricts forest management practices such as fertilization and prescribed burning. To deal with these concerns, many governments at local and state levels are initiating policies to limit sprawl and conserve rural and wildlands at the interface. The policies include regulating land use through zoning, creating urban boundaries, acquiring rural lands, and encouraging growth and development within urban centers. However, Alavalapati stresses that the implications and impacts of these policies are not always predictable, and the policies may not be cost-effective or have consistent socioeconomic benefits. Land acquisition programs, for example, face criticism due to the costs associated with the initial purchase as well as long-term management and their exclusion from the tax base of the community. Conservation easements, where only partial property rights are purchased, offer a compromise between land acquisition and government regulation and are a promising solution. Noting that public policy incentives exist for urban land uses but not for forestry, Alavalapati proposes that forestry be fully compensated for environmental benefits. In conclusion, Alavalapati emphasizes that collaborative and consensus-building actions among policy makers, resource managers, and local citizens are critical to the establishment of effective and equitable land-use and growth management policies for the wildland–urban interface.

Myszewski and Kundell present land-use planning and zoning tools that can help to protect natural resources on lands facing development, and throughout Chapter 6, they explain how and at which steps natural resource professionals can be involved in the planning process. Local planning commissions are charged with developing a land-use plan, zoning ordinances, subdivision controls, and a capital improvements program. A land-use plan is a guide that matches a community's goals with its plans for growth and development. Zoning ordinances can help implement the land-use plan by dividing the land into uses and specifying the standards for development (Figure 1.5). Myszewski and Kundell note that zoning can sometimes have unintended consequences, including the promotion of urban sprawl. However, if done properly, zoning can help a community protect its natural resources, and the authors give several examples of new types of zoning to achieve that goal. Besides zoning, communities can control growth and conserve natural resources through an increasing variety of growth management techniques such as land acquisition, conservation easements, and transfer of development rights. Myszewski and Kundell stress the need for natural resource professionals to educate the public about the issues and to actively participate in the land-use planning process.

After noting that many of the traditional federal, state, and local approaches to balancing growth with environmental conservation have failed, Porter and Marsh (Chapter 7) introduce the new paradigms of sustainable development and smart growth. The concept of sustainable development derives from McHarg's (1992) belief that development must be integrated with nature. Smart growth focuses on improving the quality of life in a community by conserving natural and economic resources. Principles include encouraging compact, mixed-use development, conserving open space, promoting infill and redevelopment in existing urbanized areas, and involving the public. These approaches are much more effective if they are conceived and implemented regionally, and Porter and Marsh cite many successful examples. Constructive solutions for rapidly growing communities such as those in the Santa Ana River Valley of California have risen from collaborative efforts and shared governance among people with diverse interests.

In order to establish effective policies and land-use plans, we must first know the ecosystems and natural

FIGURE 1.5 Zoning ordinances control the type and density of land use on a property. They can be used to manage development successfully within the interface, but if poorly planned, they can promote urban sprawl. (Photo by Larry Korhnak.)

resources in the community's landscape. Kramer introduces landscape assessment in Chapter 8 as a methodology for identifying trends and status of resources and answering questions regarding "what is there" and "what is its condition." The major tools, Geographical Information Systems (GIS) and remotely sensed imagery, help us to inventory resources and then plan and establish policies at local to federal levels. For example, as part of the Georgia Community Greenspace Program, Jackson County identified its goals for a greenspace plan, including economic development, protection of water quality and agricultural lands, and maintenance of rural character. Then, a landscape assessment identified lands with qualities supporting these goals, helping the community decide which lands should be protected and which could be developed. Landscape assessments are an important tool not only for providing information about resources but also for analyzing alternatives and communicating with stakeholders.

1.2.3 Human Dimensions and the Interface

Every chapter in this book emphasizes the need to involve people. These two chapters give us some approaches and tools for communicating with and engaging the interface's stakeholders. Whether the goal is growth management planning, providing recreation for the community, or managing invasive plants, public education and involvement are critical to program success.

Monroe begins Chapter 9 by stressing that communication, education, and public involvement are essential in the wildland–urban interface because so many people have a stake. Outreach activities are a way for agencies and organizations to keep people informed and involved. Many approaches emphasize delivering information to people, but this does not necessarily mean that people will take action. Monroe introduces some innovative marketing-oriented techniques to reach stakeholders and effect changes. Successful outreach programs encompass (1) information to increase awareness of the issues; (2) enhanced two-way communication between the organization and the public; (3) education to increase knowledge and human capacity; and (4) capacity-building activities such as strategic planning and partnerships. These categories form an important framework for outreach programs at the interface, and Monroe offers numerous examples and procedural tips to help natural resource agencies and organizations develop effective programs.

The interface provides many accessible natural areas, but the challenges to planners and land managers are the large number of people who visit these sites and the environmental degradation that can ensue. In Chapter 10, Stein emphasizes a positive, proactive approach toward recreation planning in the interface, where planners aim to provide benefits for a diversity of people and the environment (Figure 1.6). Stein describes several conceptual frameworks developed by land management agencies in response to recreational demand and shows how they can be applied in interface areas. For example, the Visitor Experience and Resource Protection framework helps managers create management zones within a recreation area that are based on the range of potential visitor activities and experiences and the resource conditions that can be accommodated. Managers need to identify and then

FIGURE 1.6 Three generations enjoy a walk in the woods. People are increasingly looking to nature as a recreational opportunity, and interface wildlands face particular pressure. Planners must attempt to provide recreational opportunities while maintaining the health of natural areas. (Photo by Larry Korhnak.)

manage for acceptable changes to the environment that are caused by recreation. Increasingly, planners and land-use managers use the Limits of Acceptable Change framework to define the change that is acceptable and develop a management strategy for this change. At the interface, planners, managers, and stakeholders need to collaboratively develop acceptable standards and incorporate them into recreation plans.

1.2.4 Conserving and Managing Forests for Ecological Services and Benefits

Forestlands provide a variety of goods and services of benefit to people, but their provision is affected directly and indirectly by urbanization. Fragmentation results in smaller, more isolated forest patches that cannot support the same suite of wildlife species; paved, impervious surfaces speed the surface runoff of water and diminish the ability of forests to supply clean water; altered fire regimes lead to less frequent but more intense fires; and increased introductions of exotic invasive species threaten the biodiversity of forestlands. The chapters in this section suggest ways that resource managers, planners, and citizens can plan and manage interface forests and landscapes with an eye toward minimizing urban effects and sustaining the ecological and social benefits of forestlands.

If our goal is to minimize the impacts of urbanization on some ecosystems in the landscape, we need to know how the ecosystems function in that landscape and which features are important to retain. In Chapter 11, Zipperer sets out to guide managers on how the ecosystem and landscape can be evaluated in land-use decisions. Planning for wildland conservation at the interface begins by evaluating site content and context in the overall landscape. The physical, ecological, and cultural aspects of the site must be assessed, and the site must also be considered in the context of other ecosystems in the landscape because they are connected — management activities on one site will influence neighboring sites or ecosystems. Zipperer explains and promotes the use of ecosystem management, a broad-scale approach to land-use management decisions that enables managers to evaluate the cumulative effects of development on natural systems and their delivery of goods and services.

Chapter 12 by Korhnak and Vince provides many solutions and some hope for preventing or mitigating urbanization's impacts on the hydrologic cycle. Because of deforestation and the increase in impervious surfaces, urban development reduces critical parts of the water cycle, including infiltration, storage, and evapotranspiration. Surface runoff of water increases, rapidly carrying greater volumes of polluted water to aquatic ecosystems. Korhnak and Vince suggest that better planning and design before development can prevent these hydrological changes. Planning should be at the watershed level, should be long term, and should seek to integrate ecological, social, and economic considerations. Key elements in watershed planning include early involvement of stakeholders, increased public education and awareness, watershed assessment, and development of a plan that is implemented, monitored, and evaluated. Korhnak and Vince identify three important strategies for managing water at the interface: (1) protection of forests and wetlands that provide critical hydrological functions; (2) reduction of pollutant supplies and impervious surfaces; and (3) use of site design and stormwater management practices to maintain the predevelopment water cycle. These strategies can be implemented only if the public is educated about the importance of forested watersheds, the impacts of urbanization, and the actions that can be taken to lessen these impacts.

Fire management is an excellent example of the complexity and critical nature of natural resource problems at the interface. People are increasingly moving into or near forests that are prone to wildfires. In Chapter 13, Long, Wade, and Beall provide a history of the relationship between Americans and wildfire: attitudes have changed from acceptance of and even dependence on fire during the early history of our country to rejection of fire and promotion of a campaign to suppress fire, and more recently to growing recognition of the benefits of this natural part of the landscape. Yet, decades of fire suppression have led to the buildup of fuel loads and with them the potential for more intense fires and threats to structures. The challenges facing citizens who live at the interface and the resource managers who manage the lands include the modified, fire-prone landscape, as well as the simultaneous need to protect structures, people, and natural resources, and the expectations of landowners for fire protection services. Fire suppression is especially challenging at the interface because it involves both wildland and structural fire fighting. Many homeowners who have moved to the interface expect the same level of fire protection as they had in urban areas, and they often do not recognize the fire-fighting difficulties or the contributions that they must make for fire prevention. Long et al. describe a variety of programs and actions to meet these challenges, including risk/hazard assessment, development of subdivision and construction standards, cross training of fire-fighting organizations, modification of wildland fuels, and, especially, community and landowner education. On small lots as well as large landscapes, carefully prescribed use of fire can be an important tool for lessening fuel loads and restoring a natural process (Figure 1.7). The authors repeatedly emphasize that interface landowners must adopt substantial responsibility; they must prepare for wildland fires by making themselves aware of their fire risk and creating and maintaining defensible home and landscape conditions.

FIGURE 1.7 A prescribed burn in Palm Coast, Florida. Prescribed burning can be an important tool for reducing wildland fuel loads, even within interface subdivisions. (Photo by Cotton Randall.)

Everyone agrees that we want healthy forests at the interface, but what is forest health? In Chapter 14, Barnard notes that forest health encompasses both the forest and the individual trees and is a composite condition of many components and processes in forest ecosystems. Forests at the interface are influenced by people and are constantly changing due to new disturbances, new introductions of species, and new management practices. They are also confronted with both indigenous (native) and nonindigenous (introduced or exotic) insects and pathogens; these two types of pests pose different threats and often need different management strategies. Barnard suggests that the only effective means of controlling nonindigenous pests is to prevent their introduction or to eradicate them at first detection. In contrast, opportunities exist for managing the impacts of indigenous insects and pathogens. Often, burgeoning populations of these pests are a symptom rather than a cause of poor forest health. Site disturbances such as air pollution and soil and water changes are examples of environmental changes that may induce tree stress and provide opportunities for insect and pathogen populations to thrive. Contrary to common beliefs, the abundance of fertilizers, water, and pesticides applied to trees in the interface may also lower resistance and predispose trees to pest attacks. Barnard offers some pest-prevention guidelines that encourage enhancement of tree and forest health before and during the development of interface lands. Education of all stakeholders, from planners and policy makers to landowners, is emphasized as the best solution for promoting and managing healthy trees and forests.

Invasive plants in forest ecosystems contribute to global environmental degradation and loss of biodiversity. In Chapter 15, Reichard points out that most of these plants are introduced to urban areas, but then move through the wildland–urban interface and into native ecosystems. The interface provides a mosaic of varied and transitional habitats conducive to invasion. Dispersal of invasive plants often occurs along corridors such as roads and waterways, but wind and birds are also significant vectors. Like Barnard, Reichard stresses that it is imperative to control the introduction of invasive species, so more resources must be committed to their early detection and eradication. Reichard also proposes that the interface can serve as an important buffer to prevent further movement of invasive plants into wildlands. Some of the best practices in the interface include planting noninvasive species, taking care to avoid accidental dispersal of invasive species, maintaining thick stands of native plants to thwart wind dispersal, and frequent monitoring to detect invading populations. An important strategy is to educate and enlist the public in the prevention, monitoring, and removal of invasive plants.

1.2.5 Conserving and Managing Forests under Different Ownerships

Forestlands in the interface have a variety of owners, from individuals to government agencies, and these owners often have different objectives and face different pressures for achieving them. This section explores some of the ways in which three types of landowners — industrial, nonindustrial private, and public — are pursuing conservation and management of forests in the interface.

Based on interviews with industrial forest managers, Munson and Haines in Chapter 16 estimate the additional costs to industry of managing forestlands in the interface. These forests are more expensive to manage than more remote forests due to greater administrative costs, higher taxes, and constrained management options. The added expenses, increased risks, and rising land prices in the interface create pressure for industry to sell their forestlands. Munson and Haines make the case for why society should promote the retention of industrial forestlands in the interface, and suggest several avenues. They argue that tax policies should encourage all types of forestland owners — industrial and nonindustrial — to invest in timber-producing forests. Another means is by conservation easements, whereby industry can continue to grow and harvest timber while the public benefits from, and pays for, the ecological and hydrological services of the forests.

Hubbard and Hoge, in Chapter 17, observe that the owners of nonindustrial private forestlands (NIPFs) in the interface face similar obstacles to timber production as industrial owners; additionally, the generally smaller parcels make many forest management practices unfeasible. Some solutions are small-scale logging systems; cooperatives, where landowners reduce their costs by sharing labor, equipment, and other resources; and forest certification, where the landowner adheres to prescribed, sustainable management practices and expects to profit from higher market prices for the products. However, many NIPF owners do not intend to market forest products and own their land for a number of other reasons, including recreation, viewing wildlife, and aesthetics. Hubbard and Hoge note that these forestland owners, and particularly those in the wildland–urban interface, often do not partake in (or even know about) traditional education and landowner assistance programs. They recommend that service providers — state forestry agencies, extension foresters, and private sector consultants — do more to understand the attributes and goals of these NIPF owners and to design and market educational programs that will assist them in maintaining healthy forests and achieving other objectives (Figure 1.8).

As urban development sprawls through a region, often the largest contiguous areas of wildlands remaining are those that are publicly owned. Moreover, due to land conversion and increased posting of privately owned forests, these lands may be the only sites available for recreation. In the last chapter, Dwyer and Chavez describe the demands on public wildlands in the interface and attempts at federal, state, and local levels to meet the intense pressures of nearby and distant users. Using case studies from the Los Angeles and Chicago metropolitan areas, they argue for adaptive and collaborative management. Public land managers must monitor and modify their management practices to ensure that objectives are being met in a rapidly changing environment. In the interface, public lands include federal, state, county, and city natural areas and parklands, and these are intermingled with privately owned forestlands. Collaboration among all the agencies and landowners administering these lands and with users and local communities is essential to protect the natural areas and to meet public needs. While many may despair at what seems to be an unstoppable spread of development across the landscape, Dwyer and Chavez show that all may not be lost: some developed lands, such as the former Joliet Army Ammunition Plant in the Chicago area, are being restored to wildlands.

FIGURE 1.8 Natural resource managers listen to a local citizen express her concerns about living in the wildland–urban interface. (Photo by Larry Korhnak.)

1.3 MEETING THE CHALLENGES

Common elements unite the many tools and approaches suggested in this book to meet the challenges of sustaining forests at the wildland–urban interface. Foremost is public education and involvement, especially at the local level. Outreach programs to increase awareness of the issues and to fully engage the public in resolving them are essential. Fire prevention educational programs can provide good models. By including the public in developing guidelines for defensible homes and landscapes, these programs can empower homeowners and communities to find solutions and assume responsibility by preparing for wildland fires. The best solutions come from this public involvement because they are supported by and tailored to the community. Planning is another common theme because it is vital in formulating new policies, making land-use decisions, and selecting management practices for specific forests and the landscape. Proactive planning, particularly at the landscape level, can lessen the impacts of urbanization on the functioning of natural ecosystems. The most successful interface efforts also involve partnerships, often including players who have not worked together in the past. Partnerships frequently result in shared governance, collaboration, and collective ownership in defining needs and finding solutions.

Natural resource managers and scientists need to play a more dominant role in solving interface problems by communicating their science-based knowledge to the public and by participating in the process of planning and policy making. They are often skilled managers of the landscape, and they need to become just as skilled at working with people. Planners and policy makers must also expand their outlook to include ecological as well as economic, political, and social considerations.

As many contributors to this book point out, there are no general or cookbook solutions to the problems at the wildland–urban interface. However, many promising approaches and tools are available that can be adapted to solve local and regional issues. We encourage and challenge readers to dive in, increase their repertoire of ideas, and create their own solutions for conserving and managing forests in this rapidly changing landscape.

REFERENCES

Dwyer, J., S. Stewart, V. Radeloff, R. Hammer, J. Fried, S. Holcomb, and J. McKeefry, 2003. Summary Statistics August 2003, in Mapping the Wildland Urban Interface and Projecting its Growth to 2030, CD-ROM, U.S. Department of Agriculture, Forest Service, North Central Research Station, Evanston, IL.

Macie, E. A. and L. A. Hermansen, Eds., 2002. Human Influences on Forest Ecosystems: The Southern Wildland–Urban Interface Assessment, General Technical Report SRS-55, U.S. Department of Agriculture, Forest Service, Southern Research Station, Asheville, NC.

McHarg, I. L., 1992. *Design with Nature,* John Wiley and Sons, New York.

Puget Sound Regional Council, 2001. Population Change in Cities, Towns, and Counties, 1990–2000, Puget Sound Trends no. D6, http://www.psrc.org/datapubs/pubs/trends/d6trend.pdf. [Date accessed: November 13, 2003.]

Sampson, N. and L. DeCoster, 2000. Forest fragmentation: implications for sustainable private forests, *Journal of Forestry* 98: 4–8.

Smith, W. B., J. S. Vissage, D. D. Darr, and R. M. Sheffield, 2001. Forest Resources of the U.S., 1997, General Technical Report NC-219, U.S. Department of Agriculture, Forest Service, North Central Research Station, St. Paul, MN.

USDA Natural Resources Conservation Service, 2003. National Resources Inventory 2001 Annual NRI: Urbanization and Development of Rural Land, http://www.nrcs.usda.gov/technical/NRI/. [Date accessed: October 22, 2003.]

Vaux, H. J., 1982. Forestry's hotseat: the urban/forest interface, *American Forests* 88: 36–46.

Zhang, T., 2001. Community features and urban sprawl: the case of the Chicago metropolitan region, *Land Use Policy* 18: 221–232.

2 The Challenge of Maintaining Working Forests at the Wildland–Urban Interface

John C. Gordon
Yale School of Forestry and Environmental Studies

R. Neil Sampson
The Sampson Group Inc.

Joyce K. Berry
College of Natural Resources, Colorado State University

CONTENTS

2.1 INTRODUCTION

If you do an Internet search on "wildland–urban interface," you get more than 2900 Web page matches. Almost all of these, including the first 20, have to do with fire risk and suppression where wildlands and cities come together. This seems to illustrate what Amatai Ezione called "the American genius for the practical." The most pressing problem of the wildland–urban interface, as seen by practical people, is the risk posed by wildland fires to urban structures. We would like to argue for a broader construction of the problem, both in terms of values at risk and potential solutions. For both scientific and political reasons, there is cause to think that all forests share a version of the risks that characterize the wildland–urban interface, in that all forests are threatened by change wrought by humans, either through direct activity or remote influences such as anthropogenic climate change.

2.2 DEFINITIONS

Public and scientific views of forests are changing, and together, science and the public are creating a better-integrated and much more complete view of their value and function (Gordon et al. 1993; Gordon 1994). Previous forest research, policy, and management strategies have usually been created and applied in pieces. Resources ("timber," "wildlife," "open space") were reified (made "real" and discrete rather than part of an integrated landscape), used, and had their use optimized largely independently. Also, most management tacitly assumed that overall land use stayed constant, that is, forests remained forests. Thus, conflicts over forest resource use most often have been resolved or approached as land allocation problems within a static total forest area. Various uses of forests were either optimized for a given area, or blended in multiple use management, in which all uses were said to be served simultaneously within a fixed area. Nowhere is the limited nature of those solutions more clearly demonstrated than at what we call the wildland–urban interface. Indeed, the term *wildlands* is ambiguous. Here, we take it to mean all land not subject to relatively intensive development for human habitation. Similarly, working landscapes can contain areas that are set aside

1-56670-602-5/05/$0.00+$1.50
© 2005 by CRC Press LLC

from the usual kinds of economic production. At the wildland–urban interface, the allocation problem is absolute. There, land currently devoted to landscape-based production of traditional marketed and nonmarketed values is either maintained or, more usually, converted to uses that preclude or greatly diminish the production of both. Thus, for the purposes of this chapter, the *working landscape* is the land within reach of near-term (a decade or less) urban conversion that is currently producing other values, including marketed ones. Working forests are the part of the working landscape covered by trees.

2.3 THE PROBLEM

The fate of the *satoyama* woodlands in urbanizing Japan provides a useful illustration of what we think is the general problem (*sato* = inhabited areas or villages; *yama* = hills or mountains). Therefore, *satoyama* woodlands were areas of forest used by a particular village or community, usually in concert with rice production and general farming. Their principal commodity function was usually the production of firewood and charcoal, but they were also used for the gathering of mushrooms and other nontimber forest products. They were often reproduced by coppicing.

> Satoyama woodlands are areas of forest used by a particular village or community. Up until the 1960s [they] ...had been used to provide domestic and commercial firewood and charcoal, manure, edible wild plants, mushrooms, timbers and other forest products. [They] were also repeatedly used for "slash and burn" agriculture/shifting cultivation.
>
> Satoyama woodlands have been destroyed by the expansion of cities, principally because over the past thirty years the woodlands have lost their commercial and forestry value (Tabata 2000).

In other words, once *satoyama* lost their utility as forests in the eyes of owners and urbanites, they were turned into something else. Only forests that are recognized as valuable and integral parts of the urban infrastructure survive for long within the reach of urban populations. Greenbelts and rural watersheds that serve urban dwellers come to mind.

Another lesson of the *satoyama* is the necessity of good information regarding wildland–urban interface forests reaching urban decision makers. Part of the reason the value of the *satoyama* woodlands was not identified was that they were not studied. When wood and charcoal were replaced by gas and oil, their usefulness was assumed to be at an end. Outreach to urban publics regarding the utility of *satoyama* did not occur because information on their other uses and values was not available in a form that was understandable and compelling to the urban public. Also, no one had the job of carrying out the needed public involvement. That this problem is not unique to Japan is demonstrated by the findings of the Meridian Institute Dialogue on Forested Lands and Taxation:

> The public and decision makers are probably not fully cognizant of the level and value of ecosystem services provided by intact forest lands. Support for federal tax changes and other policy options to address fragmentation and conversion will depend in part on public awareness of forest resource values (Meridian Institute 2001).

Had there been, during the critical time, people whose job was to talk to urban publics about *satoyama,* they might not have had much to tell. Because *satoyama* are by definition not pristine, they had not been much studied by ecologists.

> Since biologists had not recognized the importance of the satoyama environment and woodlands, which had been disturbed by human activities and were, therefore, no longer in their natural state, satoyama had not been properly researched. We do not know the biological and ecological features of satoyama, and have only now started undertaking the necessary research on it in Japan. (Tabata 2000)

This may be another characteristic problem with the wildland–urban interface working environment. The attention of scientists and conservationists is often primarily directed to those forests perceived to be pristine or unique, usually far from the wildland–urban interface. Yet, for example, according to the National Wildlife Federation, urbanization is the primary threat to 188 of the 286 species listed as threatened and endangered under the Endangered Species Act in California (Doyle 2001).

It is ironic but characteristic of forest loss at the wildland–urban interface that there is now a movement to restore *satoyama* where possible, particularly at sites close to cities. Oku and Fukamachi (2001) say, "It is our belief that during the next decade 'satoyama' will become the keyword in reference to outdoor recreation in Japan." It is likely that in Japan, as within many wildland–urban interface areas elsewhere, much effort and money will be spent to restore what might have been retained at a much lower cost with adequate planning, regulation, and incentives. The problem, then, from a professional land management point of view, is to propose solutions at the wildland–urban interface that use the tools of ecosystem management to sustain the provision of the market and nonmarket goods (clean water, for example) from the working landscape. This must be done while accommodating the seemingly inexorable march of urbanization and must incorporate the values of the expanding urban fringe. No challenge to our professional community has ever exceeded this one in scope (it is a global problem), importance (its outcome affects everyone), and complexity (it involves the service of all human values).

Urban modification of the working landscape often proceeds in a slow, largely invisible, creeping process that defies much of what we know about monitoring and managing land-use change. By the time the surveys have measured the working lands lost to other uses, the result simply indicates the outcome of years, if not decades, of change.

The process starts when urbanization begins to gnaw away at the structure of the working landscape. A farm is developed; a working forest becomes a wooded subdivision. These changes occur one at a time, with none of them appearing particularly important. But the neighbors notice, and they also notice when a mill closes, an equipment dealer disappears, or a local contractor moves away. To the working farm or forest, these changes mean higher costs, reduced market opportunity (see Section 2.4.1), and most important, a nagging uncertainty about the future of production farming and forestry. Where this creeping conversion is accompanied by local controversy over the dust or smell of agriculture, the inconvenience of large trucks on a rural road, or the visual impact of a timber harvest, the message to producers is clear: you and your activities are no longer wanted in this community.

The result is predictable. People who are fearful of the future become hesitant to make investments. Facilities do not get updated, new machinery is bypassed by making the old stuff "go a little longer," and long-term investments like tree planting are avoided. Who would spend several hundred dollars an acre to establish forests that they may be prevented from managing or harvesting?

That the march of urbanization is continuous and accelerating is well documented. Acres developed increased steadily over the last century, as has single-family house lot size (Peterson and Branagan 2000) (Figure 2.1). Conversion of rural land to developed land has accelerated over the past two decades, and conversion is most rapid near urban areas (Figure 2.2). Less well known and understood is that the rate of forestland conversion is both accelerating and goes both ways. In the 1992–1997 period, for example, about 9.9 million acres were converted from forest to other uses, and about 11.7 million acres were converted to forest from other uses (U.S. Department of Agriculture, Natural Resources Conservation Service 2000) (Figure 2.3). Most of the conversion from forest is within the wildland–urban interface and to developed land, while most of the conversion to forest is remote, from agricultural land, and reversible.

These national-level data indicate a national problem, but they do not guarantee that there is any national-level or even regionally general solution to the problem of forest loss and degradation at the wildland–urban interface. However, the advances in ecosystem and social sciences in the last two decades provide the hope of solutions tailored to specific urban areas and wildland–urban

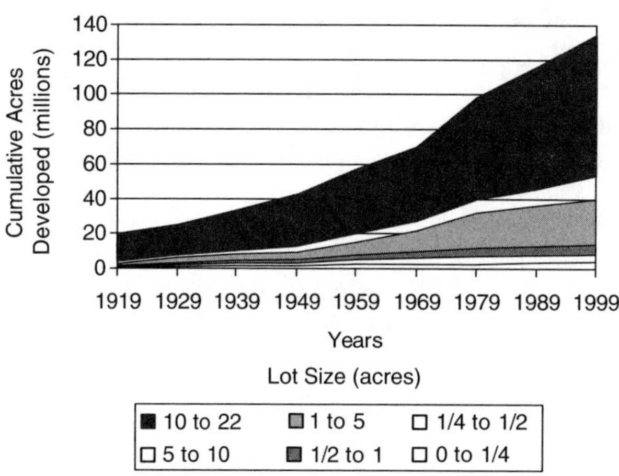

FIGURE 2.1 Amount of land developed for single-family housing by lot size, 1919–1999. (Adapted from Peterson and Branagan 2000.)

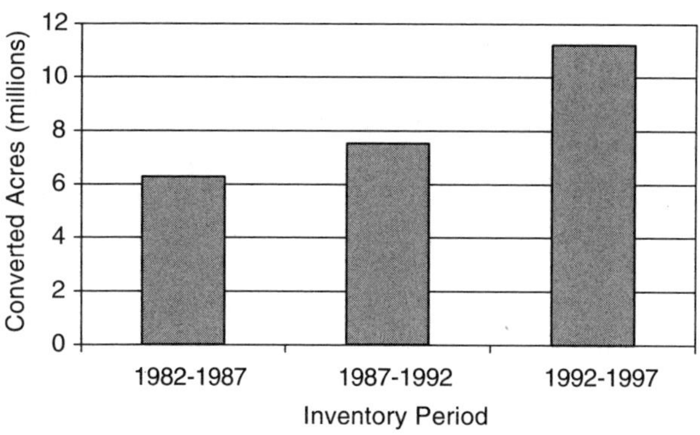

FIGURE 2.2 Conversion of rural lands to developed land in the U.S., 1982–1997. (Adapted from U.S. Department of Agriculture, Natural Resources Conservation Service 2000.)

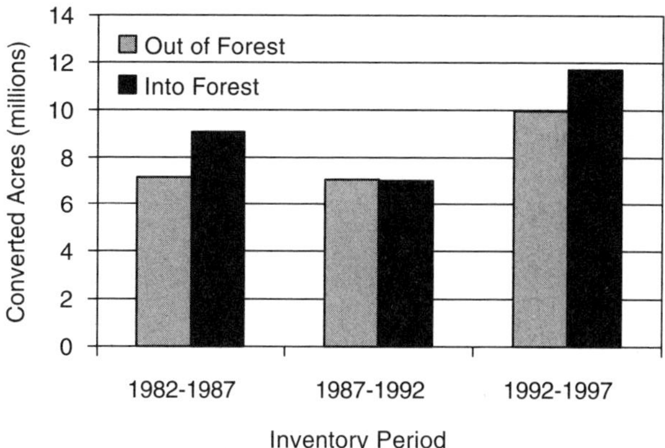

FIGURE 2.3 Forested land conversion in the U.S., 1982–1997. (Adapted from U.S. Department of Agriculture, Natural Resources Conservation Service 2000.)

interfaces. But the major mistake we can make here is to look for absolutely general solutions to wildland–urban interface problems. The nature and culture of urban areas and their surroundings are complex separately and unique in their interaction. Thus, careful local problem analysis, with broad public involvement, is probably the only universal element in any prescription for action at the wildland–urban interface.

People are now seen and studied as part of ecological systems, and urban areas recently have become legitimate objects of study for ecosystem ecologists. Ecosystems, including agricultural and wood production areas, are increasingly seen to be tightly integrated and to modify their surroundings. For example, studies of soil formation indicate that ecosystems modify their substrate much more rapidly than previously thought. Human actions increasingly condition the environment and its living content. Human-caused global warming may soon bring about easily observable changes in forest distribution and health (Wayburn et al. 2000). Air pollution (acid rain) already does. Alien invasive species have transformed forests and waters, and are accelerating in their effects (Meyerson et al. 1998). Water law and regulation have wrought great changes in wetland and riparian vegetation. All of these deliberate and accidental human influences have their greatest impact at the wildland–urban interface. Thus, urban conversion probably causes the most pervasive, drastic, rapid changes in forest ecosystem structure and function and competes in scale and cumulative effect with natural disasters resulting from hurricanes, vulcanism, and wildfire.

2.4 AN EXAMPLE

Human influence is detectable even in the areas most remote from human population concentrations. Thus, the wildland–urban interface is both a gradient on the landscape and a mental construct. A universal theoretical definition may not be possible now. We therefore present an example that we feel illustrates the problems and prospects of forest and landscape management at the wildland–urban interface in the northeastern U.S.

2.4.1 New Hampshire

New Hampshire is an excellent illustration of the fate of the working forest at the wildland–urban interface, and of the gradient nature of the problem. The state is heavily forested (in percentage terms, it is the second most heavily forested state) and contains most of White Mountain National Forest, the largest National Forest in the northeastern U.S. It is also rapidly urbanizing and its forests are being parcelized and fragmented. Because most of the state's forests are within a 3-hour drive from Boston, and a 5-hour drive from greater New York City, the whole state, despite the remoteness of its northern third, is in the wildland–urban interface (Montreal is close to this northern third). Forest cover peaked at 87 percent in the 1980s and is now estimated to be 83 percent and declining. Its average private forest parcel size has dropped from 47 acres to 37.5 in 15 years. The state's population doubled between 1950 and 1998, and it is still the fastest-growing state in New England, with a 55 percent increase in housing stock since 1980. The Society for the Protection of New Hampshire Forests (Thorne and Sundquist 2001) lists several outcomes of this rapid change with respect to working forests:

- Foresters and loggers see a strong relationship between parcel size and economical operations (Table 2.1).
- Economies of scale apply to stumpage prices paid to landowners (Figure 2.4) and to costs of preparing for timber sales (Figure 2.5). Owners of white pine stands over 500 acres receive 12 percent more in unit returns than owners of tracts of 50 acres.
- About 10 percent of the timber sales in the state are now "liquidation cuts" to make way for development (Table 2.2). Mill owners are concerned about timber supply a decade out and are restricting investment in new plants (Table 2.3).
- Forest investors are looking for parcels of 5000 acres or more, and the price of these is being bid up because of development potential. In New Hampshire, there are few parcels of this size left (Table 2.4) and none whose price does not reflect some development value, unless the development value has already been subtracted by an easement.

An interesting wrinkle in the New Hampshire wildland–urban interface story is that competition has developed between developers, who wish to focus on 10- or 12-acre building lots, and real estate brokers, who find a

TABLE 2.1
Parcel Size Below which Harvest Is Uneconomical for One-Time Marking for a Timber Sale

Parcel Size (acres)	Loggers Responding (%)	Foresters Responding (%)
5 or less	46	33
6–10	14	45
11–25	18	17
26–50	14	0
More than 50	10	6

Source: Thorne (2000). With the permission of the Society for the Protection of New Hampshire Forests.

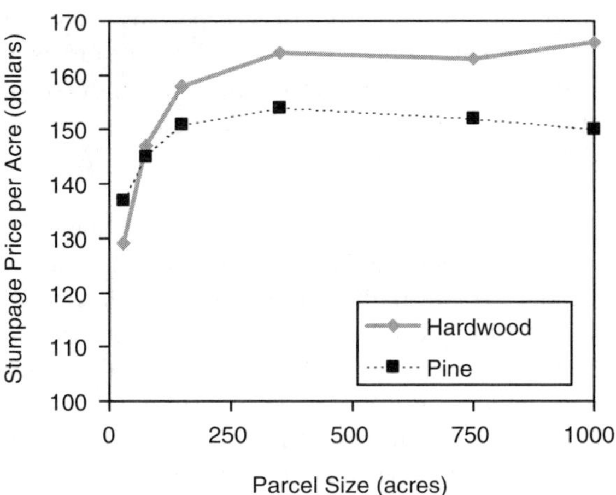

FIGURE 2.4 Average per acre stumpage price paid for white pine and northern hardwood stands harvested under forester supervision. (Adapted from Thorne 2000.)

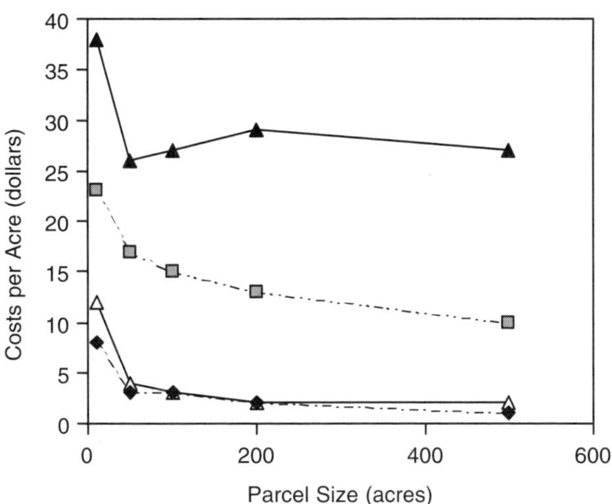

FIGURE 2.5 Foresters' average per acre costs for white pine stand harvest. (Adapted from Thorne 2000.)

premium market for residences (mostly old farms or portions of them) on 50 or more acres. In any event, the culture and politics of New Hampshire are responding in a variety of ways to its transformation into wildland–urban interface. Many old-timers declare a desire to "move north of the notch" (meaning north of Franconia Notch) to the less developed northern third of the state in an effort to avoid flatlanders. Others welcome the increased economic activity in a state that until fairly recently had a per capita income less than half of their neighbor to the south, Massachusetts. Most worry about retaining the rural character of the state, with its forested hills and classic New England villages. The threat of forest fire losses increases with the insertion of valuable residences into

TABLE 2.2
Forestland that Was Harvested for the Last Time in Preparation for Land Development

Land Undergoing Liquidation Harvest (%)	Mill Owners (%) (*N*=50)
0	28
Less than 10	32
10–25	32
26–50	4
More than 50	4

Note: Mill owners were asked to estimate how much of the acreage yielding their wood supply in 1999 had undergone a terminal harvest in preparation for land development.

Source: Thorne (2000). With the permission of the Society for the Protection of New Hampshire Forests.

TABLE 2.3
Concern about Availability of Wood Affects Investment Decisions by Forest Product Mills

Concern	Pulpchip Mill (%) (*N*=6)	Biomasschip Mill (%) (*N*=6)	Pulplog Mill (%) (*N*=10)	Sawlog Mill (%) (*N*=40)	Total (%) (*N*=47)
Yes	0	17	30	55	51
No	100	83	70	45	49

Source: Thorne (2000). With the permission of the Society for the Protection of New Hampshire Forests.

TABLE 2.4
Distribution of Private Forestland Acreage and Ownerships by Size Class, New Hampshire, 1993

Size Class, Acres	Acreage Owned	Forestland Acreage (%)	Number of Owners	Owners (%)
1–9	117,000	3	28,900	34.5
10–19	382,000	9	27,700	33
20–49	171,000	4	5,300	6
50–99	947,000	23	14,900	18
100–199	511,000	12	3,700	4
200–499	682,000	16	2,200	3
500–999	365,000	9	600	0.7
1000–4999	365,000	9	500	0.6
5000+	605,000	15	Withheld	0.1
Total	4,144,000	100	83,700	100

Source: Birch (1996).

unbroken forest, and fire protection costs can be expected to go up. It is interesting to note that the thought of forest fire threat to residences in the forest is not regarded as important or even credible in most of New England. Whereas fire is not a prominent component of forest dynamics in the Northeast, the threat is in fact real wherever urban development is dispersed in fuel-loaded forests. The state has a "current use" law that gives tax breaks to landowners that retain their land in forest or farming, but the minimum acreage is only 10. Conservation easements are increasingly popular but are relatively new, and many problems in their implementation remain to be solved. It is now safe to say that with only 26 percent of parcels over 500 acres in a status that will reasonably assure their retention as working forests, New Hampshire has a major problem and no solution in place. No comprehensive problem analysis has been carried out, although a number of organizations are working in that direction, including the Society for the Protection of New Hampshire Forests, the New Hampshire Timberland Owners Association, and the Nature Conservancy. Ironically, or perhaps realistically, much current effort is directed "north of the notch" to the area labeled the Great Northern Forest, rather than to the rapidly urbanizing south. Although the cultural and biological differences are considerable, the tragedy of the *satoyama* is being repeated in New Hampshire.

2.5 APPROACHES TO SOLUTIONS

These examples (*satoyama* in Japan and woodlands in New Hampshire), although strikingly different in geography and structure, seem to present some common themes, which might be formed into principles to underpin action at the wildland–urban interface.

It seems clear, for example, that to get, retain, and manage wildland–urban interface open space, urban publics should support rather than oppose or ignore rural producers. The very phrase open space seems to deny productive content or use beyond place holding and may therefore be counterproductive when dealing with its owners and occupants. Population numbers and affluence increasingly are making the old dichotomy between wilderness and development obsolete. Urban disturbance of wilderness is often equal to or more disruptive than production disturbance (e.g., McMansions in New Hampshire vs. timber production; ski slopes in Vermont vs. dairy farms).

Public involvement structures and methods appropriate to wildland–urban interface need to establish a continuing dialogue between the current working landscape owners/users at the wildland–urban interface and urban populations. Producing "action visions" for the wildland–urban interface is the main outcome of this dialogue. Identifying the decision makers is difficult both because of rapid changes in ownership and because of the web of overlapping laws, regulations, and political jurisdictions at the wildland–urban interface. Simply determining what these are often takes a long time. Thus, appropriate public involvement is always at odds with the crisis mentality that often surrounds rapid urban growth. Public involvement done properly is a complex, continuing process requiring constant attention and high-quality management (Figure 2.6).

The major conservation goal appropriate to the common interests of urban and rural publics is keeping forests as forests and keeping agricultural land as agricultural land, in short, to keep the working landscape within the wildland–urban interface working. This is an idea very different from preservation of nature in a pristine form. It is also very different from the creation and management of parks. But many values that natural preserves and parks serve are also served by working landscapes. Much of the biological diversity of the U.S., for example, is represented and preserved on these working lands.

The superiority of retaining working lands over the purchase of parks and natural preserves has several dimensions. The first is cost. No one, we think, would argue against the public purchase of more parks and natural areas in the wildland–urban interface. Fee ownership of open space has many advantages when viewed from the theoretical perspective of planners. However, in most urban and urbanizing areas, it is a continual struggle to gather public funds to acquire and adequately manage the

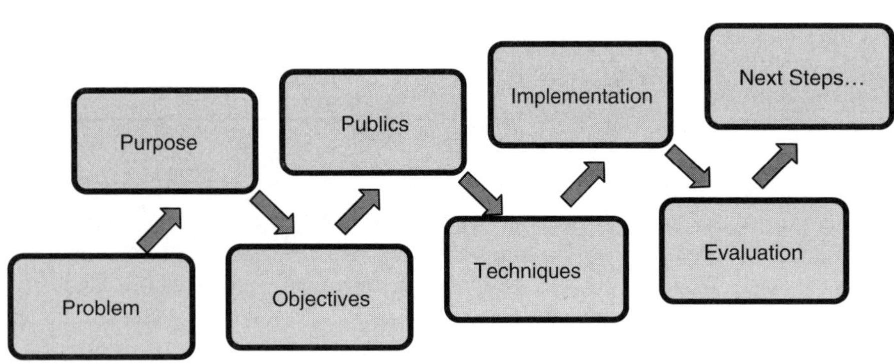

FIGURE 2.6 A framework for public involvement. All components are interconnected through process sequencing and feedback.

parks and natural areas that exist now. Some years ago, for example, the first operational definition for a park that a 14-year-old from Baltimore could think of was "a place where they find bodies."

So perhaps the first promise of working lands is that they can be secured at lower cost than lands purchased in fee. A quick (and egregiously destructive) leap from this is often made, however, that says open space should be provided at no cost to urbanites. This argument says that forest owners (and farmers) "ought to want to" protect their forests and that therefore paying them to do so is unnecessary and probably immoral, and all that is needed are clever rules and enforcement. This is the classical argument of the politically or physically strong to the weak: "you will like doing what I tell you to, or at least you had better pretend to." It is also ultimately an unnecessary argument. With the emergence of democracy and market capitalism in the forestry arena (Gordon 1999), and the assumptions democracy and capitalism imply about the universal power of individuals to speak with their votes and vote with their money, comes the notion of "user pays." It is now clear that if the politically powerful urban populations want ecological services from lands that at least appear rural, they will have to pay for them, with their money. The owners of open space clearly are not going to give up or give in without a fight. But we are back to one of the problems illustrated by the fate of the *satoyama.* If asked, most urbanites will probably say, "what services?"

Two broad kinds of ecological services are provided at the wildland–urban interface by working landscapes. Both must be maintained by a realistic approach to problem solving at the wildland–urban interface. The first kind is probably more important but is usually less visible. This is the maintenance of the basic ecological cycles that underpin all life, human and otherwise. The hydrological cycle turns in detail on the condition of watersheds and waters. Atmospheric cycles are subject to the condition of vegetation and soils; if sequestering carbon from the atmosphere is a necessary consequence of our increased emissions of greenhouse gases, it is plants and soils that will do it. A trouble with these services (and we can throw in the retention of biological diversity through habitat maintenance) is that they are mostly invisible to the casual or short-term observer. Another trouble is that the actions and conditions needed to maintain these cycles (the costs) are geographically specific (i.e., some designated owner has to create or incur them), but the benefits are broadly spread. These two conditions allow urbanites, cynically or not, to avoid payment while enjoying the benefits of the working landscape.

The second kind of ecological service is best illustrated by the right of access to something on the land. In our society, one usually thinks of recreation access (mountain biking or picnicking or hunting). In other places, gathering mushrooms or medicinal plants for personal use might be more important. The creation of visual pleasure is roughly of this kind but is a very special case, where actual access is often provided by public roads and paths, but the view is provided privately. These access services are easy, in theory, to deal with economically. Point-of-service charges for recreation, hunting leases, and even the use of road taxes to help dairy farms in Vermont are examples. Problems arise, however, when scale is considered, especially if large areas that may include many different private landowners and many categories of users are involved (long-distance winter recreation comes to mind, taking snowmobilers and skiers over many landholdings). This second kind of service is dealt with better now than the first, but certainly not everywhere successfully.

2.5.1 Actions to Secure Benefits

Improving the wildland–urban interface through the conservation of working landscapes will depend on a variety of approaches, all tailored to local and regional conditions:

- *Regulation: The role of rules.* Getting the legal structure right is a complicated task, with many specific local dimensions. For example, New Hampshire's current use law is enforced differently in different jurisdictions. The basic unit of direct democracy there is the civil town, and towns tend to guard their independence fiercely. For this and other reasons, zoning laws are a beginning and floor, if properly done and enforced, but not at all a total solution. Laws that establish freedom to farm or to harvest timber may have a role, but if they are seen to be antithetical to the financial or environmental interests of urbanites, they may be counterproductive in creating an "us and them" situation.
- *Land purchases and easements: Scale, cost, consistency.* Money for open space purchases is scarce with respect to the scale of the problem. The main principle in purchases may be cost-effectiveness through leverage. The scarce money should be spent strategically, in ways that ensure the maximum return in retained working landscape. These may include the purchase of areas of outstanding character that are widely known and perceived to be under threat, the protection of working landscapes from neighbor problems, or the acquisition of the working landscape itself, as in town forests. Easements have great promise and great complexity. They almost always imply some form

of public–private partnership and are notoriously difficult to nurture and maintain. The entity holding the easement is often a nonprofit organization without the resources to effectively monitor and enforce the easement. Easements that contain detailed management prescriptions in an inflexible legal form tend to become obsolete quickly. Also, easements that involve public money easily become hostage to unrelated political initiatives (Young 2001). But a coordinated program of purchase and easement based on solid land-use law can be a sound prescription for maintaining working landscapes (Gentry et al. 2000).

- *The attractiveness of investing in people.* Perhaps the most underinvested method of solving wildland–urban interface problems is the use of public involvement of a sophisticated and long-term kind. If the focus of our attention could be shifted from regulation and purchase to creating broad public understanding of the benefits of working lands at the wildland–urban interface, all tasks would be easier. Usually, we are reluctant to make the investments necessary. The process starts slowly and benefits are hard to document. But the practical application of the principles of social science could be the shortest route to real success. One concrete form of a possible outcome of public involvement is a landowner cooperative (Barten et al. 2001) that coordinates management over a large area of the wildland–urban interface.

2.6 CONCLUSIONS

- Without large-scale action, forests at the wildland–urban interface will continue to degrade, and this will have a negative impact on the quality of life in urban spaces; water, biodiversity, recreation, and culture all will suffer (Tyrrell and Dunning 2000). It is important to increase the intensity of action now, while some flexibility in the landscape still exists (Sampson and DeCoster 2000).
- Forests and other working lands at the wildland–urban interface are best viewed as part of the human-constructed infrastructure. The wildland–urban interface extends as far as urban influence, which, as we have argued, extends much farther in both time and space than land-use maps are likely to portray (Seville et al. 2000).
- Easements and other public–private partnerships show extraordinary promise, but need to be tried and studied more (Gentry et al. 2000).
- Forestry and farm people need to cooperate better at the wildland–urban interface. Gifford Pinchot suggested that "The farmer and the forester are like fingers on the same hand." Organizations within each community need to better coordinate their activities between communities, even though their specific goals may differ.
- Resource managers must learn to speak urban. The ultimate solution lies with urban populations. Where urban populations view the working lands of their region as a value to be saved, farms and forests will be. Where they continue to think of them as "development in waiting," those farms and forests are doomed, even where they still appear on the land.

REFERENCES

Barten, P., D. Damery, P. Catanzaro, J. Fish, S. Campbell, A. Fabos, and L. Fish, 2001. Massachusetts family forests: birth of a landowner cooperative, *Journal of Forestry* 99: 23–30.

Birch, T. W., 1996. Private Forest-Land Owners of the Northern U.S., 1994, Resource Bulletin 108, U.S. Department of Agriculture, Forest Service, Northeastern Forest Experiment Station, Radnor, PA.

Doyle, K., 2001. Paving Paradise: Sprawl's Impact on Wildlife and Wild Places in California, National Wildlife Federation, Washington, DC.

Gentry, B., M. Rosen, S. Jones, J. Gordon, and G. Dunning, 2000. Using Partnerships for Land Conservation, YFF Review 3(4), Yale School of Forestry and Environmental Studies, New Haven.

Gordon, J., 1994. From vision to policy: a role for foresters, *Journal of Forestry* 92: 16–19.

Gordon, J., 1999. The Seventh American Forest Congress: can democracy and resources co-exist? in *Forestry in the 21st Century: Dealing With the Consequences of Success,* J. Baker and J. O'Loughlin, Eds., 1997–1998 Plum Creek Lectures, University of Montana School of Forestry, Missoula, pp. 36–46.

Gordon, J., B. Bormann, and R. Kiester, 1993. The Physiology and Genetics of Ecosystems: A New Target, or "Forestry Contemplates an Entangled Bank," in Proceedings of the 12th North American Forest Biology Workshop, S. Colombo, G. Hogan, and V. Wearn, Eds., Ontario Ministry of Natural Resources, Sault Ste. Marie, Ontario.

Meridian Institute, 2001. Keeping Forests as Forests: Recommendations for Reducing Forest Fragmentation on Private Land in the U.S. through Changes in Federal Tax Policy, Meridian Institute Dialogue on Forested Lands and Taxation, Dillon, CO.

Meyerson, L. A., K. A. Vogt, G. Dunning, and J. Gordon, Eds., 1998. Invasive Alien Species, YFF Review 1(2), Yale School of Forestry and Environmental Studies, New Haven.

Oku, H. and K. Fukamachi, 2001. New Trends in Past Fuel Wood Management through Outdoor Recreation in Japan, Forestry and Forest Products Research Institute, Kyoto, Japan.

Peterson, T. and M. Branagan, 2000. Findings from the American Housing Survey: The Effect of Lot Sizes on Land Consumption in the U.S., presented at the Keep America Growing Conference "What Is Happening to the Land," Washington, DC, May 9, 2000, http://www.keepamericagrowing.org. [Date accessed: August 2002.]

Sampson, N. and L. DeCoster, 2000. Forest fragmentation: implications for sustainable private forests, *Journal of Forestry* 98: 4–8.

Seville, D., A. Jones, and D. Meadows, 2000. The Forest System Project: Exploring the Future of the Northern Forest, Sustainability Institute Report, Hartland Corners, VT.

Tabata, H., 2000. The future role of Satoyama woodlands in Japanese society, in *Forest and Civilizations,* Yoshinori Yasuda, Ed., International Research Center for Japanese Studies, Lustre Press/Roli Books, New Delhi.

Thorne, S., 2000. New Hampshire Forest Land Base Survey: Report of Results, Report for the New Hampshire Division of Forests and Lands, Department of Resources and Economic Development, Society for the Protection of New Hampshire Forests, Concord, NH.

Thorne, S. and D. Sundquist, 2001. New Hampshire's Vanishing Forests: Conversion, Fragmentation and Parcelization of Forests in the Granite State, Report of the New Hampshire Forest Land Base Study, Society for the Protection of New Hampshire Forests, Concord, NH, http://www.spnhf.org. [Date accessed: January 24, 2003.]

Tyrrell, M. and G. Dunning, Eds., 2000. Forestland Conversion, Fragmentation, and Parcelization, YFF Review 3(6), Yale School of Forestry and Environmental Studies, New Haven.

U.S. Department of Agriculture, Natural Resources Conservation Service, 2000. Summary Report: 1997 National Resources Inventory (revised December 2000), USDA Natural Resources Conservation Service, Washington, DC, and Statistical Laboratory, Iowa State University, Ames, IA, http:// www.nrcs.usda.gov/technical/NRI/ 1997. [Date accessed: March 14, 2003.]

Wayburn, L., J. Franklin, J. Gordon, C. Binkley, D. Mladenoff, and N. Christensen, 2000. Forest Carbon in the U.S.: Opportunities and Options for Private Lands, Pacific Forest Trust, Santa Rosa, CA.

Young, S., 2001. State Lawyer Pans West Branch Deal, *Bangor Daily News,* August 30, Bangor, ME.

3 An Assessment of the Southern Wildland–Urban Interface

L. Annie Hermansen
Southern Center for Wildland–Urban Interface Research & Information, USDA Forest Service Southern Research Station

Edward A. Macie
State and Private Forestry, USDA Forest Service Southern Region

CONTENTS

1-56670-602-5/05/$0.00+$1.50
© 2005 by CRC Press LLC

3.1 INTRODUCTION

Severe wildfires in Florida in 1998 demonstrated the complexities that the wildland–urban interface presents for a diverse group of people who live and work there. These fires cost millions of dollars in suppression costs, reduced tourism, and damaged timber, businesses, and homes. Entire communities had to be evacuated, and many elderly people and others afflicted with respiratory illnesses needed medical attention. Forest ecosystems were endangered.

Shortly after these fires, the chief of the USDA Forest Service conducted a review of the South and concluded that the wildland–urban interface is a key issue for the region affecting the condition, health, and management of forest resources. The Southern Research Station and Southern Region of the USDA Forest Service, in cooperation with the Southern Group of State Foresters, responded by developing a southern assessment of wildland–urban interface issues, challenges, and needs. This chapter summarizes this assessment, titled *Human Influences on Forest Ecosystems: The Southern Wildland–Urban Interface Assessment* (Macie and Hermansen 2002).

3.1.1 The Wildland–Urban Interface

The South is experiencing unprecedented population growth, resulting in rapid land-use change and profound human influences on forest ecosystems. As a result, the goods, services, and management of these forests are altered. These areas of rapid change are referred to as the wildland–urban interface. The wildland–urban interface can be defined in many ways, from a variety of perspectives. For this assessment, we defined the wildland–urban interface as an area where increased human influence and land-use conversion are changing natural resource goods, services, and management.

Wildland fires that threaten lives and property are perhaps the most obvious problems being faced by residents in the wildland–urban interface, but there are other issues of equal importance. As the number of private forest landowners in the South is increasing and parcel size is decreasing, the challenges associated with managing small-scale parcels for a diversity of management objectives are growing. Other critical management challenges in the interface include watershed management and protection, nonnative species invasions, forest health, wildlife management and conservation, recreation demand, and many more.

3.1.2 A Southern Assessment

The South is undergoing change at a rate unlike any other time in its history. Although change has been a constant since people first settled in the region, the current rate, pattern, and permanence of this change are unprecedented. Humans are influencing surrounding forests in a variety of ways. The first section of the assessment describes some of these major influences, including population and demographic changes, economic and tax influences, and land-use planning and policy issues. The second section of the assessment relates how urbanization and other human influences are changing forest ecosystem structure, function, and composition. This section also summarizes major forest resource management and conservation challenges and social changes in the interface. The third and final section presents a case study using fire to show the interdisciplinary nature of wildland–urban interface issues, and describes major themes and needs of the interface.

The assessment covers the 13 southern states from Virginia to Texas (Figure 3.1). Although many studies have looked at individual wildland–urban interface issues across the U.S., few have been conducted in the South and from an interdisciplinary perspective. While the assessment demonstrates that the rate and extent of change in the South are greater than other regions of the U.S., many of the interface issues, themes, and recommendations are applicable to other parts of the U.S. and abroad.

An extensive literature search was conducted to determine the current state of knowledge on interface issues. This indicated where gaps in knowledge still exist and facilitated the identification of wildland–urban interface research and information needs. A series of 12 focus groups in six communities experiencing rapid growth was also conducted to help refine and validate interface issues presented in this assessment (see Box 3.1).

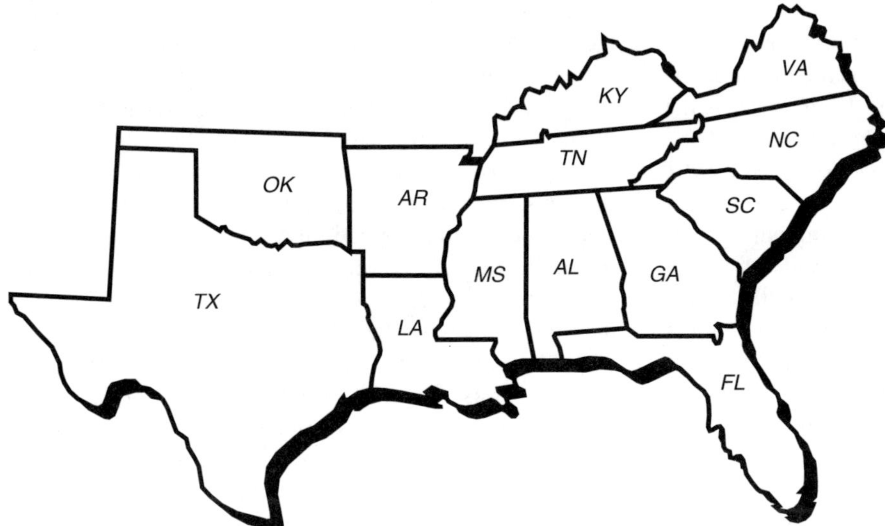

FIGURE 3.1 The 13 southern states covered by this assessment.

> **Box 3.1**
>
> **ASSESSMENT FOCUS GROUPS**
>
> In May and June 2000, 12 focus groups were conducted in six southern states to better understand the wildland–urban interface and identify issues to be addressed in the assessment. The groups helped identify new tools, knowledge, and skills needed by natural resource managers, decision makers, and others affected by the changes occurring in the interface. A diverse group of people was invited, including planners, foresters, developers, firefighters, private landowners, local and state policy makers, and many more.
>
> A facilitator asked the group members a series of questions about interface issues. Some example questions are:
>
> 1. Pretend you are a tour guide and describe the wildland–urban interface for me. What would we see, hear, and smell?
> 2. Describe factors that drive change in the interface areas you just described.
> 3. What are the key issues in the interface? What are the specific challenges you meet when attempting to manage resources in the changing wildland–urban interface?
>
> For more information on the methodology used and the results, refer to Monroe et al. (2003).

3.1.3 Purpose

The main purpose of this assessment was to provide direction for establishing a program of research and technology transfer within the USDA Forest Service, which began in January 2002 in Gainesville, FL. The Southern Center for Wildland–Urban Interface Research and Information is addressing the need for new interface research, technologies, outreach programs, and educational material for managers, landowners, local governments, and others.

3.1.4 Objectives

The five main assessment objectives were to:

1. Explore the wildland–urban interface from an interdisciplinary perspective in order to understand the complexity and connectivity of interface issues.
2. Examine factors driving change in the interface, including population and demographic trends, economic and taxation issues, and land-use planning and policy.
3. Explore the consequences of this change on forest ecosystems, resource management, and social systems.
4. Identify gaps in our knowledge of interface issues to help us identify research and information needs.
5. Promote dialogue about and heighten awareness of interface issues among practitioners, researchers, and the general public.

3.2 FACTORS DRIVING CHANGE

3.2.1 Population and Demographic Trends

The South's population is growing, moving, aging, and more culturally and ethnically diverse (Cordell and Macie 2002). Population growth in the South is increasing relative to other regions of the U.S. Between 1990 and 2000, the South's population grew 14 percent to 91 million residents and now accounts for 33 percent of the national total. The South's population is projected to increase another 24 percent to 114 million people by 2020. Births, deaths, and net immigration are major determinants of this population growth. Most significant is the net immigration into the South from other countries, which amounted to almost 6 million people between 1981 and 1990 (Cordell and Macie 2002).

An increase in the number of people from other countries is also creating a more culturally and ethnically diverse society. The ethnic makeup of the South is projected to see large shifts by 2020, most notably among the Hispanic population (Table 3.1). Non-Hispanic whites are steadily becoming a smaller percentage of the total population (Woods and Poole Economics 1997). It is important that natural resource managers be aware of these shifts because people from diverse backgrounds and age groups have different perspectives, attitudes, and values with respect to the use of forests and other natural resources (Cordell et al. 2002). For example, in one study Mexican Americans rated "doing something with your family" and "doing something with your children" significantly higher than non-Hispanic whites as favorite outdoor activities (Gramann and Floyd 1991).

Another important component of social change in the region is the aging of the population. The median age of the U.S. population has increased steadily from 18.9 years in 1850 to 32.8 years in 1990. In the South, median ages currently range from a low of just under 34.5 in Texas to a high of over 42 in Florida. This trend has important implications for forest ecosystems since forested and other natural lands are attractive retirement destinations (Cordell and Macie 2002).

More dramatic than these population dynamics is the conversion of rural and forestland to urban uses and the sprawling pattern of urban growth in the South. Increased numbers of people create more demand for housing, businesses, and transportation systems. This leads to greater urban growth and expansion of the wildland–urban interface and results in increased pressure on the forest resources found there (Cordell and Macie 2002).

Between 1992 and 1997, six of the 10 states in the U.S. with the highest levels of rural to urban conversion were in the South. Annually, more rural acreage is converted to urban uses in the South than any other region of the U.S. Metropolitan counties are accounting for about 82 percent of all population growth and today over 80 percent of the U.S. population is urban. This suggests that the urban constituency, which increasingly values forests more for noncommodity benefits than traditional forest products, will exert the greatest influence on national and state policies affecting natural resources and management of public land. Targeting the urban public for natural resource information and technology transfer programs, therefore, may have the greatest influence on the creation of public policies that support natural resource management and conservation (Cordell and Macie 2002).

Of the South's approximately 432 million acres of rural land, 78 percent is in corporate or individual private tracts. Individual private ownership is the region's primary ownership category, with 66 percent owning less than 500 acres (Figure 3.2). Also, of these ownerships the number of absentee vs. resident landowners is increasing, primarily motivated by recreation and speculation. Landowners have a variety of reasons for owning rural land (Figure 3.3), ranging from wanting to live in a rural environment to providing wildlife habitat (Teasely et al. 1999). Diverse management options for smaller tract sizes that meet a variety of landowner objectives are not currently available or must be adapted from large-scale practices to these smaller sizes.

TABLE 3.1
Projected Shift in the Ethnic Composition of the South

	Percentage of the Southern Population	
Race	**1990s**	**2020**
Non-Hispanic whites	72.4	61.0
African American	16.7	19.5
Hispanic	8.9	16.2
Asian and other races	2.0	3.0

Source: Woods and Poole Economics, Inc. (1997).

3.2.2 Economic and Tax Issues

Economic trends and tax policies considerably influence the rate of change in land use in the wildland–urban interface. Some economic and tax policies can accelerate development, while others help to shape development to meet the needs of a growing population while retaining as much land as possible in a rural condition (Moffat and Greene 2002).

The South's economy has evolved from one based primarily on agriculture and natural resource extraction to one that is diversified, including the service sector, industry, and computer manufacturing. Since 1978, nearly four of every 10 jobs gained in the U.S. were in the South, and the number of jobs has increased by 54 percent in the South compared to 38 percent for the rest of the nation (Moffat and Greene 2002). This change has helped promote the immigration and migration to the South discussed previously.

Efforts to improve the southern economy have contributed to several economic trends affecting change in the interface. For example, since local governments receive most of their funding from property and sales taxes, they have little reason to attempt to limit land development in their jurisdictions. This can lead to overzoning for development by local governments seeking to maximize their tax revenue. For example, in Loudoun County, Virginia, current zoning allows approximately 50,000 new housing

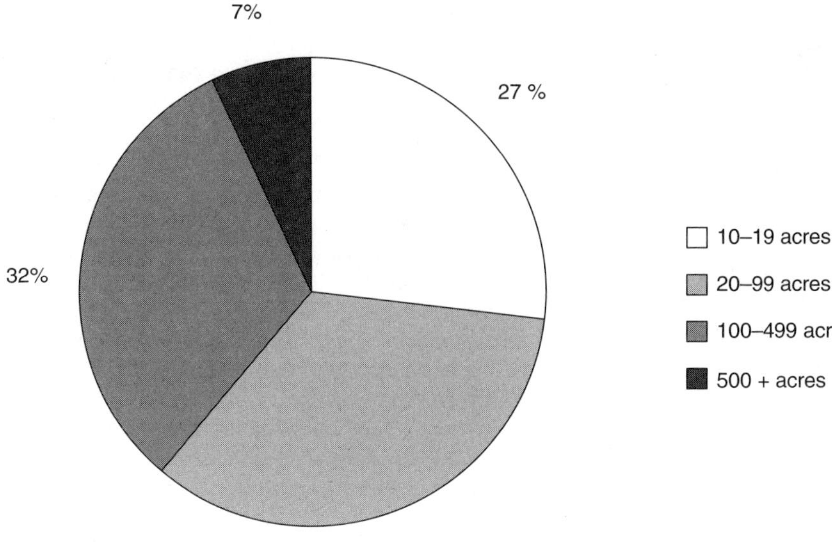

FIGURE 3.2 Percentage of individual nonindustrial landowners by the size of tract owned. (Adapted from Cordell and Macie 2002.)

units to be built, while the current demand is only about 3000 units per year (Lindstrom 1997).

As cities grow, the interface becomes more attractive to develop and inhabit. Subdivision of the land can be quite profitable for rural landowners, and often, as land values and property taxes rise, landowners may be forced to subdivide to keep any land at all. Water, sewer, garbage, fire, schools, and other services must be provided to new interface residents. Larger roads must be constructed to accommodate the increased traffic. Some people can work in the tranquility of their own homes, while others must commute longer distances. With time, these interface areas start to take on the qualities that people were trying to escape, and they therefore may seek a new interface, repeating the cycle (Moffat and Greene 2002).

Land development in the wildland–urban interface generates less revenue than municipal governments must pay to extend services to these areas. Several "cost-of-community services" studies have shown that local governments spend between 15 and 80 cents in services for every dollar of tax revenue generated by farms and forests, between 15 and 47 cents for every dollar of revenue generated by commercial development, and between \$1.04 and \$1.55 for every dollar collected for residential development! Moreover this does not include the nonmonetary values associated with maintaining land in agriculture and forests, such as reducing stormwater storage requirements that could save governments millions of dollars (Moffat and Greene 2002).

Nonindustrial private forest landowners face many economic pressures from federal and state taxes. Although many taxes affect land-use change in the interface, perhaps most notable are the federal income tax and the federal and state estate taxes. Other taxes that affect rural landowners to various degrees are state income taxes, property and yield taxes, and severance taxes. The federal income tax has the greatest economic effect of any

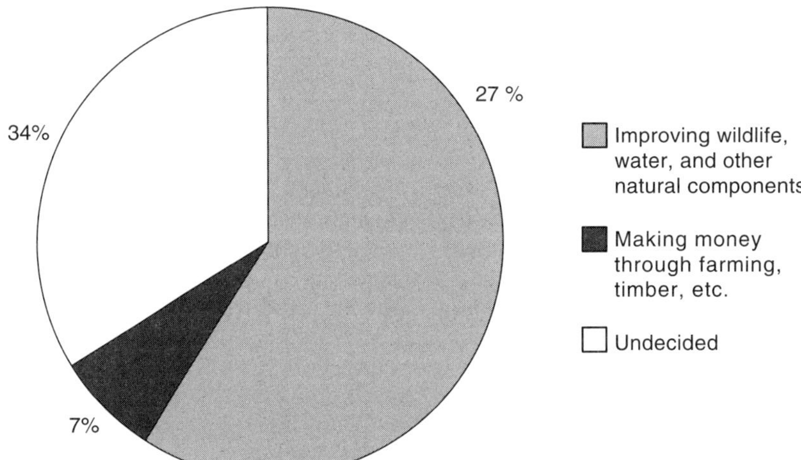

FIGURE 3.3 Percentage of nonindustrial private landowners by land management emphasis. (From Cordell and Macie 2002.)

tax on working land in the South because the rate is uniform across the region and is high compared to most other taxes (Greene 1995, 1998). This increases the costs of owning or managing rural land and therefore influences production decisions (Gregory 1972). It can create pressures for landowners to sell their land, particularly if the cost of keeping land in its present use is increasing.

Estate and gift taxation is another area of concern for landowners and foresters throughout the U.S. Federal and state death tax burdens resulting from insufficient estate planning can cause disruptions in forest management, abandonment of timber production by heirs, and fragmentation of ownerships. Greene et al. (2001) estimated that, nationwide, 2.6 million acres of forest must be harvested and 1.3 million acres must be sold each year in order to pay the federal estate tax.

Several tax and economic tools, such as conservation easements (see Box 3.2) and forest banks (see Section 3.3.2), can help landowners maintain forestlands at the interface. There are also many tax incentives, such as income averaging and permitting the immediate deduction of reforestation expenses, which could help reduce the federal income tax burden. Although these and other tools provide some assistance, they are for the most part still underutilized or of limited effectiveness without the help of policy makers to integrate and coordinate federal and state tax codes and landowner assistance programs (Moffat and Greene 2002).

3.2.3 Land-Use Planning and Policy

Land use in the wildland–urban interface is also greatly affected by current land-related public policies at federal, state, and local levels (Kundell et al. 2002). At the federal level, policies often appear to be in a tug-of-war. Some federal policies have created incentives for development and changes in land-use patterns, such as the federally subsidized National Interstate Highway System. On the other hand, there are also numerous federal policies and programs, such as the Clean Air and Clean Water Acts, that attempt to conserve and protect natural resources and contain provisions for limiting certain land uses.

Authority to guide land-use decisions lies mainly with the states, which may choose to give this control to county or municipal governments (Kundell et al. 2002). Some of the main state policies and programs affecting land use in the interface are:

- Forest practice ordinances, which act to protect environmental quality and local government investment in roads, bridges, and highway infrastructures. These ordinances are often enacted in response to local concerns over rapid land development, and range from simple tree replacement standards to comprehensive ordinances addressing natural resource issues (USDA Forest Service 2002).

Box 3.2

Existing Land-Use Planning and Policy Tools

Many tools currently exist for protecting natural resources within the interface:

- *Technologies*, such as Geographic Information Systems (GIS), can aid in planning land use and analyzing land-use trends. For example, the GIS-application CITYgreen, developed by American Forests, allows users to calculate the environmental and economic benefits of forests and trees (American Forests 2002).
- *Land-related policies* — The following growth management policies can be used in the interface (Daniels 1999):
  - *Smart growth programs*, which include a range of approaches that promote more efficient and compact urban development patterns. An example is urban growth boundaries, which encourage compact development and provide an appropriate direction for expansion of development over time.
  - *Alternative zoning ordinances,* which allow planners to design developments that better fit the land and to set aside more green space. One example is cluster developments, which are subdivisions in which development must be placed on a portion of the parcel and the rest must remain in undeveloped open space.
  - *Conservation easements,* which are voluntary legal agreements between a landowner and another party that restrict the development of a tract of land and provide tax benefits that can help to maintain land in rural uses.
  - *Purchase-of-development-rights (PDR)* programs, which enable the preservation of farm and forestland by giving the state and local governments the ability to purchase development rights (conservation easements) from landowners and restrict the land to farm, forestry, and open-space uses.
  - *Transferable-development-rights (TDR)* programs, which enable preservation of sensitive lands by moving development potential from one tract of land to another, unlike the outright retirement of development rights under PDRs.
  - *Land trusts*, which are private nonprofit organizations that can either receive donations of property, conservation easements, and money or buy property and conservation easements. Land trusts play a useful role both in working with landowners to preserve land and by acting as intermediaries between government agencies and landowners who share a common interest in keeping the land intact.

- Statewide growth management plans, which establish statewide goals and policies, create regional agencies to review and coordinate local plans, and require local governments to prepare plans that implement statewide goals. All too often, however, lack of local government cooperation prevents achievement of the plans' goals. In Florida, for example, local zoning decisions favor low-density, sprawling development, even though these practices are inconsistent with the statewide growth management plan (Nelson et al. 1995).

- State infrastructure policies, which often contribute greatly to problems with land development patterns in the interface. For example, state transportation departments can build roads without regard for local plans, and state community infrastructure funding often emphasizes new development over restoring older systems.

Local governments use zoning ordinances as the primary tool when making land-use decisions. There are many examples, however, of how local zoning policies indirectly promote growth. Often, local governments try to reduce housing density by increasing lot size. This policy actually increases land consumption, causing development to sprawl out over the landscape (Kundell et al. 2002).

Current land-use policies are largely ineffective for managing growth because they are based on traditional programs that were not designed for that purpose (Kundell et al. 2002). Zoning ordinances, for example, were designed to protect private property values and public investment. Complications also arise from the overlapping of multiple federal, state, and local jurisdictions. As a result, various levels of the government are independently making land-use decisions without any common understanding of what long-range growth management goals each government level wants to achieve. There is also no common approach for addressing environmental issues that cross jurisdictional boundaries (Kundell et al. 2002).

Fortunately, a broad array of policies, programs, and other tools exist to help guide and control growth in the interface (see Box 3.2). With the implementation of these tools, natural resource protection and management in the wildland–urban interface can be greatly improved. However, natural resource managers and the public, as well as state and local officials, need to become both more aware that these tools exist and be more willing to put them into practice.

3.3 CONSEQUENCES OF CHANGE

The effects of urban development on southern forest ecosystem goods and services are profound. Resource professionals face difficult challenges in their attempts to manage these forests and to minimize the changes to natural resources that are occurring.

3.3.1 Urban Effects on Forests

The most obvious direct effects of urbanization and other human activities on forests are the reduction of total forest area (Table 3.2) and fragmentation (Zipperer 2002), which is a deforestation process that subdivides forest cover into smaller and more isolated forest parcels. Rates of forest loss are fastest near major urban centers, along major transportation routes, and near recreational areas such as national parks (Boyce and Martin 1993).

Additionally, humans indirectly alter forest ecosystems by modifying hydrology, altering nutrient cycling, introducing nonnative species, modifying disturbance regime, and changing atmospheric conditions. These changes significantly affect forest health and modify the goods and services provided by forest ecosystems (Zipperer 2002).

Fragmentation has significant effects on biodiversity (Zipperer 2002). The loss of forested corridors can create isolated wildlife populations and consequently reduce genetic flow. This reduction can potentially lead to inbreeding and local extinctions. Fragmentation alters the physical environment and biotic communities of forests, including greater temperature fluctuations and increases in parasitism and predation. Forest patches have an increase in edge habitats, which can change the species composition by favoring edge species, such as raccoons and deer, over species that require interior conditions, such as ground-nesting birds.

Besides reducing and fragmenting forest cover, urbanization alters water flows and significantly affects aquatic habitats (Zipperer 2002). Impervious surfaces increase surface runoff, changing streambank stability,

TABLE 3.2
Tree Canopy Losses in Selected Areas in the South

Location	Forested Area Loss (thousands of acres)	Time Period	Tree Canopy Loss (%)
Atlanta metropolitan area	1747	1974–96	26
Chattanooga, Tennessee	110	1974–96	21
Houston metropolitan area	692	1972–99	8
Roanoke, Virginia	313	1973–97	9
Fairfax County, Virginia	125	1973–97	20

Note: Because measurements of canopy losses and fragmentation are scale dependent, a comparison across different studies is difficult. Analyses by American Forests (2002) were used because the same protocol was employed to analyze each region. A 30-m Landsat pixel was classified as forestland if it had at least 50 percent tree cover. The use of these analyses, however, does not imply an endorsement of techniques or models developed to obtain these values.

Source: Zipperer (2002).

water quality and quantity, and biodiversity of aquatic systems. Besides the increase in impervious surfaces, urbanization also often results in channelized streams, drained wetlands, and increases in the amounts of pesticides and nutrients found in streams. Development often occurs in the headwaters of streams and rivers, endangering local species that are extremely sensitive to adverse environmental changes.

Forests in urban landscapes differ environmentally, compositionally, and structurally from rural forests (McDonnell et al. 1997) (Figure 3.4). Forests do not need to be disturbed directly by development to be affected. Adjacency to urban land uses can create changes in forests over time, such as by exposure to nonnative species (Zipperer 2002).

In general, as one moves along a gradient from rural to urban ecosystems, species richness of plants increases, but it decreases for mammals, birds, amphibians, and reptiles (Kowarik 1990). Also along this continuum, the number of native species decreases, while nonnative species increase. These forest alterations are a result not only of urbanization but also of past and current agricultural and forestry practices (White and Wilds 1998). Altered forests are much more susceptible to the invasion of nonnative species because of modified soils and the absence of natural plant and animal predator species. Even native species in high population densities can affect ecosystem composition and structure (Zipperer 2002). Increases in the whitetailed deer (*Odocoileus virginianus*) population in the South, for example, have resulted in denuded understory vegetation, which significantly affects the breeding success of groundnesting bird species.

With increased urbanization also comes an increase in air pollutants, such as oxides of nitrogen (NO_x) and sulfur (SO_x) and tropospheric or ground-level ozone (O_3). Although these pollutants occur naturally, human activities are increasing their presence in the atmosphere (Zipperer 2002). The highest concentration of SO_x in the U.S. was found in a spruce forest in the Appalachian Highlands (Johnson and Lindberg 1992). At high concentrations, these pollutants can change ecosystem processes, damage plant tissue, and make forests susceptible to other environmental stresses (Berish et al. 1998).

Although addressed here independently, all of these urban effects act together. For example, atmospheric deposition alters nutrient availability in the soil and damages plant tissue. These effects subsequently make plants more susceptible to pests and pathogens (Zipperer 2002).

A healthy forest ecosystem is one that is free of distress syndrome, which refers to the ability of an ecosystem to recover naturally. The many direct and indirect effects of urbanization make forests vulnerable to distress syndrome. An integrative and interdisciplinary approach is necessary to address urban effects on forest health. The approach must account for the complexity of interactions among the social, ecological, and physical components of an ecosystem (Zipperer 2002).

3.3.2 Challenges to Forest Resource Management and Conservation

As previously discussed, urbanizing forest ecosystems are changing in their structure, composition, function, and processes. Additionally, forest tract size is decreasing, the number of owners is increasing, and forest management preferences are more diverse. These changes set the stage for new challenges, as well as new and innovative approaches, to forest resource management in the interface (Duryea and Hermansen 2002). Changes and challenges associated with the management of water resources, traditional forest products, fire, recreation, and wildlife are covered in this chapter of the assessment.

3.3.2.1 Water Resources

The management of water resources for quantity and quality in the urbanizing environments of the South is a complex task. A growing southern population requires increasing supplies of water; yet, more and more people are settling and recreating in primary watersheds of large cities (Minahan 2000). These increasing demands for water bring more complex issues over allocation of water for a variety of purposes, such as water-based recreation and adequate water supplies for wildlife and aquatic species habitat (Sedell et al. 2000) (Figure 3.5).

Human health concerns from polluted water sources is another important water resource issue. Municipal waste facilities in rapidly developing areas face difficulties with handling and treating increased waste loads, and sewage overflows may occur after heavy rainfalls. Septic tanks are often placed at high densities in the interface; they are extremely vulnerable to failures and are a chief contributor to fecal coliform contamination (Minahan 2000). Nonpoint source pollution, such as farm and stormwater runoff, is difficult to trace to its origin and is the cause of approximately 50 percent of water pollution problems in the U.S. (U.S. Environmental Protection Agency 1992).

Rather than concentrating on separate pollution dischargers and managing within the constraints of political boundaries, watershed management takes a "holistic" approach to managing water quality and is critical for effective water management (Rubin et al. 1993). This approach provides a framework for designing the optimal mix of land covers to minimize the effects on water resources and for coordinating management priorities across landownerships.

3.3.2.2 Traditional Forest Products

Southern forests are an important national source of timber, making up 40 percent of U.S. timberland (Faulkner et al. 1998). Management and conservation of these forests,

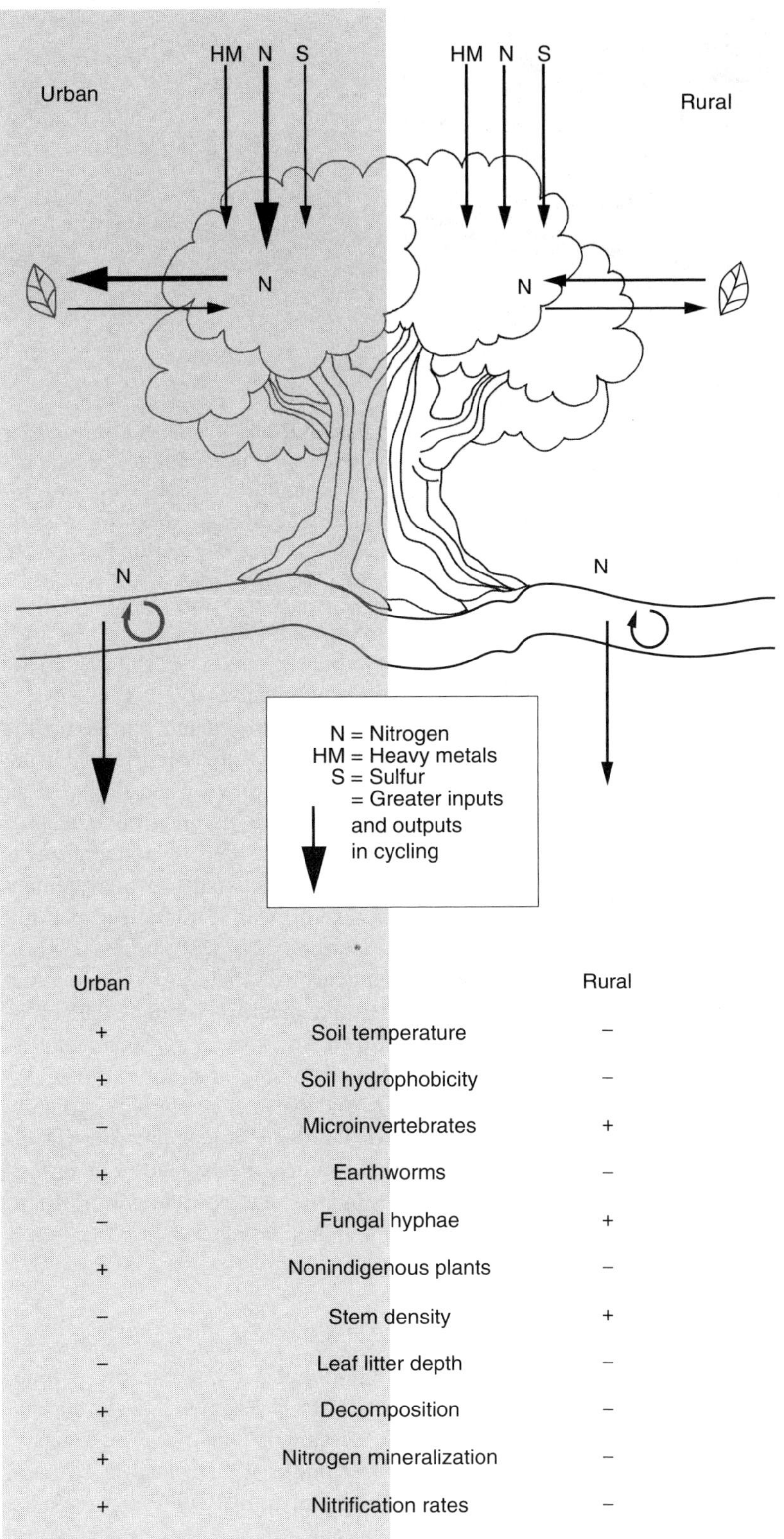

FIGURE 3.4 Generalized illustration depicting structural and functional differences of forests in urban and rural landscapes having similar physical environments and species composition. (From Zipperer 2002.)

however, are increasingly difficult in the interface. The cost of land near metropolitan counties is high, thus making the production of timber from forests on these lands expensive. This can discourage landowners from making investments in forestry in these areas. Selling and subdividing the land can be more profitable, and the rapidly changing land-use patterns characteristic of the interface may also discourage landowners from making long-term investments in forestry.

FIGURE 3.5 As human demands for water and water-based recreational opportunities increase, concerns for providing enough water for wildlife and aquatic species also grow. (Photo by Larry Korhnak.)

In the interface, closer public contact with forestry practices provides more opportunities for conflict. Opposition to the use of herbicides, prescribed burning, and other aspects of forestry operations can result in regulations that affect the quantity of timber available and the costs of transporting it. Forest management practices adapted to the special conditions and human influences of the interface are thus necessary (Duryea and Hermansen 2002). Examples include the use of partial cuts and limited use of herbicides close to public areas (Bradley 1984). Where timber production is not possible, nontimber commodity products (e.g., pine straw, medicinal plants) may be a viable alternative for landowners, especially for owners of smaller tracts.

Coordinated forest management across ownerships can help ensure healthy ecosystem function and provision of ecosystem goods and services. Partnerships among private landowners and organizations can help overcome the challenges of managing on a landscape scale (Duryea and Hermansen 2002). For example, by forming partnerships with private landowners, The Nature Conservancy's Forest Bank aims to protect the ecological health and natural diversity of working forests while ensuring long-term economic productivity (Dedrick et al. 2000).

3.3.2.3 Fire

Decades of fuel buildup and increases in the number of people in the interface have created many challenges for fire agencies across the South and the nation. The ability to use prescribed fire to enhance ecological processes is increasingly difficult, while the challenges associated with preventing and suppressing wildfires have increased. Negative public opinion regarding prescribed fire is one of the biggest obstacles that fire agencies must overcome. People may not understand the benefits of fire or may be concerned about public health and safety.

For these reasons, fire management cannot be the same in the interface as in rural areas. Different firing techniques and ignition patterns may be needed. Weather and fuel characteristics that are optimal in rural areas may not be practical in the interface due to concerns over excessive smoke production. Thus, smoke management becomes a priority because of health, safety, and liability concerns (Duryea and Hermansen 2002).

Fire protection agencies are charged to first protect human life and property, then natural resources. The problem with this is that forest fire suppression personnel do not usually have sufficient training in structural firefighting, and municipal fire departments are not typically equipped or trained for wildland fire suppression (Davis 1986). A challenge in the interface lies in combining firefighting expertise in both areas and providing cross-training opportunities (Duryea and Hermansen 2002).

3.3.2.4 Recreation

Recreation is an increasingly significant part of southern lifestyles. However, opportunities for recreation, particularly on nonindustrial private forestlands, are decreasing. This puts considerable pressure on public land managers to provide recreational opportunities for a diverse spectrum of users and to maintain the quality of natural resources on the site (Duryea and Hermansen 2002). This is especially true when these lands are close to large urban centers where recreational opportunities in inner cities are declining.

As previously discussed, the South holds an increasingly diverse population. Recreation managers must consider the needs and expectations of the different groups using wildland–urban interface recreation sites. They must also possess the ability to communicate with diverse user groups who have different perceptions and values regarding land management.

3.3.2.5 Wildlife

The most significant wildlife management challenge in the interface is conserving, managing, and restoring wildlife habitat (Duryea and Hermansen 2002). Fragmentation of forests has created many unconnected patches of habitat. Utilizing corridors to connect small forest patches to larger reserves is especially valuable for wildlife. Site history, adjacent land-use types, and current influences should be taken into account when developing wildlife plans (Nilon and Pais 1997).

In the interface, some residents enjoy closer contact with wildlife, while others find this a nuisance. Balancing desires of residents to be close to wildlife with their desire to avoid nuisance and human health problems, such as transmission of Lyme disease from deer ticks, is a major wildlife challenge in the interface (Duryea and Hermansen 2002).

Nonconsumptive uses, such as wildlife viewing (Figure 3.6), are increasingly popular while consumptive uses, such as hunting, are declining (Cordell 1999). Being aware of local public attitudes toward wildlife management and conservation and incorporating both consumptive and nonconsumptive uses into management strategies are important for wildlife managers in the interface (Duryea and Hermansen 2002).

3.3.3 Social Consequences of Change

Social consequences are at least of equal importance to the environmental consequences of change in the interface. Economic, political, and community and landowner consequences of change are covered in this chapter.

3.3.3.1 Economic Consequences

As the rural forest transforms to an urban one, the economics of land management change (Hull and Stewart 2002). The perceived benefits of trees change and perhaps increase, as do the costs of planting and maintaining them (Dwyer et al. 2000). Decisions over whether and when to harvest trees are more complex in the interface because of community members' concerns about environmental quality and forestry practices, such as large-scale clear-cutting.

It also appears that urbanization reduces timber supply, increases harvesting costs, and decreases the profitability of timber production (Barlow et al. 1998; Wear et al. 1999), although much still remains to be known. Some studies suggest that nontimber commodities, such as fruits and medicinal herbs, can generate more money per acre in the interface than do rural lands growing traditional forestry crops. This money can help supplement family income and even help to retain land in forests that might otherwise be subdivided (Hull and Stewart 2002). Jobs follow people to the interface, providing additional opportunities to supplement household incomes.

3.3.3.2 Political Consequences

Interface forests differ from rural forests in the number and complexity of political issues affecting them (Hull and Stewart 2002). Land-use decisions tend to be more contentious and attract more attention than those in rural areas. Public participation also tends to be more abundant and diverse. Typically, with new owners and neighbors, decision making processes become more formalized. New owners may give more emphasis to environmental concerns than do longtime residents, although they may not have the personal community contacts that can help to influence land-use decisions. They do, however, tend to have more contact with national and regional organizations and insist on more formal procedures than long-term residents (Smith and Krannich 2000). This can have long-term benefits for a community, although decision making becomes more complex (Hull and Stewart 2002).

3.3.3.3 Community and Landowner Consequences

New forestland owners often have little contact with the professions that traditionally offer management advice, turning more toward garden care professionals and landscape architects for information, and possess different values and management objectives from longterm residents. Traditional methods of providing forestry advice, such as forest management plans, may not be effective for these new owners (Hull and Stewart 2002).

Community quality of life is also affected by settlement of interface forests. Increased development in the interface can bring increased access to health care, education, and jobs. Being closer to nature and farther from urban stressors is an accepted benefit of living in the interface, although increasing population density can generate the very qualities that were supposedly left behind in the urban environment and can encourage migration to even more remote areas. Finding an acceptable balance between these social costs and benefits is an ongoing challenge (Hull and Stewart 2002).

3.3.3.4 Needs of Natural Resource Professionals

Natural resource professionals need many new skills and tools to remain effective in the changing environment of the interface. They need new methods for communicating with landowners and distributing forestry advice. They need new skills, such as techniques for managing forests on small scales. They must also work effectively with the

FIGURE 3.6 Birdwatching is an increasingly popular outdoor recreation activity. (Photo by Larry Korhnak.)

TABLE 3.3
A Selected History of Wildland–Urban Interface Fires in the U.S.

Location	Year	Structures Lost (number)	Area Burned (acres)
Pine Barrens, NJ	1963	383	1,83,000
Laguna, CA	1970	382	1,75,425
Sycamore, CA	1977	234	805
Panorama, CA	1980	325	23,600
Palm Coast, FL	1985	99	13,000
Burke County, NC	1985	76	2,000
Onslow County, NC	1986	0	73,000
Monterey County, CA	1987	31	160
Nevada County, CA	1988	90	33,500
Sisters, OR	1990	22	3,300
Paint Cave, CA	1990	641	4,900
Oakland Hills, CA	1991	2,900	1,500
Chelan County, WA	1992	32	2,400
Craven County, NC	1994	0	24,600
Millers Reach, AK	1996	344	37,336
Poolville, TX	1996	141	16,000
State of Florida	1998	330	5,00,000
Juniper, CA	1998	44	6,000
St Lucie, FL	1999	43	759
Colbert County, AL	1999	20	3
Los Alamos, NM	2000	235	47,650
Russell County, AL	2000	6	4
Chambers County, AL	2001	2	30
Talledega County, AL	2001	1	347

Source: Monroe (2002).

large number of stakeholders with diverse values and interests (Hull and Stewart 2002).

Building partnerships is an important aspect of interface management. Natural resource professionals must become involved with the institutions that influence the management and development of interface forests. They need mechanisms that encourage and enable crossboundary ecosystem management. Additionally, resource professionals need a new language and conception of forestry that incorporates the understanding and concerns of the new owners and neighbors of interface forests (Hull and Stewart 2002).

3.4 MAJOR THEMES AND NEEDS

3.4.1 Fire

Fire is but one of many important issues in the interface; yet, it is the one that attracts the most attention. Although fire was already mentioned in Section 3.3.2, here wildland–urban interface fire is used as a case study to reinforce the concepts brought up throughout the assessment and demonstrate how ecology, resource management techniques, economics, public policy, and demographics influence efforts to manage and protect both people and natural resources. Fire concerns cannot be resolved solely from a natural resource perspective. Using information and perspectives from each discipline, we can come up with better solutions to wildland–urban interface challenges (Monroe 2002).

At one time, wildland fire was not a problem in the South. To the contrary, it was considered a normal event, or fire was set intentionally by Native Americans and early European settlers to improve wildlife habitat and clear land for cultivation. It is no longer possible, however, to let fires run their course. There are small towns, timberlands, vacation homes, and ranchettes in what were formerly wildland areas. The protection of human lives and investments necessitated the purposeful exclusion of fire from the South. Although this strategy successfully protected many lives and structures, it created huge fuel loads and increased the risk of catastrophic fires. Also, as more people live in the interface, the chances of a fire being ignited have increased. A selected history of wildland–urban interface fires in the U.S., provided in Table 3.3, demonstrates the scope and breadth of the problem.

3.4.1.1 Ecological Structure and Function

Many southern forest ecosystems have developed adaptations to fire. Longleaf pines, for example, have a thick bark that insulates and dissipates heat, a grass stage to protect the bud, and a substantial root system that allows it to grow quickly above the ground fire zone (Myers and Ewel 1990). The purposeful exclusion of fire in the South has led to huge fuel buildups and the encroachment of plant species not tolerant of fire. This has resulted in high-intensity fires that can kill even large, mature trees, despite these adaptations. Frequent fires can help maintain healthy southern forests by providing ecosystem services such as releasing nutrients, scarifying seeds for germination, and releasing natural fertilizers such as ash and carbon (Brennan et al. 1998).

3.4.1.2 Natural Resource Management

Prescribed fire is one resource management tool that temporarily reduces heavy fuel loads in the interface and helps maintain healthy, diverse forests (Figure 3.7). However, there are concerns about air quality and public safety from the smoke produced during the fires. Where prescribed fire is not an option, alternatives such as mechanical reduction and herbicide treatment have been explored. These options, however, may not provide the same benefits for forest health as prescribed fire. Additionally, herbicide use may be even less acceptable to the public than fire (Monroe 2002).

3.4.1.3 Demographics

Interface residents are quite diverse — retirees, exurbanites, vacation, or weekend residents — representing a variety of ethnic and cultural backgrounds. Thus, not surprisingly opinions and attitudes about wildland fire vary from group to group. For example, English- and Spanish-speaking Florida residents who were recently exposed to wildland fires differed in their knowledge and perception of fire risk (Loomis et al. 2000). New residents who move to the interface may not be aware of the wildland fire risk. It is therefore important for natural resource managers to understand whom they are talking to when communicating about fire in the interface (Monroe 2002).

3.4.1.4 Economics

It is quite expensive to suppress and recover from interface fires. Because of the high housing densities in the southern interface, agencies must suppress fires at great cost, and any wildland fire is likely to put interface homes at risk. One complex of fires near Orlando, Florida, in 1998 cost over $5 million in suppression in less than 3 weeks. Additionally, the cost of conducting prescribed fires is much higher in the interface than in the wildlands because more preparation and public contact are needed (Greenlee et al. 1999).

3.4.1.5 Land-Use Planning and Policy

There are a variety of policies and recommendations about wildland fire suppression, the use of prescribed fire, zoning, firewise landscaping, and building construction. For example, Flagler County, Florida, which had a countywide evacuation during the 1998 wildland fires, enacted an ordinance requiring brush mowing and selective thinning of mature pine trees (Flagler County Ordinance No. 98-14). Because of liability issues related to smoke, several Florida counties have adopted ordinances requiring that prospective homebuilders be told about the use of prescribed fire in nearby state-owned natural areas (Wade and Brenner 1995). Florida's Certified Prescribed Burning Program [Florida Statute 590.125 (3)(b)] requires written prescriptions for each burn and protects the burner from liability unless gross negligence is proven. Successful interface policies will be ones that the public supports (Monroe 2002).

FIGURE 3.7 Prescribed fire is a fuel reduction method that can help reduce the risk of catastrophic wildfire in the interface. (Photo by Larry Korhnak.)

3.4.2 Assessment Themes

Throughout the course of this assessment, many themes emerged, crossing disciplinary boundaries. These themes were narrowed down to four principal areas that helped us to identify corresponding research and information needs in the interface.

3.4.2.1 Wildland–Urban Interface Issues Are about People

Wildland–urban interface issues are about people and their relationship with and effect on natural resources. Public perceptions, values, and attitudes affect land use and ultimately determine future forest management strategies and policies in the interface. Research can help us better understand and predict the intricate and complex relationship between people and natural resources.

3.4.2.2 Public Policy Plays an Important Role in Creating and Solving Interface Problems

Some public policies act to protect and conserve natural resources in the interface, while others provide incentives for urban development. Some policies may conflict with one another. Obtaining an understanding of and becoming involved with the various policies and decision making processes unique to the interface are critical for natural resource professionals. Natural resource professionals must also become involved by providing the best available scientific information to policy makers.

3.4.2.3 Interface Issues Are Interdisciplinary

This assessment demonstrates the crosscutting nature of interface issues and the need for interdisciplinary approaches in solving the complex issues within the interface. Building relationships across various disciplines and professions improves opportunities for addressing interface issues.

3.4.2.4 Issues Involve Multiple Ownerships, Jurisdictions, and Scales

Many natural resource management and conservation challenges are associated with multiple ownerships, jurisdictions, and issues related to scale. As land is subdivided, the increase in landowners and decrease in tract size present the need for a wider variety of management options to meet multiple objectives. There is also a lack of management techniques to address a variety of tract sizes. Multiple jurisdictions in the same region can implement different and often conflicting policies that complicate land use and management of forest resources. Ecological concerns often exist at landscape or watershed scales but may only be addressed at much smaller scales. These challenges are addressed most effectively when efforts are coordinated across the landscape and multiple stakeholder involvement is sought.

3.4.3 Research and Information Needs

3.4.3.1 Explaining and Adapting to Human Influences on Forest Ecosystems

The effects of land conversions, forest fragmentation, pollution, and nonnative species on forest ecosystem structure, function, composition, and processes need to be better understood. Research in these areas would help us to understand these effects of urbanization and to develop management techniques for multiple small-scale ownerships. Modeling and long-term monitoring that assesses these urban effects on ecosystems are also needed.

3.4.3.2 Identifying the Influences of Public Policy on Forest Ecosystems and Their Management

The relationships among public policy, land-use change, and resulting effects on forest ecosystems are still poorly understood. Research in this area could help us understand the roles, strengths, and weaknesses of various policies that affect natural resource management and conservation in the interface. We also need information about environmental quality indicators to identify long-term threats and prioritize environmental needs.

3.4.3.3 Identifying and Reducing Risk to Ecosystems and People in the Wildland–Urban Interface

Fire, invasive species, groundwater contamination, and other environmental changes can present risks for human and forest communities. Controlled experiments, historical studies, modeling, and long-term monitoring are needed so that we can better understand, predict, and avert risk.

3.4.3.4 Understanding and Communicating Public Attitudes, Values, and Perceptions

Knowledge of the diverse public preferences, values, and attitudes with respect to resource management and conservation is an important element to any natural resource program. Research in this area would help us understand how differences in age, ethnicity, and cultural backgrounds influence public use and management of forests (Figure 3.8). Natural resource managers and others could then use this information to develop effective communication strategies, education programs, and outreach messages.

FIGURE 3.8 Differences in age, ethnicity, and cultural backgrounds influence public use and management of outdoor recreation areas. (Photo by Larry Korhnak.)

Demographic research could help monitor and forecast urban expansion, economic development, and resulting human influences on the landscape.

3.5 CONCLUSION

The unique conditions, challenges, and needs of the wildland–urban interface call for an integrated and adaptive approach to natural resource management. New research is required so that we can better understand changing demographics, land-use patterns, and the resulting effects on forest ecosystems and their management. A greater public understanding of the complex relationships between people and natural resources is needed, and hence innovative approaches must be developed for disseminating information to the new and diverse landowners in the interface. The Southern Center for Wildland–Urban Interface Research and Information in Gainesville, Florida, will help meet these needs by providing new interface research, technologies, outreach programs, and educational material for managers, landowners, local governments, and others. Equipped with this knowledge, people can address interface issues and make informed decisions that will affect the future sustainability of wildland–urban interface forests.

ACKNOWLEDGMENTS

We would like to thank the members of the assessment steering committee, the Southern Wildland–Urban Interface Council, for their help in planning and advising this project from its inception. We also thank the planners of and participants in the focus groups.

REFERENCES

American Forests, 2002. Regional Ecosystem Analysis, www.americanforests.org/resources/rea. [Date accessed: April 5, 2002.]

Barlow, S.A., I.A. Munn, D. A. Cleaves, and D. L. Evans, 1998. The effect of urban sprawl on timber harvesting: a look at two southern states, *Journal of Forestry* 96: 10–14.

Berish, C. W., B. R. Durbrow, J. E. Harrison, W. A. Jackson, and K. H. Riitters, 1998. Conducting regional environmental assessments: The Southern Appalachian experience, in *Ecosystem Management for Sustainability*, J. D. Peine, Ed., Lewis Publishers, New York, pp. 117–166.

Boyce, S. G. and W. H. Martin, 1993. The future of the terrestrial communities of the southeastern U.S., in *Biodiversity of the Southeastern U.S.: Upland Terrestrial Communities*, W. H. Martin, S. G. Boyce, and A. C. Echternact, Eds., John Wiley, New York, pp. 339–366.

Bradley, G. A., Ed., 1984. *Land Use and Forest Resources in a Changing Environment: The Urban/forest Interface,* University of Washington Press, Seattle.

Brennan, L. A., R. T. Engstrom, W. E. Palmer, S. M. Hermann, G. A. Hurst, L. W. Burger, and C. L. Hardy, 1998. Whither Wildlife Without Fire? Transactions of the 63rd North American Wildlife and Natural Resources Conference 63; pp. 402–414.

Cordell, H. K. (ed.), 1999. *Outdoor Recreation in American Life: A National Assessment of Demand and Supply Trends,* Sagamore Publishing, Champaign, Il.

Cordell, H. K., G. T. Green, and C. J. Carter, 2002. Recreation and the environment as cultural dimensions in contemporary American society, *Leisure Sciences,* Special issue 24(1): 13–41.

Cordell, H. K. and E. A. Macie, 2002. Population and Demographic Trends, in Human Influences on Forest Ecosystems: The Southern Wildland–Urban Interface

Assessment, E. A. Macie and L. A. Hermansen, Eds., General Technical Report SRS-55, U.S. Department of Agriculture, Forest Service, Southern Research Station, Asheville, NC, pp. 11–35.

Daniels, T., 1999. *When City and Country Collide: Managing Growth in the Metropolitan Fringe,* Island Press, Washington, DC.

Davis, J. B., 1986. Danger zone: the wildland–urban interface, *Fire Management Notes* 47: 3–5.

Dedrick, J. P., T. E. Hall, R. B. Hull, and J. E. Johnson, 2000. The forest bank: an experiment in managing fragmented forests, *Journal of Forestry* 98: 22–25.

Duryea, M. D. and L. A. Hermansen, 2002. Challenges to Forest Resource Management and Conservation, in Human Influences on Forest Ecosystems: The Southern Wildland–Urban Interface Assessment, E. A. Macie and L. A. Hermansen, Eds., General Technical Report SRS-55, U.S. Department of Agriculture, Forest Service, Southern Research Station, Asheville, NC, pp. 93–113.

Dwyer, J. F., D. J. Nowak, M. H. Noble, and S. M. Sisinni, 2000. Connecting People with Ecosystems in the 21st Century: An Assessment of Our Nation's Urban Forests, General Technical Report PNW-GTR-490, U.S. Department of Agriculture, Forest Service, Pacific Northwest Research Station, Portland, OR.

Faulkner, G., J. Gober, J. Hyland, K. Muehlenfeld, S. Nix, P. Waldrop, and D. Weldon, 1998. Forests of the South, Southern Forest-Based Economic Development Council, Southern Group of State Foresters, www.southernforests.org/publications/forest-of-the-south/contents.htm. [Date accessed: April 5, 2002.]

Gramann, J. H. and M. F. Floyd, 1991. Ethnic assimilation and recreational use of the Tonto National Forest, Texas Agricultural Experiment Station, College Station, Contract report prepared under Cooperative Agreement PSW-89-0015CA between the Texas Agricultural Experiment Station and the Wildland Recreation and Urban Culture Research Project of the U.S. Department of Agriculture, Forest Service, Pacific Southwest Research Station.

Greene, J. L., 1995. State Tax Systems and their Effect on Nonindustrial Private Forest Owners, in Managing Forests to Meet People's Needs, Proceedings of the 1994 Society of American Foresters/Canadian Institute of Forestry Convention, September 18–22, 1994, Anchorage, Alaska, SAF Publ. SAF-95-02, Society of American Foresters, Bethesda, MD, pp. 414–419.

Greene, J. L., 1998. The Economic Effect of Federal Income Tax Incentives in Southern Timber Types, in Meeting in the Middle, Proceedings of the 1997 Society of American Foresters National Convention, SAF Publ. SAF-98-02, Society of American Foresters, Bethesda, MD, pp. 231–241.

Greene, J. L., T. Cushing, S. Bullard, and T. Beauvais, 2001. Effect of the Federal Estate Tax on Rural Land Holdings in the U.S., in Proceedings of Global Initiatives and Public Policies: 1st International Conference on Private Forestry in the 21st Century, Auburn University, Forest Policy Center, Auburn, AL.

Greenlee, J. M., F. McGarrahan, and T. Namlick, 1999. Wildfire Mitigation in the 1998 Florida Wildfires, Federal Emergency Management Agency, After Action Report, FEMA-1223-DR-FL.

Gregory, G. R., 1972. *Forest Resource Economics*, Ronald Press Company, New York.

Hull, R. B. and S. I. Stewart, 2002. Social Consequences of Change, in Human Influences on Forest Ecosystems: The Southern Wildland–Urban Interface Assessment, E. A. Macie and L. A. Hermansen, Eds., General Technical Report SRS-55, U.S. Department of Agriculture, Forest Service, Southern Research Station, Asheville, NC, pp. 115–129.

Johnson, D. W. and S. E. Lindberg, 1992. *Atmospheric Deposition and Nutrient Cycling: A Synthesis of the Integrated Forest Study,* Springer-Verlag, New York.

Kowarik, I., 1990. Some responses of flora and vegetation to urbanization in central Europe, in *Urban Ecology: Plants and Plant Communities in Urban Environments*, H. Sukopp, Ed., SPB Academic Publishers, The Hague, pp. 45–74.

Kundell, J. E., M. Myszewski, and T. A. DeMeo, 2002. Land Use Planning and Policy Issues, in Human Influences on Forest Ecosystems: The Southern Wildland-Urban Interface Assessment, E. A. Macie and L. A. Hermansen, Eds., General Technical Report SRS-55, U.S. Department of Agriculture, Forest Service, Southern Research Station, Asheville, NC, pp. 53–69.

Lindstrom, T., 1997. Land use planning in Virginia: The truth about the Dillon Rule, www.heyhom.com/Sprawl/1997/Vol1/DillonRule/dillon.html. [Date accessed: April 5, 2002.]

Loomis, J. B., L. S. Bair, P. N. Omi, D. B. Rideout, and A. Gonzalez-Caban, 2000. A Survey of Florida Residents Regarding Three Alternative Fuel Treatment Programs, Joint Fire Sciences Program Report, www.nifc.gov/joint_fire_sci/floridafinal.pdf. [Date accessed: April 5, 2002.]

Macie, E. A., and L. A. Hermansen, Eds., 2002. Human Influences on Forest Ecosystems: The Southern Wildland–Urban Interface Assessment, General Technical Report SRS-55, U.S. Department of Agriculture, Forest Service, Southern Research Station, Asheville, NC.

McDonnell, M. J., S. T. A. Pickett, and P. Groffman, 1997. Ecosystem processes along an urban–rural gradient, *Urban Ecosystems* 1: 21–36.

Minahan, K., 2000. The Upper Etowah River Watershed: Our Land, Our Water, Our Future: A Guide for Local Residents, Policy Makers, and Resource Agencies, Calhoun, Georgia, The Upper Etowah River Alliance, The Limestone Valley Resource Conservation and Development Council.

Moffat, S. O. and J. L. Greene, 2002. Economic and Tax Issues, in Human Influences on Forest Ecosystems: The Southern Wildland–Urban Interface Assessment, E. A. Macie and L. A. Hermansen, (eds.), General Technical Report SRS-55, U.S. Department of Agriculture, Forest Service, Southern Research Station, Asheville, NC., pp. 37–52.

Monroe, M. C., 2002. Fire, in Human Influences on Forest Ecosystems: The Southern Wildland–Urban Interface Assessment, E. A. Macie and L. A. Hermansen, Eds., General Technical Report SRS-55, U.S. Department of Agriculture, Forest Service, Southern Research Station, Asheville, NC, pp. 133–150.

Monroe, M. C., A. W. Bowers, and L. A. Hermansen, 2003. The Moving Edge: Perspectives about the Southern Interface, General Technical Report SRS-63, U.S. Department of Agriculture, Forest Service, Asheville, NC.

Myers, R. L. and J. J. Ewel, Eds., 1990. *Ecosystems of Florida*, University of Central Florida Press, Orlando.

Nelson, A. C., J. B. Duncan, C. J. Mullen, and K. R. Bishop, 1995. State and regional growth management approaches, in *Growth Management Principles and Practices*, Planners Press, Washington, DC, pp. 19–36.

Nilon, C. H. and R. C. Pais, 1997. Terrestrial vertebrates in urban ecosystems: developing hypotheses for the Gwynns Falls watershed in Baltimore, Maryland, *Urban Ecosystems* 1: 247–257.

Rubin, D. K., M. B. Powers, H. Carr, and D. B. Rosenbaum, 1993. A "whole" lot of planning going on, *Engineering News Record* 231: 38–44.

Sedell, J., M. Sharpe, D. Dravnieks Apple, M. Copenhagen, and M. Furniss, 2000. Water and the Forest Service, FS-660, U.S. Department of Agriculture, Forest Service, Washington Office, Washington, DC.

Smith, M. D. and R. S. Krannich, 2000. "Culture clash" revisited: newcomer and longer-term residents' attitudes toward land use, development, and environmental issues in rural communities in the Rocky Mountain west, *Rural Sociology* 65: 396–421.

Teaseley, R. J., J. C. Bergstrom, H. K. Cordell, S. J. Zarnoch, and P. Gentle, 1999. Private lands and outdoor recreation in the U.S., in *Outdoor Recreation in American Life: A National Assessment of Demand and Supply Trends*, H. K. Cordell, Ed., Sagamore Publishing, Champaign, Il, pp. 183–218.

U.S. Department of Agriculture, Forest Service, 2002. Urban tree ordinances, http://www.urbanforestrysouth.usda.gov/ ordinances/index.htm. [Date accessed: April 5, 2002.]

U.S. Environmental Protection Agency, 1992. Protecting the Nation's Wetlands, Oceans, and Watersheds: An Overview of Programs and Activities, EPA 840-S-92-001, U.S. Environmental Protection Agency, Office of Water (WH556F), Office of Wetlands, Oceans, and Watersheds, Washington, DC.

Wade, D., and J. Brenner, 1995. Florida's Solution to Liability Issues, in Fire Issues in Urban and Wildland Ecosystems: The Biswell Symposium, General Technical Report PSW-GTR-158, U.S. Department of Agriculture, Forest Service, Pacific Southwest Research Station, Walnut Creek, CA, pp. 131–137.

Wear, D. N., R. Liu, J. M. Foreman, and R. M. Sheffield, 1999. The effects of population growth on timber management and inventories in Virginia, *Forest Ecology and Management* 118: 107–115.

White, P. S. and S. P. Wilds, 1998. Southeast, in *Status and Trends of the Nation's Biological Resources,* M. J. Mac, P. A., Opler, and C. E. P. Haecker, Eds., U.S. Department of the Interior, Geological Survey, Washington, DC, pp. 117–129.

Woods and Poole Economics, Inc., 1997. *1997 Complete Economic and Demographic Data Source (CEDDS)*, Woods and Poole Economics, Washington, DC.

Zipperer, W., 2002. Urban Influences on Forest Ecosystems, in Human Influences on Forest Ecosystems: The Southern Wildland–Urban Interface Assessment, E. A. Macie and L. A. Hermansen, Eds., General Technical Report SRS-55, U.S. Department of Agriculture, Forest Service, Southern Research Station, Asheville, NC, pp. 73–91.

Part II

Planning and Managing Growth

4 Economic Values at the Fringe: Land Use and Forestry in the Wildland–Urban Interface

David W. Marcouiller
Department of Urban and Regional Planning, University of Wisconsin-Madison

CONTENTS

4.1 INTRODUCTION

The rapid urbanization of America during the latter part of the 20th century has created significant social, economic, and environmental demands on land and land-based resources. Although these demands and the change they bring about exist throughout America, they are particularly acute within the regions encompassing the wildland–urban interface surrounding major metropolitan areas. Residential, commercial, and industrial land needs have pushed outward from urban core areas like spreading wildfire, the extent of land-use change being largely determined by the winds of economic growth. These rapid changes in land use on the fringe have created a confusing array of forest management issues. Those responsible for managing forest resources are increasingly faced with a myriad of complex issues with which they are only partially equipped to deal.

The basis for this rapidly changing land use can be explained in several ways. Perhaps, the most useful explanatory approach lies within the theory and application of neoclassical market economics. The "invisible hand" of Adam Smith helps us understand the phenomenon of economic growth and provides a straightforward theory of resource endowments, firm decision making, and household utility. This explanation revolves around the central concepts of supply, demand, and efficient

1-56670-602-5/05/$0.00+$1.50
© 2005 by CRC Press LLC

market operation. The intent of this chapter is to provide an economic explanation for understanding rapid land-use change on the wildland–urban interface.

Indeed, the role of economics at the wildland–urban interface supplants other theoretical constructs associated with forested land-use change. There are no silvicultural or environmental theories that explain forest clearing for residential purposes. These are clearly anthropocentric phenomena explained by human-driven economic pressures. To understand land-use change, one must delve into the social sciences that explain macrointeractions between humans and their environment. This chapter is written to help the reader distinguish and separate alternative economic concepts of land and land-based resources such as productive use value, speculative development value, and amenity value. An appreciation for the more comprehensive set of values associated with land and land-based resources more completely captures the essence of this rather dramatic phenomenon of rapid land-use change.

Urban core areas of the U.S. and Canada provide markets and trading centers. These centers of dense human population serve as a huge mass of buyers and sellers, all interacting through the exchange of goods and services. Given competitive human tendencies for efficiency to maximize economic returns and minimize costs, metropolitan regions have grown dramatically during the past century. Urban core areas enjoy the relative advantages of efficient transportation networks, close proximity between willing buyers and sellers, and the natural tendencies of regional economies to agglomerate. This growth is further fueled by public policies that reward entrepreneurism and competition, characteristics of more conservative political regimes that have fostered relatively unfettered economic growth in the U.S. and Canada during the past decade.

Land as a primary factor of production and the resources found on and within it have provided a key basis for expansion of human civilizations throughout the ages and across the world. Although Leopold (1989) recognized that the abuse of land resources occurred due to the dominant view that land is a commodity belonging to its owner, it is rather naive to expect that in the current U.S. and Canadian political environment that serves to promote less-fettered economic market forces, society would characterize land as anything but something that is bought and sold. Only through conscious public and private planning does land use take on attributes of being something other than a private commodity. This is particularly important in understanding the changes taking place on forested lands within the wildland–urban interface and the implications for managing forests in these regions.

The unintended consequences of unfettered market forces have left their mark on regions surrounding metropolitan areas (Figure 4.1). Forests have been converted to residential neighborhoods, hills have been leveled to build highways, and these highways have eliminated open space to provide efficient transportation corridors. These corridors provide avenues for continued growth and development, and so on. Previously sleepy rural communities have been transformed by newcomers seeking relatively inexpensive land upon which to build their homes, relocate their businesses, and conduct exchange. This transformation has been a hallmark of late-20th-century North American economic prosperity. The nature of change in the wildland–urban interface has been both rapid and complete.

Silviculture, as a productive use of land, is superseded by other values associated with land and land-based resources. As will be shown, this is due to several economic aspects associated with land productivity, the value of timber products relative to other goods and services, transportation costs, and a host of other economic variables. In the wildland–urban interface, the economics associated with growing trees for timber have been dwarfed by the economics associated with open spaces, ready for development of new residential and commercial/industrial buildings. The

FIGURE 4.1 Urban market forces create commercial, industrial, and residential development land demands that supersede demands for land as a productive input for commodity outputs such as agriculture or forestry. The rate of this "exurbanization" process is determined largely by the rates of economic growth.

inherent value of forests within this rapidly growing region is dominated by the amenity values associated with healthy, high-quality, well-placed shade trees.

This rapid growth within once-rural communities has implications for a broad array of changes in economic, social, and environmental structures. Furthermore, there are differential impacts on people in these areas depending on initial endowments of resources. For instance, those who own land can stand to enjoy significant windfall benefits associated with land exchange and/or development. On the other hand, those without substantial endowments face steadily rising land prices, housing values, and tax bills with little more than fixed asset and income levels with which to afford increased costs of living. Although this chapter deals primarily with the microeconomics associated with land and land-based resources, it is important to remain aware of the broad scope of change being dealt with on the urban fringe.

This chapter is organized into four additional sections. First, I outline the basis for land-based resource values and the traditional valuation of land with focus on the concept known as land "rent." The next section deals with techniques developed to estimate land and resource values with a specific interest in incorporation of land and resource value into regional planning and policy making. The third section details tradeoffs that occur between economic perspectives and environmental perspectives, with particular interest in how forests are managed at the wildland–urban interface. Finally, I conclude with a section on policy tools and planning approaches that can affect how land-use change occurs.

4.2 THE VALUATION OF LAND AND LAND-BASED RESOURCES AT THE WILDLAND–URBAN INTERFACE

The value we place in land and land-based resources can, and does, take many forms. Using land for growing trees or for producing row crops represents one aspect of land value, namely, the use of land as an input to the production of commodities. To assess this value and distinguish it from other land-use values, we need to develop a conceptual approach that is spatially sensitive. In other words, we need a basis for evaluating land that directly accounts for the premium paid for land use based on distance to an urban core. This regional, or spatial, approach is fundamental to understanding why certain land uses are found in one location and not in another. This section outlines an early, and very simple, spatial approach useful in distinguishing land use throughout and across the wildland–urban interface.

4.2.1 The Spatial Economics of Alternative Land Uses

The economic story of land and land-based resources for production at the wildland–urban interface begins with the thinking of a well-known German academic named Johann Heinrich von Thünen, who lived from 1783 to 1850 in rural Germany. Von Thünen (1966) was the first to develop a basic analytical model of the relationships among markets, production, land use, and distance to trading centers. Land use, in von Thünen's world, was thought to be determined by the relative costs associated with transporting different commodities to the central marketplace (this and other topics are fully described in a classic text on land economics by Barlowe 1986). The most productive activities compete for the closest land, while less productive activities locate farther and farther away from the central place. Thus, distance to central place, the needs of human sustenance, and land productivity (both fertility and type of production) provide key elements associated with explaining alternative land uses.

Von Thünen perceived a rather obvious and regularized pattern of settlement throughout the countryside of rural Germany and used this perception to develop a generalized conceptual model of alternative land uses in his region. This perception provided the basis for explaining differing land-use types as a function of distance from a central market center. Our understanding of the spatial array of alternative land uses in the wildland–urban interface in modern-day North America follows a pattern strikingly similar to von Thünen's view of the countryside. The regular progression of land-use types and change in land-use patterns can be explained by their relative productive uses and transport costs.

4.2.2 The Concept of Land Rent

Humans, like all organisms on the planet, have necessary requirements to prosper, which include food, water, and shelter. The provision of these basic staples lies at the heart of the way we combine our labor and meager capital resources with land through production of commodities. As this story progresses, we will interject a fourth staple of life that was beyond von Thünen's grasp and represents improved standards of living — and is critical to bring this explanation up to modern times. This fourth staple involves quality of life or amenities and is increasingly important in locational decisions of firms and households. Forest use for amenity values and speculative land values will inevitably dictate land use in an expanding wildland–urban interface.

Settlement patterns and central business districts originate at crossroads of transportation infrastructure. In industrializing Germany during the late 17th century and early 18th century, this involved the intersection of navigable streams and cart or footpaths. Today, we see this agglomeration of market activities where railroads, highways, and ports exist. For illustration, Figure 4.2 outlines a stylized village where the distance from central market and infrastructure are key determinants of land use.

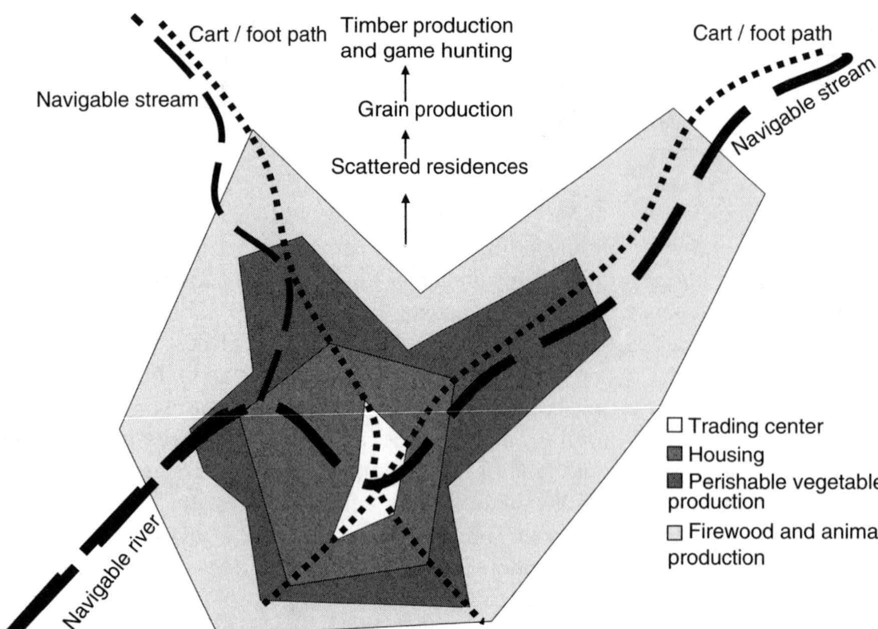

FIGURE 4.2 A von Thünen perspective of the stylized array of alternative land uses relative to a market center in rural Germany during the 1700s.

At transportation crossroads in 18th-century Germany, one could find retail and service sector businesses engaged in market activities. Just outside of this central place were the homes of merchants, service workers, and local residents. Because refrigeration had yet to be invented, von Thünen realized that the transport of goods sold in the central place required production in close proximity to the market. Thus, just outside of central marketplaces and just beyond residents' housing, he noticed a pattern of agricultural production. In particular, he noticed the production of those products that were highly perishable, like tomatoes, potatoes, carrots, and cabbage.

Another key need in sustaining the food needs and industrial development activities of human settlements is fuel production for cooking and industry as well as for preserving the perishable milk and cheese from dairy cows. This relatively lower level of site productivity was found just outside of the region where perishable food was produced. This change in land use with increasing distance to market center progresses to include the production of grains (swiddle agriculture) and forestry activities associated with timber production. The latter was required for construction materials to build homes, shops, and factories.

The economics behind settlement patterns and alternative uses of land focuses on the production activities of people, both individually and in teams (today's small-scale entrepreneurs and business interests). To simplify the explanation, it is useful to characterize production from the standpoint of costs. These costs involve the costs of growing or making salable goods, marketing goods to potential buyers, and transporting these goods to market for final sale. In a simple example, we can assume that the costs of production and marketing of goods remain relatively the same, regardless of the distance to market. What changes with distance are the costs of transporting the good or service to the market. Figure 4.3 outlines the concepts behind production cost. Given a fixed market price determined outside the region, distance affects transportation costs and identifies the relative profit associated with producing commodities at different locations. If production and marketing costs remain relatively stable with distance, the difference between market price and costs of delivering a good to final market (production, marketing, and transportation costs) identifies what is commonly referred to as *land rent* (specifically *von Thünen land rent*). The amount of land rent for production of a commodity decreases as the distance to the market center increases.

Note from Figure 4.3 that it becomes unprofitable for producers to engage in the production of commodity Y beyond the distance where the total cost of production exceeds the market price. This zone of production around a central business district has a limit identified as the *zero-rent margin*, or the point where land rent equals zero. Beyond this margin, the land rent associated with the production of the good or service is negative (e.g., costs now begin to exceed the returns from its sale).

If we now simply focus on land rent from the market center to the zero-rent margin, we can begin to build the picture of decreasing profitability (or land rent) in producing goods as we move farther and farther away from the market center. This notion of decreasing land rent by commodity with increasing distance is outlined in Figure 4.4. Note that this is simply a transposition of the land rent found in Figure 4.3 up to the zero-rent margin.

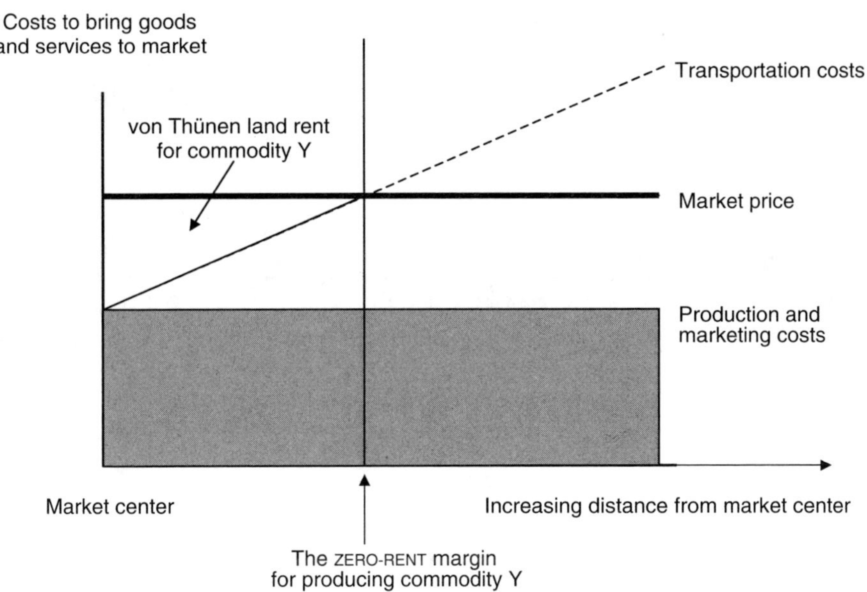

FIGURE 4.3 Typical costs of production include fixed production and marketing costs plus variable transportation costs. In the simple von Thünen approach, transportation cost increases with increasing distance, leading to a reduction in land rent as production takes place farther and farther from the central market.

It is important to note that different commodities will experience different costs of delivery to final market. For example, producing tomatoes (a highly perishable commodity) will be much more sensitive to distance from market than producing wheat or timber. Also, if we consider the fact that land for housing or industrial/retail uses necessarily requires proximity to market centers and supersedes the productive use of land for agricultural or silvicultural purposes, a gradient of land uses becomes evident, as shown in Figure 4.5. Here, we need to apply the notion that residences, offices, and factories require a highly intensive use of land (typically measured not in acres but in square feet) that is inextricably tied to the market center. An excellent example of this is the phenomenon of the skyscraper, which requires very little land per square foot of office space.

Thus, land uses will be arrayed spatially from the market center. In Figure 4.5, the region that surrounds the market center outward to the point labeled "a" will be dominated by land use for production of commodity X (e.g., housing and space for commercial/industrial activity). Note that land use will begin to change at point "a" as people realize that higher land rents can be obtained by switching from the production of commodity X to commodity Y. The concentric ring delineated by the distance from point "a" to point "b" will be dominated by the production of Y. This is simply due to the relatively higher "profits" or land rents that can be obtained by producing commodity Y. Likewise, land uses will switch at point "b" to the production of commodity Z because now land rents are higher with production of the alternative commodity.

There exists a seemingly limitless array of alternative land uses, each with a unique distance decay function that specifies the "premium" associated with proximity to the market center. If we were to superimpose each land use and its respective rent function with reference to distance

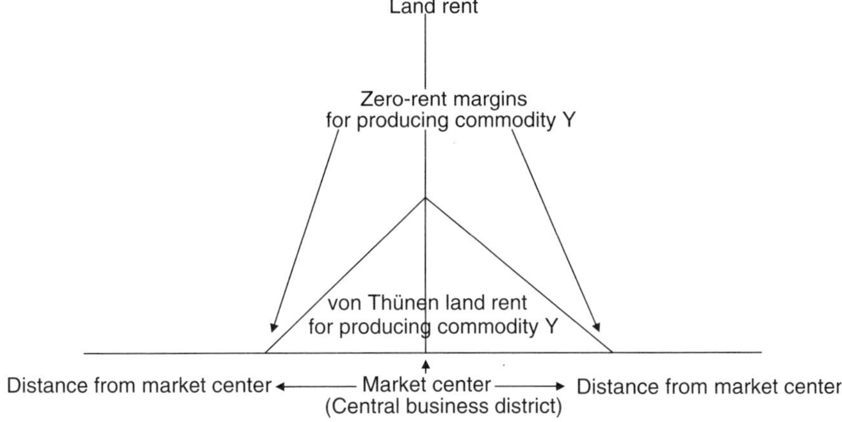

FIGURE 4.4 Von Thünen land rent by location of a single land use relative to the market center.

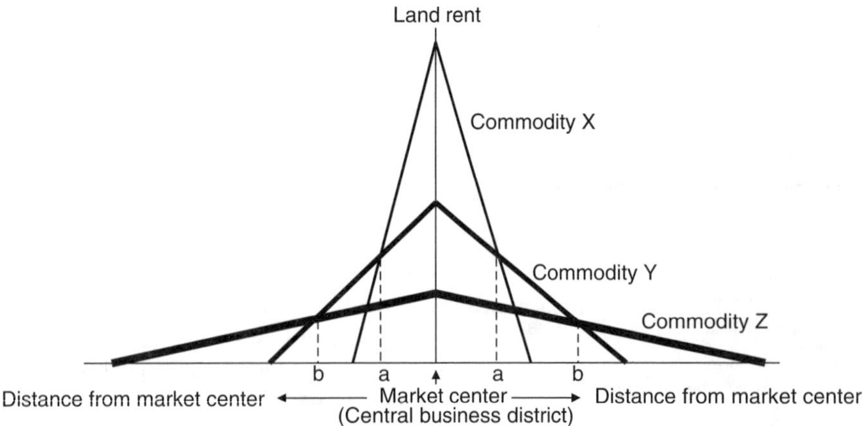

FIGURE 4.5 Von Thünen land rents producing three commodities relative to the market center. Note that a is the point where land-use change occurs between production of commodity X and Y, while point b is where land-use change occurs between production of commodity Y and commodity Z.

to market center, we would end up with a smooth distance decay cone that declines with increasing distance to market center. This is commonly referred to as a *bid-rent cone* that arrays alternative land uses by their respective land rents as a function of distance to the market center. This bid-rent cone is graphically presented in Figure 4.6.

Although American cities have progressed through several variants of this conical representation of land rent (e.g., urban decay in some cities during the 1960s and 1970s created a depression in the center of the cone with beltway highways surrounding cities determining maximum land rents), this simple representation helps us understand the relative boundaries associated with the spatial array of land-use activities surrounding cities. It helps us understand and spatially define the limits of urban sprawl at any given moment in time. This is particularly true given the notion that commuting distances, like perishable commodities, have an outward limit. For instance, few people are willing to commute for more than an hour to get to work. Given modern interstate highways, this limit is roughly identified at 40–60 mile from the urban center. Of course, this varies with highway congestion, itself a function of time of day that commuting takes place, size and efficiency of transportation networks, and type, or mode, of travel.

Alternatively, another early approach to land rent was developed by David Ricardo, an 18th-century economist, and has been aptly named Ricardian land rent (Barlowe 1986, p. 137). Ricardo developed the concept of land rent through the realization that land cannot be assumed to be of equal productivity (i.e., the fertility of soils is highly variable with topography and other determinants). Ricardian land rents arise because of different qualities of land. Although Ricardian land rent does capture fertility differences and fully utilizes the cost structure of production, it remains aspatial. That is to say, it does *not* deal with fundamental aspects of distance. These spatial gradients are critical to explain the development of urban areas and land-use alternatives on the urban fringe. Spatial explanations are necessary to develop the basic implications for alternative land uses and the distance decay of land rent away from urban core areas.

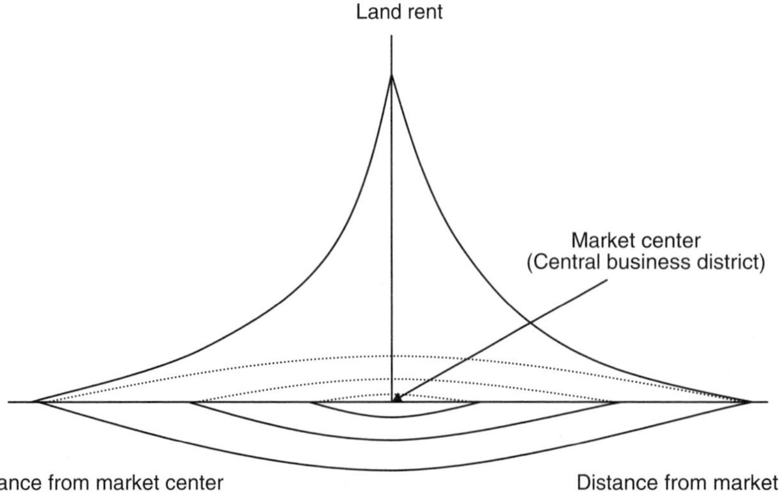

FIGURE 4.6 A typical bid-rent cone for land surrounding an urban area.

4.2.3 The Economic Basis of Urban Sprawl

Given this economic conception of urban influence on alternative land uses as we progress away from the city center, we now have a rather convenient and useful tool upon which to examine the issues associated with urban sprawl. In order to understand land-use change in the wildland–urban interface, it is important to realize that market forces sort out alternative land uses by their respective "highest and best" uses (i.e., rational economic competitors will have a tendency to locate the production of outputs that use land in a manner that minimizes the relative costs of land). Agriculture and forestry, as land uses, are lower in value than residential, commercial, and industrial uses. Furthermore, the bid-rent gradient associated with agriculture and forestry is nearly flat relative to urban land uses. Thus, we can superimpose a flat bid-rent gradient for forestry within the urban bid-rent cone as shown in Figure 4.7. Given static conditions, land-use change will occur from urban uses (residential, commercial, and industrial) to rural uses (agriculture, forestry, open space) where the bid-rent function yields the highest and best use (denoted in Figure 4.7 as the edge of the city).

Now, for the sake of argument, let us assume that economic growth occurs in this urban area. Economic growth often stimulates labor markets, resulting in lower unemployment and higher levels of income. With growth, better jobs become plentiful, new residents are attracted to the region, more residents create further demands for residential, commercial, and industrial lands, and so on. In a nutshell, each respective bid-rent surface experiencing higher demands shifts upward. Although agriculture and forestry experience some higher demands, their rate of change is outstripped by land uses characterized by more elastic distance decay functions. The edge of the city now moves outward (denoted in Figure 4.7 as the new edge of the city). The distance between the old city edge and the new city edge can now be referred to as "sprawl." Certainly, this oversimplification neglects a host of additional locational issues but provides us a rudimentary economic explanation of urban sprawl.

Other additional locational issues associated with land rent gradients involve publicly owned goods such as infrastructure and amenities. Recently, economists have been working to incorporate the presence (or absence) of amenities into estimates of land rent (cf. Lancaster 1966; Kohlhase 1991; Earnhart 2001). Similar to market centers, amenities can also be an important initiator of increased land rents. To more fully explore this notion, we must first briefly recognize some important aspects of

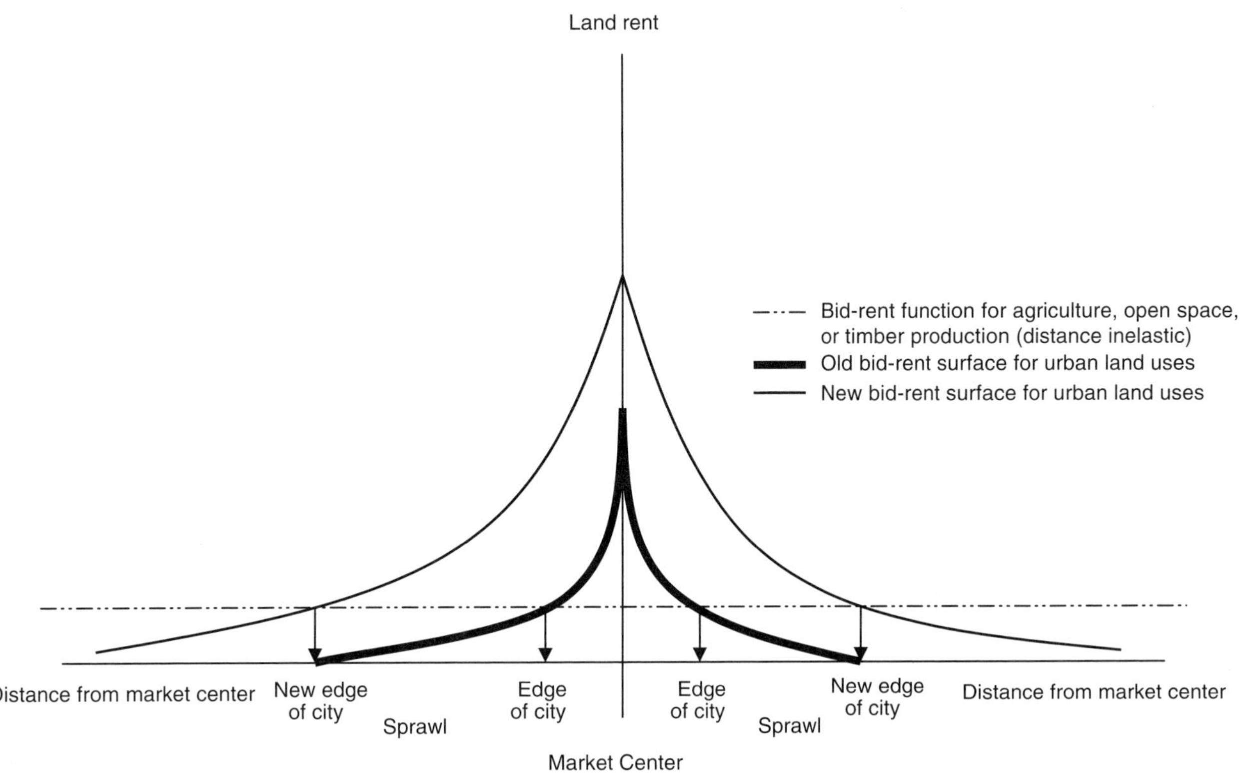

FIGURE 4.7 An economic perspective of urban sprawl based on growth-induced bid-rent functions. Additional economic growth shifts the bid-rent surface for urban land uses upward because of increased demands for residential, commercial, and industrial lands, and the edge of the city moves outward.

how contemporary economic thinking characterizes the amenity basis for land, land-based resources, and the use of land surrounding cities.

4.2.4 Alternative Values for Land and Land-Based Resources

Much of the explanation of change in land use along the fringe of urban America is due to recent increases in the value we place on nonmarketed goods and services associated with land. That is to say, the driver of suburban land-use change is land use for nonagricultural and nonsilvicultural purposes. It is important to understand that productive uses of land are just one of a wide variety of uses to which we put land. Economists have been addressing the variety of land uses and their respective values and have provided us a useful categorization scheme (Hodge and Dunn 1992; Hodge 1995). An overview of total economic value can be found in Figure 4.8.

4.2.4.1 Use Values: Direct

Note from this figure that what might be termed "productive" uses (i.e., those uses that rely on site productivity, endowments of minerals, or location as a basis for activity) are categorized at the far left as *direct use values.* These include commodity production for row crops, livestock, and timber. Also, extraction of mineral endowments provides another example of direct use value. Estimating direct use values associated with the production of these commodities is relatively straightforward and involves the quantity of commodity produced multiplied by the market price of the commodity at the time of sale. The value of land in this case is a function of its ability to produce commodities.

Managers of commodities falling within this category attempt to maximize returns and minimize costs of production given site productivity and some predetermined time frame associated with sustained production. For example, forest managers interested in a sustainable harvest of timber into perpetuity will account for a forest stand's rotation age (largely determined by site productivity) and some objectively based preference for the time value of money to determine both the type and level of timber extraction. From an economic perspective, the "profitability" of using land for timber production is based on initial growing stock levels, wood prices, and the direct costs of extraction (including land rent largely determined by stand accessibility and distance to market). From a strictly timber perspective, alternative and less market-oriented goods are absent from this calculus.

Another example of direct use value of land is the development of land for residential, commercial, industrial, and infrastructural uses. Indeed, these are direct applications of land value, but in most circumstances, the portion of total value related to land is small and is generally not related to the productive capacity of soil or the endowment of mineral wealth located beneath the site. The key element of land associated with these types of direct use values has to do with available space in a relevant location for direct development. Relevant locations are often wrapped up in tradeoffs between proximity to market centers, infrastructure, and amenities or quality-of-life measures. Once again, the estimation of value in this circumstance is reflective of the amount of land involved and the market-based price of land with

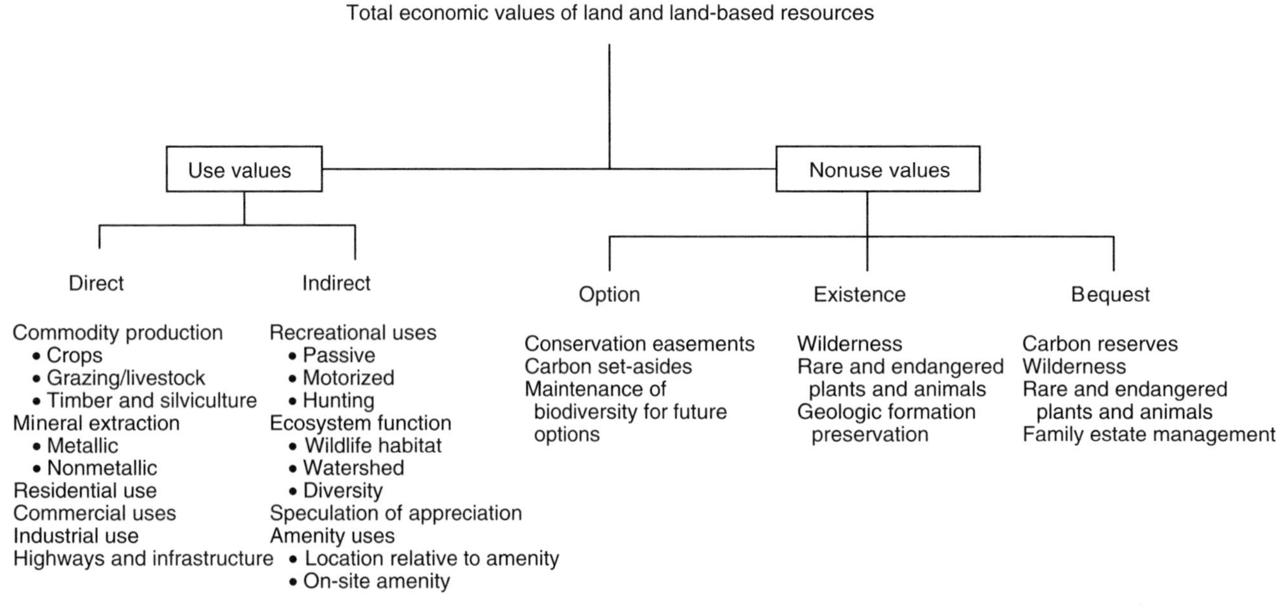

FIGURE 4.8 An economist's outline of total economic value: land and land-based resources.

respect to the competitive aspects of available land and demand for land. In regions surrounding urban centers, stiff competition for land and limited supply often lead to dramatically inflated land prices and resulting values that often greatly exceed those values derived from using the land for commodity production. This is particularly so with respect to prices for land that contains or is near concentrations of amenities. In other words, the "highest" and "best" use of land in urban areas is for commercial, residential, and industrial purposes. These purposes supersede the use of land for production of commodities.

4.2.4.2 Use Values: Indirect

In addition to these direct use values, however, less direct forms of land exploitation lead to what is termed *indirect use value.* Indirect use valuation is more complex given the various intervening mechanisms that relate to economic activity within local communities. Included within this category are recreational land-use values. Estimating the value of land used for recreation has followed several alternative methods. A straightforward approach involves estimation of traveler expenditures for items purchased in local markets (see Box 4.1). Additional approaches to estimate the value of recreational use include the development of stated and revealed preference models (further discussion can be found later in this section).

Another indirect use value associated with land falls within a category broadly referred to as amenity uses. Closely related to recreation, this form of land use relates to the hedonic, or pleasurable, aspects of land and land-based resources. Scenic resources associated with the outdoors are key elements of amenity use and include several different types that can be categorized as sociocultural amenities and natural amenities. Sociocultural amenities include historic sites and special events such as festivals. Natural amenities include regional features such as lakes, forests, mountains, and other recreational sites. The presence of sociocultural and natural amenities and their significant positive relationships with regional economic growth have been shown in recent studies (English et al. 2000; Deller et al. 2001).

Yet another example of indirect use value is for the maintenance of operable ecosystems (typically referred to as ecosystem function). Specific examples of this type of land use include the tacit incorporation of wildlife habitat or watershed concerns within silvicultural prescriptions. Increasingly, these ecosystem function roles of land and land-based resources take on important land management considerations and have provided the primary objective of public (local, state, and federal) agencies and private not-for-profit institutions in situations where critical ecosystems need to be maintained.

Box 4.1

Determining the Indirect Use Value of Forests for Recreation

Nonmarketed goods and services are an increasingly important output of forests. This is particularly so within the wildland–urban interface and in rural areas close to major cities. A primary nonmarketed service of woodlands is represented by forest-based outdoor recreation. Under the commonly encountered conditions of open access and/or public ownership, there are neither operating markets within which forest-based outdoor recreation is traded nor mechanisms in place to discover prices for these forest service outputs.

Forest-based recreation is only indirectly connected to markets. This indirect connection to markets can be represented by the activity of tourism-type establishments found within or around forests. Thus, the economics of forest-based recreation represents indirect use value and can be approached using the characteristics of forest-based recreation that affect this indirect market. Namely, how recreationists spend money can be used to infer the value they place on forests for recreation.

In a recent study of Wisconsin forests (Marcouiller and Mace 1999), the indirect use value of woodlands for recreational purposes was estimated using an expenditure approach. Using surveys of forest-based recreationists, researchers collected data on important characteristics of recreational spending patterns. Examples included characteristics of the visitors themselves (from where did they originate, where and how did they recreate, and other demographic characteristics) and how these visitors spent money. This latter set of spending characteristics included the extent (how much), type (sectoral distribution), and location (where spending took place) of spending.

Survey results suggested several avenues that more fully outline the alternative values we hold for forests. In addition to attitudinal and recreational use attributes, survey results suggested that spending by forest-based recreationists exceeded $5.5 billion per year in Wisconsin, roughly $2.5 billion of which was spent locally within 25 mi of where the recreational use took place (within or close to forested lands). Spending varied by user group and is shown in Figure 4.9. Although motorized users had the highest aggregate amounts of recreation-related spending, passive users (campers, hikers, and silent sports enthusiasts) spent the most locally (within or close to forested lands).

Individually, primary user types provided meaningful and statistically significant differences that helped to disaggregate the rich detail of recreational

—*Continued*

Box 4.1 — (continued)

use. The recreational types assessed included hunters, motorized users, and passive users. Overall, each group exhibited very different characteristics that extended beyond spending patterns, including unique perceptions about forestland use, attitudes about economic development, and sociodemographic characteristics.

While this study identified gross patterns of indirect use value for forest-based recreation in Wisconsin, there remain many unanswered questions. How do changing social and demographic characteristics of the North American population affect alternative values for forests? How do telecommunications and the dynamic aspects of economic distance bear on forest use in suburban, exurban, and very rural regions? There is a clear need to continue further research to provide a more comprehensive understanding of total economic values associated with forests in the wildland–urban interface.

4.2.4.3 Nonuse Values: Option, Existence, and Bequest

Perhaps the most challenging aspects of both management and valuation of land and land-based resources are the relatively less tangible aspects of economic value associated with providing resource service functions for current existence, future options, and transference of resources among generations. These are classified in Figure 4.8 as *nonuse values* and represent an important, albeit less tangible, aspect of land and land-based resources. Examples of these values from a land-use practice perspective include the basis for preservation efforts to set aside tracts of land for wilderness, bioreserves, and carbon sequestration efforts.

From a valuation method perspective, the stated preference models discussed in the following section can provide evidence and estimates of people's demand for these services. It is important to note that estimates of nonuse value can, and often do, greatly surpass levels of use value. There are many explanations for this, but one important point of reasoning comes from the notion that nonuse values are often societal in nature and reflect the overall value placed on resources by society as a whole, while use values are more specific to individuals, private concerns, and focused interest groups.

4.2.5 Characterizing Nonmarket Goods and Services

An overview of land and land-based resource values would not be complete without a discussion that distinguishes, in a very crude way, a significant problem associated with land-based resource valuation. This dilemma often crops up with land and land-based resources due to the rather ill-defined nature of property rights with respect to the array of goods and services demanded from land. Certainly, property rights with respect to the land itself are relatively straightforward as a result of standard land-surveying techniques, record keeping, and the historical development of real estate institutions. Much less clear, though, are the rights that people have over what can be done with land.

Property ownership has often been described as encompassing a "bundle" of rights. This bundle can extend to include mineral rights for endowments found beneath the land, air rights above the land, and the surface use rights for agriculture, forestry, or alternative uses such as residential, commercial, and industrial purposes. Toward the end of this chapter, we will develop a discussion of land-use tools. Many of these tools use this notion

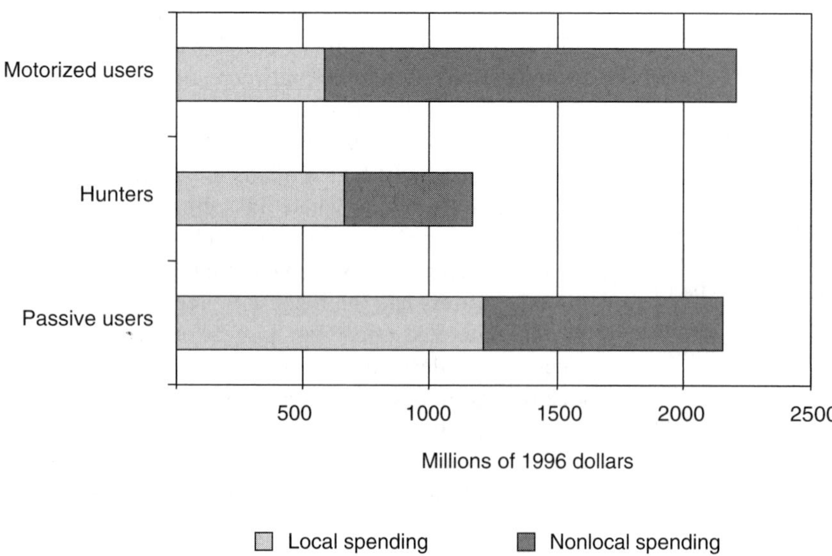

FIGURE 4.9 Spending by forest recreationists in Wisconsin varied by the type of recreationist and the location of spending. Local spending (defined as within 25 mi of the forest-based recreational use) was highest for those recreationists who could be categorized as passive (campers, hikers, and silent sports enthusiasts). (From Marcouiller and Mace 1999.)

of a bundle of rights, especially the idea that regulations can affect various aspects of this bundle. For instance, land-use zoning specifies the nature of surface rights with respect to land development. With zoning, the freedom to decide about these development rights rests with the publicly elected decision makers of the jurisdiction within which the land falls. This notion of separable rights builds on the unique characteristics of land and land-based resources that we will discuss now as *rivalry* and *exclusion*.

The economic values we place on land and land-based resources are, in large part, determined by their characteristics as goods. Property rights affect the ability of a resource to be considered a good because property rights deal with ownership. Ownership of a resource allows the owner to exclude use to those who are not willing to pay. Thus, excludability is an aspect that often presents problems with valuing land-based resources. Another aspect that affects a resource's consideration as a good has to do with how competitive the environment is with respect to using the resource. We sometimes refer to this as the level of "rivalness" that a good exhibits. The Latin root of this word is *rivalis,* which literally means "one using the same stream as another." A rival good is one in which the addition of another user acts to diminish the original user's value. A nonrival good is one that all can enjoy without diminishing the other user's value. Once again, in characterizing resource value, it is important to bear in mind that land-based goods vary in their level of rivalness.

Different types of land and land-based goods can be identified with respect to their levels of exclusivity and rivalry. As shown in Figure 4.10, the type of good is largely determined by the extent to which the good is excludable and rival. By the very essence of increased demands on a limited supply of land, many land values are sought competitively (characterized as rival in nature). In other words, for many of the uses we put to land, additional use has a diminishing effect on others' value. For example, an additional recreationist in a forest is likely to diminish the value of others using the forest. In general, people view more solitary experiences with land in higher regard; thus, values for fewer users are often higher than that with crowded use. As shown in Figure 4.10, most land-based recreational uses view the land itself as a "common property resource."

Certain final products rely on exclusion within a competitive environment. For instance, timber harvested from a forest is considered a *private good.* The value we place on this type of good, however, is typically a function of the costs of processing, delivery, and marketing, not so directly related to the land itself. *Club goods* are sometimes applicable with land resources but, again, relate to the packaging of an experience to an individual. An example of this type of good might be the value an individual places on an exclusive lease for hunting. Finally, there are some values that primarily relate to what was discussed earlier as nonuse values. These are typically referred to as purely *public goods* where exclusion is not possible and use is not rival. For instance, the values people place on the simple existence of untouched wilderness or in certain endangered species is an example of a purely public good. Another's value for this simple existence will not affect an individual's own value (nonrival). Also, an individual's existence value is not something from which the individual can be excluded.

Once again, most of our indirect-use values associated with land and land-based resources view the land itself as a *common property resource.* Publicly owned parks and forests can be used to illustrate the characteristics of a common property resource. Because these public lands are open

Level of rivalry	Level of exclusivity: Exclusive	Level of exclusivity: Nonexclusive
Rival	Private land-based goods • Timber production • Amusement parks • Mineral extraction	Common-property goods • Lakes and river systems • Public forests • Public parks
Nonrival	Club goods • Resorts • Golf courses • Heli-skiing	Purely public goods • Aesthetics • Sunsets • National defense

FIGURE 4.10 Characteristics of land and land-based resources.

to the public, no one can be excluded from using them (although user fees may be charged). Eventually, however, as more and more people use these lands, congestion becomes a problem. Consequently, each individual's enjoyment of the property will be reduced. This is in stark contrast to the items that we purchase in the marketplace.

Common property resources exist in between purely public goods and private goods. The nonexcludable nature of many land-based resources makes it nearly impossible for market forces (the purchase and sale) to operate due to the presence of what are commonly referred to as *free riders*. Free riders create a situation where no one is willing to pay for the use of the resource if other people are able to have access to it without paying. This is only partially the case with the problem of free riders. Indeed, people may be willing and able to pay, but there exists no market mechanism to extract payment.

As we have seen, in the case of private goods, the market determines who gets to use a resource and who does not by granting use to the highest bidder. But in the case of many land-based resources, this only works when a market allows trading to take place. Without trading, the market is not able to place a value on uses that are not producing any sort of saleable good. Thus, we have to rely on a combination of market valuation as well as nonmarket valuation of land use to determine which uses are the most valuable to society. Nonmarket valuation techniques, used for many years to determine the value of alternative uses of land and land-based resources, can be divided into two primary categories: stated preference models and revealed preference models.

4.2.5.1 Stated Preference Approaches

This aptly named category of methods for estimating the demand for nonmarket goods and services relies on surveys of people who are involved in some aspect of the nonmarket good. It allows people to directly state their preference for a nonmarket good or service. The key elements of interest in these types of approaches typically involve a two-step process. These two steps involve developing a reasonable market for trading a nontraded good and then querying people to estimate what they would be willing to pay for receipt of this good or service. Contingent valuation and contingent ranking are two specific methods that fall into this category of estimation techniques (a good overview of these methods can be found in Boyle and Bishop, 1988).

To be sure, there are many possible problems and biases associated with this type of approach. Those who have been applying this method have worked to minimize several key types of bias. Examples of problems with these types of surveys include *strategic bias* (or the ability of a respondent to strategically affect the outcome of the research), *hypothetical bias* (or the inability on the part of the respondent to understand the contingent market being set up), and *starting point bias* (or the inability of the respondent to respond accurately given the absence of any reasonable benchmark upon which to conceive of value). Although beyond the scope of this chapter, specific correction mechanisms have been instituted to allow use of stated preference methods as a reasonable approach for assessing the demand structure of nonmarket goods.

4.2.5.2 Revealed Preference Approaches

Once again, this is an aptly named category of research methods that attempts to uncover evidence of people's value for nonmarket goods through how they spend money in "proxy" markets. In other words, these approaches try to develop an indirect estimate of people's revealed preference for nonmarket goods or services by examining how they spend money in the marketplace for closely related market-based goods or services. Examples of specific techniques within this category include hedonic price models and travel cost models. Again, given the size limitation of this chapter, a complete discussion of these approaches remains beyond our scope. Interested readers can obtain a more thorough detailing of nonmarket goods valuation from benchmark texts in natural resource economics (cf. van Kooten 1993; Lesser et al. 1997; Kahn 1998).

We now turn our attention to a discussion that places valuation of land and land-based resources within the political and decision making framework of how land-use planning takes place within regions surrounding rapidly growing cities. In this way, nonmarket goods and services can be held up against market-based land uses and the complex array of stakeholders involved in weighing the benefits and costs of decisions about alternative land uses in the wildland–urban interface.

4.3 AFFECTING LAND USE IN THE WILDLAND–URBAN INTERFACE THROUGH PLANNING

Directing the course of land use within the region surrounding urban areas is the primary responsibility of land-use planners. Land-use planners are typically public servants of local units of government (cities, towns, countries, and regional planning commissions) or private consultants working for themselves or for larger engineering firms. Where resources are available, these local units of government will employ a cadre of planners to perform the necessary analytical and process-oriented tasks. Consultants often perform a significant number of the specific tasks associated with land-use planning. This is particularly true in smaller governmental units that do not have the resources to support a planning staff and in situations where specialized expertise is required.

Land-use planning involves a myriad of actors, and requires a significant amount of stakeholder involvement, technical analysis, and political decision making. An important aspect of land-use planning at the wildland–urban interface involves both fiscal and project assessments for costs and benefits. Economists and their analyses provide but one input into a very complex and politically motivated landscape. In the end, it is often the case that politics remains the principal driver of local decision making regarding land and land use. The planner's role is to provide locally elected officials with the best information possible in order that decisions can be made objectively and rationally. The ability of land-use planning analysts to identify winners and losers among an array of alternative land-use decisions is meaningful only to the extent these affected people and groups are involved in the political process. That is to say that estimates of costs and benefits in evaluating specific projects provide only one input into how decisions are made.

There are several categories of evaluative tools used by planners to assess the economic tradeoffs associated with land-use decisions. These analytical tools (broadly defined) include project-level cost–benefit analysis, fiscal impact assessment, economic impact analysis, and an array of accounting tools. Although viewed through various academic lenses as being distinct lines of research and inquiry, each has a very specific role in assisting planners with analysis of alternatives with respect to land-use decisions.

Cost–benefit analysis encompasses numerous analytical approaches that are applied to evaluate costs and returns at the project level. Ultimately, these approaches are dynamic, in the sense that they account for changes in both resources and the valuation of money over time. Key elements of cost–benefit analysis necessarily employ specific measures to capture how we value goods and services across various time frames. This temporal valuation approach raises important issues involving appropriate discount rates and the compound nature of time preferences with respect to economic value. Interested readers are referred to classic texts by Mishan (1976) and Pearce (1983) for full descriptions, caveats, and applications of cost–benefit analysis.

Given the complex nature of benefits derived from land and land-based resources, there is a need to incorporate stated and revealed preference approaches into benefit estimation. This is particularly true when land-use alternatives have differential impacts on the availability of nonmarket goods and services. The techniques of nonmarket valuation are slowly beginning to be applied but are dependent upon the skills available and the specific problems being addressed. As these techniques become more widely known and as the values they represent become more important, routine cost–benefit analysis to evaluate alternative land uses will inevitably represent a more comprehensive array of weighting mechanisms.

Fiscal impact assessment deals with decisions about land-use alternatives from the point of view of local government revenues and costs. Common questions that are analyzed in fiscal impact assessment include how one development option differs from another with respect to the local tax base. How does the potential for increased property taxes offset the need for increased services provided by the local unit of government? Public policy tools of local governments used to encourage certain forms of development are often determined through fiscal impact assessment, examples of which include the determination of development impact fees, levels of tax increment financing, and programs designed to offer property tax relief. Again, the full details and limitations of specific fiscal impact analysis techniques are beyond the scope of this chapter. Interested readers are referred to a classic text in fiscal impact modeling by Burchell and Listoken (1978).

Economic impact analysis provides an assessment of the gross benefits associated with land-use alternatives with particular interest in how changes in land use affect the overall economic activity of the region. Economic impact analysis incorporates specific tools that can be both descriptive and inferential. Descriptive techniques of economic impact analysis provide useful measures of economic activity that provide context for decision making. These include industry advantage measures such as location quotients, shift-share analysis, and export-base multipliers. Inferential techniques of economic impact analysis attempt to use available data on past and present economic activity to estimate current impacts of change and forecast future activity resulting from some policy. Examples of these techniques include input–output analysis, social accounting matrix analysis, computable general equilibrium analysis, and an array of econometric forecasting methods. Again, a complete description of economic impact analysis is beyond the scope of this chapter, but interested readers can learn more from classic texts in this topic written by Pleeter (1980) and Miller and Blair (1985).

It is important to note that there are several dilemmas associated with economic assessments in land-use planning within the wildland–urban interface that reinforce earlier statements made about the complexity of land-use values. Comprehensive assessments of land-use value are complex in the sense that nonmarket (indirect use values and nonuse values) goods and services of land-based resources are both critically important to capture and elusive to empirically estimate. There are several aspects of nonmarket goods and services that we do not fully understand, which make meaningful and comprehensive assessments nearly impossible at this point. To be sure, nonmarket valuation techniques are part of cost–benefit analysis, and measures of consumer and producer surplus help us understand societal welfare differences. But our ability to compare and integrate these quantitative values within a decision making framework remains somewhat limited. Furthermore, much of the interindustry analysis that makes up economic impact analysis remains at the market-based goods and services level. For instance, when we assess employment and income impacts associated with a land-use alternative, these remain at the level of earnings derived from labor and productive land/capital usage. Developing usable empirical measures to incorporate societal welfare change at the household level is a continual challenge to both academics and practitioners of economic development.

4.4 LAND-USE PLANNING TOOLS

There are often interests to maintain certain land uses within the wildland–urban interface that would not be represented were market forces to predominate. Again, this is the dilemma of land-use planners and, as they say, dilemma leads to innovation! During the recent past, increasing interest in providing an array of tools has led to some interesting approaches to land use along the wildland–urban interface. To a large extent, the progression of tools reflects the need to maintain land uses for production of nonmarket goods deemed socially desirable in a way that mimics market forces. These economic tools provide incentives for people to manage land in certain ways. Again, given totally unfettered market forces, these land uses would be superseded by other less socially desirable forms of land use.

Economic land-use tools are policies or programs that regulate land use or create incentives that encourage or assist individuals in exchanging rights in land, consistent with a set of broader land-use policy objectives. Economic incentive policies provide financial rewards (or penalties) for undertaking specified actions that support (or undermine) societal goals for land use. What follows is an outline of toolbox categories that provides a useful perspective to current land-use planning at the wildland–urban interface. First, public programs and regulatory policies that attempt to alter land use in accordance with publicly defined goals and objectives are outlined. These, then, lead to some quasi-public/private interaction tools that also affect land use but do so in accordance with more market-oriented goals and objectives.

4.4.1 Public Regulations

4.4.1.1 Zoning

Zoning (and related development ordinances) is a typical tool used by urban planners to appropriate land uses. Simply stated, zoning delineates areas, or zones, where certain activities are allowed to take place. In addition, zoning regulations and ordinances often specify details about the physical design of residential, commercial, and industrial parcels that must be adhered to by the owner in order for development to take place. Although important in directing urban form, zoning has limited value in addressing the whole-scale changes brought about by urban growth because it prespecifies highest and best uses of land. Zoning and ordinances rarely address the profitability of individual land uses within a given zone. In regions surrounding cities, zoning is often an after-effect of decisions about land-use change. One exception to this is the urban growth boundary (discussed in the next paragraph).

This said, it should also be noted that with minimal exception, zoning has not been widely applied throughout rural America. This is probably due in large part to political resistance to overt regulations that dictate how land can be used in rural areas. Also, it is important to note that although zoning can be implemented in concert with economic factors, rarely is zoning considered an economic tool for land-use planning. Rather, it provides a regulatory basis for development within already urbanized areas.

4.4.1.2 Urban Growth Boundaries

In response to dramatic urban growth and an overriding interest in maintaining open space and rurality outside of city limits, a very tight form of urban zoning control known as an *urban growth boundary* (UGB) has seen limited implementation. This unique form of land-use control is most widely known from the Portland, OR, example. Instituted during the early 1970s, this tool entails drawing a line around the expected boundary of a city and regulating development both within the delineated region and outside the region according to publicly determined land-use objectives. As part of Oregon's Statewide Planning Program, a UGB was defined for all 241 cities in Oregon. Since the 1970s, other areas (such as the San Francisco metro region, Charleston, South Carolina, and Knoxville, Tennessee, to name a few) have looked into applications of this regulatory tool.

Urban growth boundaries were developed with multiple objectives in mind. Most advocates of UGBs point to their success in providing limits to the extension of costly public services and facilities, the preservation of land for agricultural purposes, greater certainty for people who own, use, and invest in land at the edge of cities, and better coordination between city and county land-use planning. Critics often recognize these benefits but also point to the negative impacts that UGBs have by way of placing artificial upward pressures on land values within the UGB and artificial depression of land values outside of the boundary. Many argue that the tight controls of UGBs work against several socially desirable aspects of urban growth, such as the provision of affordable housing, both within the urban core and within the wildland–urban interface surrounding cities. Empirical work (Knapp 1985; Lang and Hornburg 1997; Phillips and Goodstein 2000) suggests mixed results and the notion that it is too early to draw conclusions about the efficacy of UGBs as a viable land-use tool.

4.4.2 Publicly Provided Incentives

The public, or societal, will is often administered through government incentive programs. A wide variety of programs have been established to address land management and land-use practice across the U.S. and throughout Canada. In addition to the federal government, state/provincial and local governments have a variety of incentive programs that target land use, broadly defined. Categories of publicly provided incentives programs include rental payment programs (e.g., the Conservation Reserve Program), use value assessments (reduction in property taxes that act to influence landowner decisions to retain forestlands), subsidies (e.g., Stewardship Incentives Program and state-administered cost-share programs), land purchase programs (e.g., federally administered Land and Water Conservation programs), and impact fees. With the exception of impact fees and certain state or municipal incentive programs, land management incentives are primarily targeted at rural land in general, not necessarily focused toward land and/or land use within the wildland–urban interface.

Incentives programs, whether they provide up-front rental payments, tax breaks, or cost-share payments, perform two essential tasks. First, they provide an incentive to landowners by lowering the landowners' relative costs of production. For instance, if, in return for meeting program objectives, the government provides financial assistance to landowners in the form of tree seedlings, technical expertise, annual rental payments, or tax breaks, landowners will be more willing to maintain land as a managed forest. This lowered cost of production may now allow woodland use for producing timber to better compete with alternative land uses from the landowner's financial perspective.

The second essential task of any incentive program is to translate societally determined wants and needs into land management action. For example, prior to landowners being given governmental financial assistance, program involvement often entails a landowner's agreeing to legally binding contracts or other enforcement mechanisms that specify how land will be used and managed. In this way, societally determined goals can be injected into how land is used.

Listed here as an incentive to logical development in the wildland–urban interface, *development impact fees* intend to shift the full cost of development onto those who develop and consume land within the wildland–urban interface. Development at the urban fringe often results from a failure to equitably translate the true costs of infrastructure (Brueckner 2000) to those who benefit from its existence. It is important to note that new residential developments at the urban fringe (and elsewhere) often create significant additional costs for service provision and facility development that are publicly offered by towns, municipalities, and other smaller units of government. Examples include sewer and water provision, roads, and sidewalks. Average cost pricing forces everyone in an urban area to pay for expensive extensions to existing systems. These resulting subsidies from urban center to development at the fringe provide incentives for outward expansion. Impact fees are fees assessed to developers in order to capture more fully the true (or marginal) costs of development.

Development impact fees are now a fairly standard item in most urban centers and attempt to assess the average costs of providing these services and facilities, as opposed to allowing developers to enjoy paying only the marginal costs associated with site-specific provisions. This is particularly important with sewer and water because any use of excess capacity within a system represents an incremental step toward very high fixed-cost facility upgrades. Simply stated, if developers use up all the excess capacity, the city will be forced to build a new sewage treatment plant or water treatment facility. Fees and regulations imposed on developers at the urban fringe can have important effects on the rate of development (Skidmore and Peddle 1998; Mayer and Somerville 2000), often resulting in a dampening of development pressure and creating more equitable relationships between cities and developers.

4.4.3 Quasi-Public Market-Oriented Partnerships

Several contemporary land-use planning initiatives have been developed that represent an attempt to split the various rights and responsibilities of land ownership. These are often initiated in a collaborative way between landowners, private special interest groups, and local units of government — thus they are included here as "quasi-public" partnerships. Also, land trusts have emerged as an increasingly popular quasi-public/private partnership, often intent on maintaining or preserving certain land uses. Land trusts are typically private, nonprofit organizations acting to acquire land or easements for protective purposes (Whittaker 1999).

These contemporary partnerships apply specific land-use planning tools that rely on the ability to separate rights to own land from the rights to development land. There are two categories of land-use tools that assign various land-use rights for the purpose of attaining open space or conservation demands. These include transferable development rights (or TDRs) and purchase of development rights (or PDRs). With both TDRs and PDRs, it is important to note that the tools rely on development rights that are independent of land ownership. The two approaches involve severing the right to develop land from the right to exclusive ownership.

4.4.3.1 Tradable (or Transfer of) Development Rights

TDRs are a useful land-use tool for attaining previously identified goals both to maintain open space and to foster more highly competitive urban development. TDR agreements provide a market within which development rights can be traded. For instance, if land-use planning has targeted one area as logical for development and another for maintenance of more natural landscapes and open space, a TDR structure can allow landowners in the more restricted zone to sell development rights to landowners in a development zone. The development zone landowner might be required to buy some extra development rights in order to develop the property or to increase the density of development. Thus, TDRs allow development to take place in one area while providing incentives for landowners to make decisions that are more in concert with societally determined wants and desires in another.

A key element of the TDR approach is that it relies on stable and well-recognized long-term plans of a community. The identification of sending areas (restricted) and receiving areas (areas for development) need to be clearly and unequivocally identified through an overall planning initiative (e.g., comprehensive planning) that is accepted, implemented, and under close control. The manner in which development rights are communicated often takes the form of permanent deed restrictions that are clearly specified and legally binding. These deed restrictions permanently separate the rights of owning land from the rights associated with development of that land. Often, these development restrictions are specified in the deed as *conservation easements*.

Although TDRs have been around for the past 30 or so years, some debate still brews over their efficacy. Clearly, though, developing markets for transferring development options presents a more market-driven approach to land-use planning. It has had the effect of making development restrictions more palatable to developers while permanently maintaining land as open space within the wildland–urban interface (Thorsnes and Simons 1999; Nickerson and Lynch 2001; Plantinga and Miller 2001).

4.4.3.2 Purchase of Development Rights

Another approach to affecting development within the wildland–urban interface is to provide financial incentives to landowners for their rights to develop land. Again, the essence of this approach relies on separating out the rights to own land from the rights to develop that land. Now, though, instead of trading the rights to develop for other parcels located within a development zone, these rights are purchased, hence the term "purchase of development rights," or PDRs. These payments for development rights can be in the form of outright payments to landowners or in the form of long-term tax breaks in return for restrictive deed language. Landowners agree, for a fee, to irrevocably restrict the deed to their land so as to prohibit certain uses.

The PDR approach has been successfully applied in several communities throughout the U.S. during the past 20 or so years and also represents a primary tool used by land trusts. Experience, however, has shown that wide-scale application of PDRs is an extremely costly endeavor for local, regional, or state units of government. With increasingly tight fiscal conditions and the devolution of an array of social programs, very few units of government can afford to make payments or forgo the tax revenue to implement PDR programs. Land trusts continue to utilize the PDR approach and do so only through effective solicitation and application of member contributions.

4.5 CONCLUSION: TRADEOFFS BETWEEN ECONOMIC GROWTH AND FOREST CONSERVATION

Economic growth has led to an array of land-use changes within the wildland–urban interface due to increased demands for residential, commercial, and industrial uses within agglomerated economies surrounding cities. The strongly positive correlation between increased demands for land and distance to the urban core helps identify the notion that distance, not soil productivity, plays a primary role in determining land use within rapidly growing regions. Hence, land use that is driven by soil productivity (such as timber production) will logically be superseded by more profitable uses, particularly in the wildland–urban interface of rapidly growing cities.

The essence of growth management is to balance the need for economic growth with other less market-oriented objectives of society. The maintenance of open space in these rapidly growing regions is often the basis for farmland or forestry conservation initiatives. Given rapid economic growth, the market alone will not support these uses. Indeed, many of the open space attributes demanded by society are of a nonmarket nature. The values of these nonmarket goods and services are difficult to quantify, but economists have begun to develop models and approaches to assist in decision making. To make the issue of land use in the wildland–urban interface more complex, land is also unique from the standpoint that the rights and responsibilities associated with managing or developing land are often distinct from the rights and responsibilities to own land.

Land-use planning is key to rational decisions about land use in the wildland–urban interface. Clearly, objective empirical assessments are critical to making good decisions. From an economic perspective, this analysis takes the form of fiscal impact assessment, cost–benefit analysis, and economic impact modeling. Several tools are available to the land-use planner that help in implementing land-use decisions. These include programs that offer both carrots (incentives) and sticks (regulations) to accomplish goals set out by policy makers. The arguments associated with regulation-based market inefficiencies has led to a couple of specific tools that work to create markets within which public land-use decisions can be made in concert with private decisions about development.

Even the best analysis-based toolbox, when implemented, suffers from some inherent spillover effects between land and the socioeconomic activities of urban areas. These unintended consequences of land-use planning efforts often create a spiraling effect on land values. Increased prices for land lead to inevitable difficulties for lower-income individuals and families and frequently work against public policies to foster affordable housing. Thus, land-use planning is not simply a problem with market efficiency. There is a fundamental need to address both efficiency *and* equity concerns of land use as the distribution of benefits from growth often accrues in a disproportionate manner to a minority of the stakeholders involved.

In closing, it is important to point out that the economic growth witnessed during the past decade has exceeded most predictions and that downturns in growth are inevitable as markets adjust to outside influences. Forestry and agriculture are, in many respects, residual land uses that will be strongly influenced by changing rates of growth. This is most noticeable where growth is most rapid. Until now, this rapid growth has been most obvious throughout urban America, especially in the wildland–urban interface surrounding cities. Rural America, however, is also beginning to see dramatic economic growth take place, particularly on sites directly adjacent to natural amenities such as waterfronts, vistas, and key natural landscapes. Many of the same economic forces, analytical techniques, and land-use tools discussed in this chapter also have direct application to rapidly growing amenity-based rural regions that are far away from urban centers. Our ability to balance economic growth with other societally determined wants and needs provides the challenge associated with managing growth. Sound economic, social, and environmental planning is necessary in order to maximize societal benefits while ameliorating the societal costs.

REFERENCES

Barlowe, R., 1986. *Land Resource Economics: The Economics of Real Estate,* Prentice-Hall, Englewood Cliffs, NJ.

Boyle, K.J. and R.C. Bishop, 1988. Welfare measurements using contingent valuation: a comparison of techniques, *American Journal of Agricultural Economics* 70: 20–28.

Brueckner, J., 2000. Urban sprawl: diagnosis and remedies, *International Regional Science Review* 23: 160–171.

Burchell, R.W. and D. Listoken, 1978. *The Fiscal Impact Handbook: Estimating Local Costs and Benefits of Land Development,* Center for Urban Policy Research, New Brunswick, NJ.

Deller, S.C., T. Tsai, D.W. Marcouiller, and D.B.K. English, 2001. The role of amenities and quality of life in rural economic growth, *American Journal of Agricultural Economics* 83: 352–365.

Earnhart, D., 2001. Combining revealed and stated preference methods to value environmental amenities at residential locations, *Land Economics* 77: 12–29.

English, D.B.K., D.W. Marcouiller, and H.K. Cordell, 2000. Linking local amenities with rural tourism incidence: estimates and effects, *Society and Natural Resources* 13: 185–202.

Hodge, I., 1995. *Environmental Economics: Individual Incentives and Public Choices,* Macmillan Press, Basingstoke, U.K.

Hodge, I. and C. Dunn, 1992. *Valuing Rural Amenities,* OECD Publication, Paris, France.

Kahn, J.R., 1998. *The Economic Approach to Environmental and Natural Resources,* Dryden Press, Fort Worth, TX.

Knapp, G.J., 1985. The price effects of urban growth boundaries in metropolitan Portland, Oregon, *Land Economics* 61: 26–35.

Kohlhase, J., 1991. The impacts of toxic waste sites on housing values, *Journal of Urban Economics* 30: 1–26.

Lancaster, K.J., 1966. A new approach to consumer theory, *Journal of Political Economy* 78: 311–329.

Lang, R.E. and S.P. Hornburg, 1997. Planning Portland style: pitfalls and possibilities, *Housing Policy Debate* 8: 1–10.

Leopold, A., 1989. *A Sand County Almanac,* illustrated by C. W. Schwartz, Oxford University Press, Oxford, U.K.

Lesser, J.A., D.E. Dodds, and R.O. Zerbe, Jr., 1997. *Environmental Economics and Policy,* Addison-Wesley Publishers, New York.

Marcouiller, D.W. and T. Mace, 1999. *Forests and Regional Development: Economic Impacts of Woodland Use for Recreation and Timber in Wisconsin*, Monograph G3694, Board of Regents of the University of Wisconsin System, Madison.

Mayer, C. and T.C. Somerville, 2000. Land use regulation and new construction, *Regional Science and Urban Economics* 30: 639–662.

Miller, R.E. and P.D. Blair, 1985. *Input–output Analysis: Foundations and Extensions*, Prentice-Hall, Englewood Cliffs, NJ.

Mishan, E.J., 1976. *Cost–benefit Analysis,* new and expanded edition, Praeger Publishing, New York.

Nickerson, C.J. and L. Lynch, 2001. The effect of farmland preservation programs on farmland prices, *American Journal of Agricultural Economics* 83: 341–51.

Pearce, D.W., 1983. *Cost–benefit Analysis,* 2nd ed., St. Martin's Press, New York.

Phillips, J. and E. Goodstein, 2000. Growth management and housing prices: the case of Portland, Oregon, *Contemporary Economic Policy* 18: 334–344.

Plantinga, A.J. and D.J. Miller, 2001. Agricultural land values and the value of rights to future land development, *Land Economics* 77: 56–67.

Pleeter, S., 1980. *Economic Impact Analysis: Methodology and Applications,* Martinus Nijhoff Publishers, Boston.

Skidmore, M. and M. Peddle, 1998. Do development impact fees reduce the rate of residential development? *Growth and Change* 29: 383–400.

Thorsnes, P. and G.P.W. Simons, 1999. Letting the market preserve land: the case for a market-driven transfer of development rights program, *Contemporary Economic Policy* 17: 256–266.

Thünen, J.H. von, 1966. *Isolated State* (an English edition of *Der Isolierte Staat,* edited by P. G. Hall), Pergamon Press, New York.

van Kooten, G.C., 1993. *Land Resource Economics and Sustainable Development: Economic Policies and the Common Good,* University of British Columbia Press, Vancouver.

Whittaker, M.S., 1999. Preserving open space on the rural–urban fringe: the role of land trusts, in *Contested Countryside: The Rural Urban Fringe in North America,* O. J. Furuseth and M. B. Lapping, Eds., Ashgate Publishing Company, Brookfield, VT.

5 Overview of Policies Influencing the Wildland–Urban Interface

Janaki R.R. Alavalapati
School of Forest Resources and Conservation, University of Florida

CONTENTS

5.1 INTRODUCTION

Human-induced factors are causing significant changes in the wildland–urban interface. The interface includes the edges of both large cities and small communities, where forests or other rural lands are adjacent to urban centers, areas where homes and other structures are intermixed with forests or other land uses, and islands of undeveloped lands within urban areas. From 1970 to 1990, the population of the U.S. increased by 36 million (24.2 percent), but the density of the urban population actually decreased by 23.2 percent because land in urban areas increased by 21 million acres, a 60 percent increase in total area. During the same period, about 400,000 acres of rural land per year were converted into urban areas (Garkovich 2000), and much of this land was forested. Alig et al. (2000) noted that about 14 million acres of nonindustrial private forests (NIPF) were lost to urban use between 1952 and 1997. Along with the decrease in forest area, the number of forest landowners is on the rise, suggesting an increase in forest fragmentation. Each year, about 150,000 new forest landowners are added to the list (Sampson and DeCoster 2000), many of whom may be families who are buying small parcels of forest and moving into rural areas to improve their quality of life (Egan and Luloff 2000).

While critics denounce the pace of urban sprawl and associated forestland loss and some prefer the use of regulations to prevent it, the American dream of home ownership and the value placed on private property rights favor urban development (Staley 2000; Garkovich 2000). In a 1998 poll conducted by *Desert News*, 83 percent of Utah residents either strongly or somewhat agreed that open spaces should be set aside now for future generations. Yet, 88 percent also believed that private-property owners should be allowed to do what they want with their own land subject to the zoning restrictions (Staley 2000). Reconciling this paradoxical situation and ensuring the

1-56670-602-5/05/$0.00+$1.50
© 2005 by CRC Press LLC

sustainability of forests at the wildland–urban interface is a complex task for which neither easy nor perfect solutions exist. This chapter reviews selected policies that affect the interface and the conservation and management of its forests. It begins with a summary of the effects of urban sprawl on forest management, the economy, and the environment. Next, it explores policies influencing the wildland–urban interface and forest management. Finally, it suggests approaches to influence policy development and draws conclusions relating to the interface.

5.2 IMPACTS OF URBAN SPRAWL ON THE USE AND MANAGEMENT OF FORESTS

Loss of forestland and forest fragmentation (which occurs when areas of contiguous forest landscapes are broken into small, isolated tracts surrounded by human modified environments; Figure 5.1) will have a series of negative impacts on the environment and the economy (Stevens et al. 1999). First, a decrease in timberland may cause a decrease in timber supply, and the forest products industries that contribute to regional economies will not be sustained in the long run. Barlow et al. (1998) found that proximity to urban land uses, higher population densities, and proximity to urban centers led to lower harvesting rates on forest plots. Also, Wear et al. (1999) concluded that the transition between rural and urban uses of forestlands in Virginia occurs where the population density is between 20 and 70 people per square mile, and population effects are likely to reduce commercial timber inventories between 30 and 49 percent. This suggests that loss of forestland and fragmentation will cause socioeconomic problems in many forest-dependent communities. Furthermore, with no sign of a reduction in the demand for timber products and growing restrictions on timber production in national forests, a decrease in domestic timber supply may create the need for imports to meet the domestic demand, thereby increasing timber harvest in other regions. This means that urban sprawl and forestland loss in the U.S. may have implications for tropical deforestation. Second, it is likely that residential sprawl into rural areas will make forest management more difficult because of nuisance complaints that are often brought against some activities (Garkovich 2000). For exurbanites, significant concerns can include pesticide over-spray, fertilizer use and runoff, prescribed burning, and noise from activities such as timber harvesting and transporting. Exurbanites also often possess strong preservationist attitudes that contrast with the conservation and utilitarian views about forest resources held by the original residents (Alig et al. 2000). Blahna (1990), for example, found that exurbanites in Michigan were more likely than original residents to support preservation-oriented policies. In northern Wisconsin, seasonal residents who live there primarily for recreation were more likely than original residents to support land-use controls (Green et al. 1996). Third, fragmented forests may be less healthy and more vulnerable to insect and disease attack and a higher incidence of exotic, invasive species (see Chapters 14 and 15 for more detailed discussion). For example, Florida has 67 exotic plant species that are invading and disrupting native communities (referred to as Category I species) (Florida Exotic Pest Plant Council 2003). Furthermore, many scientists note that forest habitat fragmentation is one of the greatest threats to wildlife survival and biodiversity worldwide (Rochelle 1999). Fourth, urban sprawl may actually cost local governments more than they are able to recover through property taxes (Garkovich 2000). Heimlich and Anderson (2001) noted

FIGURE 5.1 Forests are broken into smaller, unconnected parcels by land development and this fragmentation has important ecological and economic consequences. (Photo by Larry Korhnak.)

that residential development requires $1.24 in expenditures for public services for every dollar it generates in tax revenues, whereas farmland or open space requires only 38 cents for each tax dollar. Fifth, according to a survey conducted by Pew Charitable Trusts, sprawl tied with crime as two of the most pressing concerns for Americans at the local level (Sierra Club 2000). Finally, conversion of forests to urban use may be an irreversible action. It is difficult to restore urban land to its original state and as a result, future generations are deprived of the "option values" associated with forestlands. Also, both current and future generations will lose the biodiversity, ecosystem services, and esthetic benefits associated with forestlands.

5.3 POLICIES INFLUENCING THE WILDLAND–URBAN INTERFACE

A variety of factors are affecting the wildland–urban interface and the sustainability of its forests. For example, an increase in real income causes urban households to demand larger houses; advances in transportation technology and the expansion of roads and freeways into the suburbs reduce the time and cost of travel and encourage peripheral locations for business and households; and improvements in information technology have reduced the needs for business to be concentrated in urban centers (Anas 1999). Although these factors combined with population growth are forcing urban boundaries to expand, federal, state, and local governments across the nation are initiating policies to limit sprawl, conserve rural lands, and address wildland–urban interface issues. These policies include zoning regulations to stabilize urban boundaries, subsidizing infrastructure development in old urban centers to attract residents into urban areas, acquiring rural lands for preservation using tax dollars, buying conservation easements or influencing donation of conservation easements through tax policies, compensating rural landowners for generating environmental benefits, charging households the full cost of residential services, and other policies limiting regulations on forestry practices. Although the overarching objective of these policies is to limit urban sprawl and conserve farmlands and forestlands, their socioeconomic impacts may not be uniform. Policies that are cost-effective and attractive from a private property rights perspective, for example, may not be effective in limiting sprawl and conserving rural areas.

5.3.1 ZONING POLICIES: OREGON'S URBAN GROWTH BOUNDARY POLICY (1973)

In 1973, Oregon adopted a policy stipulating affordable housing within urban growth boundaries and the creation of protective zones outside the growth boundaries. The policy was intended to protect 25 million acres of farmlands and forestlands by establishing growth boundaries with a 20-year build-out capacity around all cities (Figure 5.2) (Libby 2001). According to this policy, families who wished to set up homes outside the boundary had to provide their own services, such as sewage. The consequences of this policy in Portland, for example, were many (Anas 1999; Libby 2001). First, the growth boundary limited development. Although Portland's population grew by 50 percent since the 1970s, its land area increased

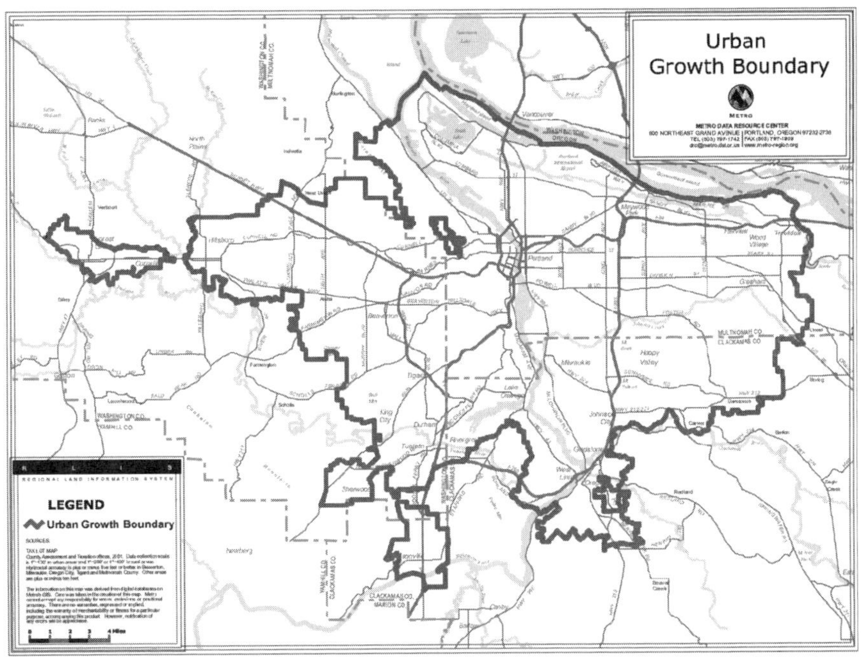

FIGURE 5.2 Urban growth boundary of Portland, OR, in 2002. The line separates developed land from agricultural and forested lands. Every 5 years, a regional council reviews the land supply within the boundary and, if necessary, expands the boundary to meet a requirement of a 20-year supply of land for residential development. (Courtesy of Metro.)

by only 2 percent (Bullard et al. 2000). Second, in some places within the growth boundary, prices of land increased by 400 percent and home prices went up by 40–80 percent. However, Phillips and Goodstein (2000) note that upward pressure on housing prices by the growth boundary was relatively small. They suggest that, while land prices increased due to the growth boundary, the upsurge in housing prices was more likely due to conventional market dynamics. Third, rental prices within the growth boundary increased by 40–80 percent, and many families with low income moved into surrounding rural towns. Also, the increase in rental prices may have increased the business cost of companies, which in turn may have had a depressing effect on the wage rate. With an increase in rental prices and a decrease in wage rate, the welfare of some families living within the boundary may have declined (Anas 1999). Fourth, landowners within the boundary made windfall gains unlike those who owned land outside the growth boundary. This policy favored households who sold their farms and forestlands for urban development and imposed costs on those who continue farming and forestry businesses by curtailing the prospects of future development on their lands.

Despite some limitations, growth boundary policies are generally favored over other policies such as taxes on households moving into semiurban areas because they are easy to enforce and their impacts on limiting sprawl are immediate. An implicit assumption behind these policies is that their overall societal benefits outweigh the societal costs. However, the downside is that benefits and costs of these policies are not distributed evenly across society and thus raise distributional equity concerns. Furthermore, in the long run, zoning policies may be less effective because they regulate rather than guide or accommodate real economic forces (Libby 2001).

5.3.2 Subsidizing Infrastructure to Urban Centers: New York's Smart Growth Economic Competitiveness Act (SGECA) of 1999

Many policies are designed to promote growth into areas with existing infrastructure. They include location-efficient mortgages for homes that are closer to public transportation or the center of the city, taxing buildings in existing communities at a rate lower than land, providing public transportation facilities through light rail, commuter trains, and high-speed buses, and subsidizing infrastructure in urban centers. Several Pennsylvania cities, for example, have formulated an innovative policy called a "split-rate tax," which specifies reduced taxes on buildings and increased taxes on land, encouraging development in existing communities and discouraging sprawl (Downs and Stanley 1995). Using a rail line that provides transportation to over 9000 commuters, the city of Englewood, CO, has attempted to redevelop a dying mall into a town center with movie theaters, shopping, residences, and the new town hall (Colorado Public Interest Research Group 1998). Along these lines, New York State has initiated the SGECA to provide subsidies for the development of infrastructure in urban centers.

The SGECA of 1999 directs future state infrastructure investments toward existing population centers and/or existing public infrastructure (Anas 1999). The SGECA attempts to limit urban sprawl by targeting subsidies to the central area and by denying them or reducing them to the periphery. Therefore, prices of properties in urban centers may increase relative to those of properties in the semiurban areas. (However, the increase in prices depends on the demand for land for urban development.) This is expected because subsidies to urban centers would reduce the cost of development relative to semiurban areas. As a result, there will be more development in and around urban centers and there may be an expansion of the urban area.

Anas (1999) casts doubts on the long-term success of the SGECA policy as follows. First, it is not certain that infrastructure development in urban centers will attract new residents, because people's preferences to live in urban centers may not be correlated with the number of facilities. Factors such as perceived crime in urban centers, quality of education, and tranquility in rural settings may have a significant impact on people's choice. Second, with growing affluence and technological developments in communication and transportation, it is unlikely that development in the semiurban area will be discouraged. Third, any program that depends on subsidies may not be financially sustainable. In the long run, an increase in prices of urban properties and an associated rise in rental values may cause families to move out of urban areas.

5.3.3 Land Acquisition Policies: A Case Study From Florida

Public support for acquiring and preserving environmentally sensitive lands is growing across the nation. In 1998, 72 percent of 240 ballot initiatives were passed authorizing \$7.5 billion (Garkovich 2000). Florida's efforts to acquire and conserve land lead the nation. In 1990, Governor Martinez called for a blue ribbon commission to look at what Florida needed to do in response to the conversion of about 19 acres per hour of forest, wetlands, and agricultural lands to urban uses. The commission estimated that about three million acres of rural lands would be converted to other uses by the year 2020, placing in jeopardy Florida's freshwater aquifers, unique ecological diversity, open space, recreational lands, and many endangered and threatened animals and plants. In response, Governor Martinez proposed a \$3 billion land preservation fund, based on \$300 million in yearly

FIGURE 5.3 Forestland purchased and protected under Florida's Preservation 2000 program. (Photo by Larry Korhnak.)

bonded funds over 10 years, to implement "Preservation 2000," the most ambitious land acquisition program in the nation. Under this program, by 1998 the state had acquired over one million acres for preservation from urban and other development (Figure 5.3). In 2000, the government reauthorized this program in the name of Florida Forever and approved the spending of $3 billion to acquire land and/or buy conservation easements during the next 10 years (Florida Department of Environmental Protection 2002). Many local governments in Florida have also initiated land acquisition and/or conservation easement programs similar to the state program. For example, in 1998, a referendum was passed by Alachua County residents to raise up to $29 million through property taxes to fund land acquisition.

Although land acquisition programs such as Preservation 2000 have made significant strides in preserving lands for current and future generations, they are facing many obstacles and criticisms.

- These programs are just too expensive for the state and counties to purchase enough land for conservation.
- The acquisition of private lands for public preservation causes a reduction in the tax base and government revenue.
- Acquired lands need maintenance, which implies that annual appropriations have to be made through tax dollars for their management. Furthermore, some people question the cost effectiveness of public management of acquired lands.
- Land acquisition programs and the associated public ownership of lands may not be consistent with the private ownership principle that many Americans cherish.

Because of these concerns, local, state, and federal governmental agencies are exploring innovative solutions, such as conservation easements, to acquire only partial property rights and leave lands in the hands of private owners for further management. In the following section, the role of conservation easements, the fastest-growing method of limiting urban sprawl and conserving farmlands and forestlands (Anderson 2001), is discussed.

5.3.4 Conservation Easements to Limit Urban Sprawl and Protect Rural Farmlands and Forestlands

The Uniform Conservation Easement Act (UCEA), which was approved in 1981 by the National Conference of Commissioners on Uniform State Laws, provides a framework to protect natural and historic resources by removing obsolete common-law defenses that might obstruct such efforts (Gustanski and Squires 2000). Diehl and Barrett (1988, p. 2) noted that "Conservation easements occupy an appealing niche in the array of land protection techniques — halfway between outright public or nonprofit ownership, at one extreme and government land-use regulation at the other…easements are tailored to the protection requirements of the particular property and to the desires of the individual landowner. Easements keep property in private hands and on the tax rolls and also carry a lower initial price tag than the outright acquisition. Easements are more permanent and often restrictive than land use regulation, which can shift with the political winds."

Some of the main difficulties associated with conservation easements are how they are qualified for federal income tax deduction, how they are treated by individual states, and how they are incorporated into the existing framework of real property law (Squires 2000). Florida has tailored its conservation easement policy to the political, social, and natural environment in which it exists (Anderson 2001). Florida enacted three pieces of legislation — Conservation Easement Enabling Legislation, Conservation Easement Assessment Legislation, and the Rural and Family Lands Protection Act (RFLPA), which was passed in 2001 — to encourage conservation easements. While the first describes the details of a conservation easement, its parties, and its enforceability, the second provides guidelines for the tax assessment of property encumbered by conservation easements. Florida Senate Bill 1992, or RFLPA, authorizes the Florida Department of Agriculture and Consumer Services to make payments for conservation easements, rural land protection easements, resource conservation easements, and agriculture protection agreements. The premise behind this program is that good stewardship of rural lands will further public goods, such as wildlife habitat and watershed protection.

Since 1964, the Internal Revenue Service (IRS) has allowed a federal tax deduction for the charitable contribution of a conservation easement. Congress continued to create incentives for conservation easements with the passage of the Taxpayer Relief Act of 1997, which included the modified version of the American Farm and Ranch Protection Act (AFRPA) that deals with estate taxes with respect to land subject to a qualified conservation easement. As of 2002, a taxpayer may gain federal tax benefits from the granting of a conservation easement in three ways. Section 170 (h) of the IRS code allows for an itemized charitable deduction in income tax. To be eligible for an income tax deduction under this section, "a contribution of a qualified real property interest" must be made "to a qualified organization exclusively for conservation purposes" (see Box 5.1). Section 2055 (f) allows for a reduction in the value of the land equal to the rights restrained by the conservation easement when assessing estate taxes. Finally, section 2031 (c) allows the conservation easement donor an additional exclusion from estate tax, beyond the exclusion allowed under section 2055 (f). To qualify for estate tax benefits, the easement must satisfy one of the conservation purposes as detailed in Conditions to Qualify for Estate Tax Benefits, except for historic preservation; the decedent or a family member must have owned the land on which the conservation easement was granted for 3 years prior to death; and commercial recreational activities must be discontinued, while commercial timber and crop production activities are permissible.

Sections 170 (h) and 2031 (c), along with 2055 (f) of the IRS code, create the backbone for conservation easement incentives. With more landowners becoming land rich and money poor, these financial incentives allow the landowners to retain their ownership while relieving much of the tax burden. Because of the potential of conservation easements to address wildland–urban interface problems, more research efforts are being devoted to them. Gustanski and Squires (2000) described the creation and use of conservation easements and provided details of conservation easement-enabling laws. Bick and Haney (2001) addressed the tax and financial aspects, how to determine the suitability of a conservation easement, and how to design an easement deed that addresses the needs of donor or seller. Anderson (2001) examined the legal aspects of conservation easements, focusing on selected southeastern states of the U.S. Further research relating to the comprehensiveness, exclusiveness, duration, and transferability of conservation easements is urgently needed.

Box 5.1

Conditions to Qualify for Estate Tax Benefits

A "qualified organization" must: (1) be organized in the U.S.; (2) operate exclusively for charitable purposes; (3) have no part of the organization's earnings benefit any individual or shareholder; (4) not be disqualified for tax exemption under section 501 (c) of the IRS code; (5) not be involved in the political campaign of any candidate for public office; and (6) normally receive the majority of its financial support from the government or public contributions.

"Conservation purposes" include: (1) the preservation of land areas for outdoor recreation by, or the education of, the general public; (2) the protection of a relatively natural habitat for fish, wildlife, plants, or an entire ecosystem; (3) the preservation of open space (including farmland and forestland) where the preservation is (a) for the scenic enjoyment of the general public or (b) pursuant to a clearly delineated benefit, or (c) will yield a significant public benefit; and (4) the preservation of a historically important land area or a certified historic structure.

Source: Anderson (2001).

5.3.5 Paying for Public Goods and Charging the Full Cost of Services to Control Sprawl and Conserve Rural Lands

If forests and farmlands are valued solely on the basis of market outputs such as timber and grain, the opportunity cost of their conversion to urban development will be low. In reality, the value of environmental and amenity services such as clear water, clean air, and biodiversity that rural lands generate is significant, and people are increasingly aware of these values (Poe 1999). For example, it is estimated that 20 years from now, traditional agricultural commodities in an Australian farm business will account for only 55 percent of revenues, and ecosystem goods and services, such as carbon sequestration, salinity control, and biodiversity, will account for the rest (Daily et al. 2000). Chichilnsky and Heal (1998) estimated that the provision of adequate clean water to New York City by forests in the Catskill Mountains was equivalent to a capital investment of \$6–8 billion and an annual \$1–2 billion in operating costs for a plant to carry out the same service. However, efforts to transform these nonmarket goods into market goods have been limited (see Buttoud [2000] and Glück [2000] for details on the full value of forests). If landowners are compensated for these services, such as through government transfers, they may not only produce more of these benefits but also bring more lands under farming or forestry. Daily and Ellison (2002) documented innovative ways to pay for ecosystem services that promote conservation efforts.

On the other hand, local, state, and federal governments subsidize residential services through various policies that promote sprawl. Tax policies that permit the deduction of nominal mortgage costs encourage suburban home buying by middle- and high-income households. Below-cost pricing of sewer and electricity services, new schools, increased police and fire protection, and public infrastructure such as roads and freeways provide an incentive for households to move into suburban areas. The fragmented nature of metropolitan governance causes suburban municipalities to compete with each other for the provision of public infrastructure, thereby contributing toward more suburbanization. Furthermore, it is not uncommon to see some communities close schools in existing neighborhoods and open new schools in semiurban areas. Between 1970 and 1990, Minneapolis-St. Paul built 78 new schools in the outer suburbs and closed 162 schools, some in good condition, within city limits (Sierra Club 1998). A study of the costs of sprawl in Washington State concluded that school costs were the number one hidden cost of sprawl in the state. In particular, the study estimated that, for the Issaquah School District, the cost of providing education was about \$18,600 for each new single-family home. The impact fees paid by developers to recoup the cost of providing services and structures, however, ranged from \$1100 to \$6140. Taxpayers paid the additional \$12,000 per household (Mazza and Fodor 2000). If developers and residents are charged the full and fair cost of bringing schools to new communities, and these communities are properly planned, the sprawl cycle may be broken.

The framework presented in Figure 5.4, which follows the well-known von Thünen model (see Chapter 4 for a detailed explanation), illustrates how land use for urban development and forestry is being influenced by below-cost pricing of residential services and lack of compensation for public goods such as environmental services. The *X*-axis measures the distance away from the urban center, while the *Y*-axis reflects the market price for land. In the figure, it is shown that the demand for urban use (line MS) decreases rapidly as the distance from the urban center increases because commuting costs rise and access to urban infrastructure declines. On the other hand, the demand for forestry use (line NT) does not decrease as rapidly as the demand for urban use. In particular, it is shown that beyond a certain distance from the urban center (R), it is no longer profitable to convert land to urban use, and it is more profitable to use the land for forestry.

Many government policies, such as the homestead exemption, affect the demand for urban use. Florida's constitution provides a \$25,000 homestead exemption from a property's assessed value if the homeowner holds a legal

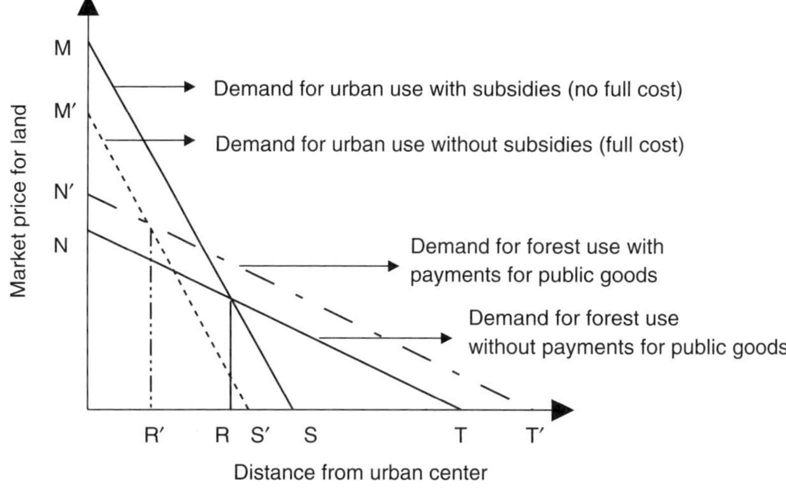

FIGURE 5.4 Effects of charging the full cost of residential services and paying for ecosystem services on demand for land for urban and forestry uses.

title to the property, maintains it as his/her permanent residence, and is a legal resident of the state as of January 1 of each year. Other policies including subsidized access to public infrastructure (e.g., roads, sewer and water systems, and recreation areas) and tax incentives on mortgage interest payments affect urban use. In particular, credit subsidies to help young families own their new single-family homes, a secondary mortgage market to increase funding for private mortgages in the suburbs, income tax deductions for mortgage interest, and property taxes that absorb some of the private cost of home ownership (Rusk in Libby 2001) are big boosters of the demand for urban development. In the absence of these policy incentives, the demand for urban land use would be much lower than the demand with policy incentives (line M′S′ vs. MS in Figure 5.4). On the other hand, the lack of incentives for societal benefits, such as biodiversity, water quality, carbon sequestration, and aesthetics produced by forestlands, cause the demand for forestland, line NT, to be lower than the demand when forestland owners are compensated for those benefits, line N′T′. With compensation payments, forestry would become more profitable and fewer forestlands would be converted to urban development (see Box 5.2 for a case study that explains how payments for carbon and biodiversity increase the value of forestlands). With no subsidization for urban development and with fair compensations for environmental benefits associated with forestry practices, the distance from the urban center at which it becomes more profitable to use the land for forestry decreases (R′ vs. R) (Figure 5.4). However, in contemporary society, we have many incentives for residential development and few for forestry, and as a result, land under urban use is increasing.

Box 5.2

Effect of Environmental Benefit Payments on Forestland Values

Alavalapati et al. (2002) estimated the profitability of a longleaf pine (*Pinus palustris*) forest (Figure 5.5) under two scenarios: (1) marketable timber and (2) timber, carbon sequestration, habitat for the endangered red-cockaded woodpecker, and other amenity benefits. They found that the value of land based on timber output alone is about $500 per hectare. If carbon sequestration benefits are considered at the rate of $50 per ton of carbon, the longleaf pine forestland value will increase to about $1350 per hectare. If red-cockaded woodpecker habitat benefits are considered along with timber and carbon benefits, the land values will increase further by about $300 per hectare, thereby making the total value about $1650 per hectare. In other words, the results show that carbon and biodiversity benefits will increase the value of longleaf forestland by more than 300 percent. Furthermore, if aesthetics and other amenity benefits are considered, the land value will increase even further. This increase in land values and associated profitability will provide incentives for landowners to practice forestry on their lands and even bring other lands into forestry.

5.4 OTHER POLICIES AFFECTING THE FORESTS AT THE WILDLAND–URBAN INTERFACE

A variety of policies beyond urban growth management policies influence forest management at the interface. These include forest fire management, water and air quality, and tax policies. A brief discussion of each follows.

5.4.1 Forest Fire Management Policies

Decades of fire suppression and the exclusion of fires from fire-dependent ecosystems are causing the accumulation of unprecedented levels of forest fuels (Hesseln 2001). In 1995, the Federal Wildland Fire Policy was revised to use wildfire to protect, maintain, and enhance forest resources and, as nearly as possible, allow wildfires to function in their natural ecological role (USDI/USDA 1995). However, as the number of people living in semiurban areas increases, the wildland–urban interface is becoming one of the most complex areas for fire managers to administer (Cortner et al. 1990; also see Chapter 13 for a detailed discussion of the challenges). Residents in the interface often oppose the use of prescribed burning for reasons including diminished aesthetics, liability, aversion to perceived risk, and smoke intrusion (Hesseln 2000). In particular, the growing public concern over smoke management has supported regulations about when, where, and how a forest may be burned (Figure 5.6). As such, the window of opportunity to safely conduct prescribed burns is narrow, and additional restrictions on smoke management narrow the window even further. This vicious cycle continues to affect the forests and residents in the interface.

Many state governments have attempted to address problems associated with prescribed burning. In Alabama, for example, the Prescribed Burning Certification Program has been initiated with an objective of reducing the liability associated with prescribed burning. Section 9-13-273 (a) of the law states: "No property owner, or his or her agent, conducting a prescribed burn in compliance with this article, shall be liable for damages or injury caused by fire, or resulting smoke, unless it is shown that the property owner or his or her agent failed to act within that degree of care required of others similarly situated." The 1999 Florida Legislature supplemented a previous bill with much greater

FIGURE 5.5 Longleaf pine forestlands once covered millions of acres in the southeastern U.S. Restoration of these forestlands is increasing, and if landowners were compensated for the ecosystem services provided by these forests, even more land would be returned to longleaf pine. (Photo by Larry Korhnak.)

FIGURE 5.6 Prescribed burning is an effective tool for reducing fuel loads in forests, but drifting smoke can raise public health and safety concerns and limit its use, especially in the wildland–urban interface. (Photo by Larry Korhnak.)

liability protection for certified burners. Although these policies assure greater care during prescribed burning at the interface, without public support and acceptance it is difficult for forestland owners to undertake prescribed burning. The public should be made aware that unmanaged forests cause an increasing danger over time for their property and lives (Beebe and Omi 1993). Information campaigns could change the widely held misconceptions about the hazards of prescribed fire (Winter and Fried 2000), and Chapter 13 gives many examples of effective programs.

5.4.2 Estate Taxes

A recent national survey questioned nonindustrial forest landowners who were involved in the transfer of a forestland estate between 1987 and 1997 (Greene et al. 2000). About 44 percent of these owners sold timber or land to pay part or all of the federal estate tax that was due, and about two thirds of the sales occurred because other estate assets were inadequate to pay the tax. Extrapolating from these results, Greene et al. (2000) estimated that in order to pay federal estate taxes, roughly 2.6 million acres of forestland are harvested and about 1.4 million acres are sold each year. Of the acres sold, approximately one fourth is converted to other uses.

A 1999 study conducted by the Mississippi State University, College of Forest Resources, and the USDA Forest Service examined the effects of the federal estate tax on Mississippi Forestry Association members (Mississippi Society of American Foresters 2000). Between 1987 and 1997, 192 landowners had been involved in an estate transfer. Forty-five percent of these estates exceeded the $600,000 exemption then in effect ($1 million in 2002), and almost 10 percent responded that timber was harvested or land was sold to pay federal estate taxes. The Mississippi Society of American Foresters concluded that federal estate tax policies present a great burden to many individuals who inherit forestland, leading to the premature liquidation of privately owned forests and/or forcing the sale of all or portions of the inherited land.

5.4.3 Water Quality Policies

In response to the Federal Water Pollution Control Amendments of 1972, also known as the Clean Water Act, the timber industry, along with individual states, developed a series of Best Management Practices (BMPs). Forestland owners associations and the timber industry maintain that their practices, through voluntary BMPs and the Sustainable Forestry Initiative, have made great strides in limiting nonpoint source pollution. Partly, this may be the reason why the Environmental Protection Agency (EPA) removed forestry from its list of sources of water impairment. As such, forestry practices are exempted from a permit requirement.

Recently, the EPA has been revising its interpretation of the term *point source* with respect to discharges associated with silviculture. As a result, it is possible that operations such as nurseries, site preparation, reforestation and subsequent cultural treatment, thinning, prescribed burning, pest and fire control, harvesting, surface drainage, or road construction and maintenance may be considered as sources of water impairment. If the EPA implements new total maximum daily load water rules requiring landowners to seek a federal permit for all forestry activities, then timber producers will perceive that their cost of production will increase. This might lead some landowners to divert their lands to alternative uses such as urban development.

5.5 POLICY DESIGN AT THE WILDLAND–URBAN INTERFACE

When creating policies that affect the wildland–urban interface, emphasis must be placed on collaborative and consensus-building processes. The National Governors Association has officially adopted the "Enlibra" doctrine, a philosophical framework, for addressing environmental issues and problems (Hirschorn 2000). It focuses on the use of collaborative processes as the principal mode for effective public participation and relies on eight interdependent principles (see Box 5.3). Enlibra's principles provide a useful framework when designing and initiating policies for the wildland–urban interface. A key ingredient that makes this approach unique and effective is recognition of the need to address multiple goals and obtain multiple benefits. This need is often blocked by the highly polarized and confrontational natural resource debates. Enlibra could bring diverse stakeholders together to focus on maintaining and improving the sustainability of natural resources at the interface and the quality of life of the local communities. Competing priorities for growth and measures to address the negative impacts of urban sprawl can be identified, and a shared vision of the desired future of the community can be developed using this collaborative process. Therefore, it is suggested that resource planners

Box 5.3

Enlibra Principles for Collaborative Stewardship

National standards, neighborhood solutions: assign responsibilities at the right level. As the Federal government sets national standards, it should consult with the states, tribes, and local governments as well as other concerned stakeholders to access data and other important information. When the objectives and standards are set, local governments should be allowed the flexibility to develop their own plans to achieve them and to provide accountability. Plans that consider local environmental, economic, and social factors have more public support and involvement and can reach the standards more efficiently.

Collaboration, not polarization: use collaborative processes to break down barriers and find solutions. Because of the highly polarized nature of environmental issues, unproductive confrontations tend to occur. "Successful environmental policy implementation is best accomplished through balanced, open, and inclusive approaches at the ground level, where interested stakeholders work together to formulate critical issue statements and develop locally based solutions to those issues."

—*Continued*

Box 5.3 — (continued)

Reward results, not programs: move to a performance-based system. Federal, state, and local policies should encourage "outside-the-box" thinking in the development of strategies to achieve desired outcomes. Solving problems, rather than just complying with programs, should be rewarded. Governments should also reward innovation and take responsibility for achieving environmental goals.

Science for facts, process for priorities: separate subjective choices from objective data gathering. Separate interests often hold polarized positions on environmental issues that interfere with reconciling the problem at hand. A public, collaborative process should be used to reach agreement on the underlying facts, as well as the range of uncertainty surrounding the issue. This process should include respected scientists and peer-reviewed science for quality control, and decision makers should use ongoing scientific monitoring information to adapt management decisions.

Markets before mandates: replace command and control with economic incentives, whenever appropriate. Market-based approaches and incentives may lead to more rapid compliance with environmental programs by rewarding environmental performance and encouraging innovation.

Change a heart, change a nation: ensure environmental understanding. Beginning with the nation's youth, people need to understand their relationship with the environment. Citizens need to understand that a healthy environment is critical to the social and economic health of the nation. Governments can reward those who meet their environmental stewardship.

Recognition of benefits and costs: make sure environmental decisions are fully informed. Economic factors, like the externalities imposed on those who do not participate in key transactions, and noneconomic factors such as equity within and across generations should be fully considered in the assessment of options.

Solutions transcend political boundaries: use appropriate geographic boundaries for environmental problems. The natural boundaries of the problem should be considered to identify the appropriate science, markets, and cross-border issues, as well as the full range of affected interests and governments that should participate and facilitate solutions.

Source: Hirschorn (2000).

and decision makers understand Enlibra's principles and consider them in addressing land growth management issues.

5.6 WORKING WITH LOCAL CITIZENS AND POLICYMAKERS

The success of a policy depends on how it is developed and implemented. Sound policies may not achieve their intended objectives unless they are developed and implemented in appropriate and innovative ways. Effective communication between resource managers and the residents of communities within the wildland–urban interface is a key component for successful management of forests in the interface (see Chapter 9). Local citizens must perceive resource managers as honest, trustworthy, and credible (Opio 1999). Community members must be made aware of forest management policies and practices and their relationship with the environment. By creating an atmosphere where local citizens, decision makers, industries, and resource managers can come together to share concerns and ideas, we can achieve consensus. A success story, Sunwood Lakes Green Belt Management (see Box 5.4), illustrates the importance of working with local people for successful management of forests at the interface.

Box 5.4

SUNWOOD LAKES GREEN BELT MANAGEMENT

In 1994, a resident addressed the Sunwood Lakes Homeowners' Association Board, in Olympia, WA, regarding the need to manage the forest to maintain forest health and to reduce future threats to life and property resulting from crowding and disease. Following the meeting, the board approached Jim O'Donnell, a resource manager who had a successful experience working with landowners with nontraditional forest management objectives, and sought his help in developing a Forest Stewardship Plan. In the development of the plan, the resource manager worked with the board in defining and ranking the owners' objectives for the forested green belts in Sunwood Lakes. Homeowner safety from fires and falling trees, forest health, aesthetics and recreation, wildlife habitat, and lake water quality protection were identified as the objectives. The board accepted the plan in 1995.

Shortly after acceptance of the plan, a set of demonstration activities and a homeowners' association tour were planned for some of the green belt areas. These included low-density commercial thinning activity immediately followed by placement of a variety of nest boxes and forage plantings to encourage wildlife habitation. Harvest activities and timber hauling were limited to normal school and business hours to minimize the impacts to the community. Timber-yarding was done by draft horses to minimize ground disturbance and damage

—Continued

Box 5.4 — (continued)

to residual trees and to further reduce noise and engine exhaust impacts to the community. Contrary to the expectations and fears of many (at one time, the board rejected a suggestion of daylight thinning in a community picnic area), the tour attendees' reactions to the demonstration activities were very positive. Following the demonstration-site presentations, similar activities continued in other areas until the end of the school year, when all harvesting activities were suspended to reduce the risk of accidents involving unsupervised children. Once some initial sites had been treated, most homeowners were very supportive of the operation. Many requested that the consultant also remove large trees on their lots in order to expose their homes to daylight and reduce the risk of windfall and fire.

The *Olympian* newspaper published two articles on Sunwood Lakes green belt management, featuring photos of the horses working close to homes. Local government agencies selected Sunwood Lakes for an Urban Forest Stewardship Field Day with the objective of introducing this innovative approach to resource managers and other community members. After visiting Sunwood Lakes green belts, other homeowners' associations hired O'Donnell to develop stewardship plans for their green belts.

Source: Meacham (2000).

5.7 CONCLUSIONS

Policy issues related to urban sprawl and forest management in the wildland–urban interface are complex. In this chapter, a range of policies, including zoning regulations, subsidizing infrastructure development in old urban centers, land acquisition programs, conservation easements, paying for ecosystem services, charging the full cost of residential services, and fire, estate tax, and water quality policies relating to the interface are reviewed. I have not attempted to provide a comprehensive treatment of each policy, but rather have suggested that each policy will have its own strengths and weaknesses, and the choice of an appropriate policy must depend on the nature of the resource, location, and societal preferences. A policy approach that has proven effective in one place may do a poor job in another because of the differences in socioeconomic and/or physical features of the regions. A number of factors must be recognized and considered in choosing one policy alternative over another.

- Intended and unintended consequences of a policy action must be assessed. Often, we run into a situation where the unintended consequences of a policy may outweigh its intended positive benefits. For example, the equity concerns that surfaced from growth boundary policy in Oregon may be serious. Policies prohibiting prescribed burning may have resulted in the accumulation of fuel loads, thereby posing a greater threat to homes at the interface.
- Short- and long-run impacts of policy alternatives must be examined by considering both market and nonmarket values. Policies that may be effective in the short run may not be effective in the long run. As noted earlier, subsidizing infrastructure in old urban centers may keep more households within urban centers for a short period of time, but in the long run this might increase the demand for urban living and thus may cause an expansion of the urban boundary or outmigration into semiurban areas. Consideration of nonmarket costs and benefits in policy development is a challenging task, but the gains from efforts to include them would be significant.
- The issues of the wildland–urban interface are multisectoral. Assessment of policy alternatives from a single-sector perspective assumes that changes in the interface do not affect other sectors of the economy. A variety of trade-off or multiplier effects may occur in response to interface policy actions. For example, conservation easement policies may not only conserve farmlands and forestlands but also promote biodiversity and related recreation and tourism activities. An increase in recreation and tourism activities will translate into additional employment and income in rural and urban communities. Failure to capture these spin-off effects may underestimate the benefits of conservation easements.
- Policies must be socially compatible and developed in collaboration with local communities. Innovative and entrepreneurial skills of local communities provide unique solutions to complex problems of the wildland–urban interface.

We must recognize that our belief and value systems influence the choice of policy alternatives. A century ago, policies such as the Homestead Act reflected our preferences for urban development and economic growth. Now, land acquisition programs and conservation easements reflect our preferences for the preservation of rural lands for current and future generations. It is hard to forecast what our preferences will be in the face of growing population pressure and technological advances and how they will shape our rural landscape. Given this uncertainty, we need to invoke the "precautionary principle" to guide our decisions relating to the wildland–urban interface.

ACKNOWLEDGMENT

Research assistance from Kristina Stephan and Enrique Anderson is greatly appreciated.

REFERENCES

Alavalapati, J.R.R., G.A. Stainback, and D.R. Carter, 2002. Restoration of the longleaf pine ecosystem on private lands in the U.S. South: an ecological economic analysis, *Ecological Economics* 40: 411–419.

Alig, R.J., B.J. Butler, and J.J. Swenson, 2000. Fragmentation in Private Forest Lands: Preliminary Findings from the 2000 Renewable Resources Planning Act Assessment, in Fragmentation 2000: A Conference on Sustaining Private Forests in the 21st Century, September 17–20, 2000, Annapolis, MD, L.A. DeCoster, and R.N. Sampson, Eds., Sampson Group, Alexandria, VA, pp. 34–47.

Anas, A., 1999. The cost and benefits of fragmented metropolitan governance and the new regionalist policies, *Planning and Markets* 2, http://www-pam.usc.edu/. [Date accessed: June 30, 2002.]

Anderson, E.R., 2001. Conservation Easements in the Southern U.S.: A Look at Issues Arising out of Laws, Relationships, and Land Uses, Unpublished Technical Report, University of Florida, Gainesville.

Barlow, S.A., I.A. Munn, D.A. Cleaves, and D.L. Evans, 1998. The effect of urban sprawl on timber harvesting, *Journal of Forestry* 96: 10–14.

Beebe, G.S. and P.N. Omi, 1993. Wildland burning: The perception of risk, *Journal of Forestry* 91: 19–24.

Bick, S. and H.L. Haney, Jr., 2001. *The Landowner's Guide to Conservation Easements*, Kendall/Hunt Publishing Company, Dubuque, IA.

Blahna, D.J., 1990. Social bases for resource conflicts in areas of reverse migration, in *Community and Forestry: Continuities in the Sociology of Natural Resources*, R.G. Lee, D.R. Field, and W.R. Burch, Jr., Eds., Westview Press, Boulder, CO, pp. 159–179.

Bullard, R.D., G.S. Johnson, and A.O. Torres, Eds., 2000. Introduction, *Sprawl City: Race, Politics, and Planning in Atlanta*, Island Press, Washington, DC.

Buttoud, G., 2000. How can policy take into consideration the "full value" of forests? *Land Use Policy* 17: 169–175.

Chichilnsky, B.D. and G. Heal, 1998. Economic returns from the biosphere, *Nature* 391: 629–630.

Colorado Public Interest Research Group (CoPIRG), 1998. http://www.coPIRG.org/.

Cortner, H.J., P.D. Gardner, and J.G. Taylor, 1990. Fire hazards at the wildland urban interface: what the public expects, *Environmental Management* 14: 57–62.

Daily, G.C. and K. Ellison, 2002. *The New Economy of Nature: The Quest to Make Conservation Profitable*, Island Press, Washington, DC.

Daily, G.C., T. Söderqvist, S. Aniyar, K. Arrow, P. Dasgupta, P. Ehrlich, C. Folke, A. Jansson, B. Jansson, N. Kautsky, S. Levin, J. Lubchenco, K. Mäler, D. Simpson, D. Starrett, D. Tilman, and B. Walker, 2000. The value of nature and the nature of value, *Science* 289: 395–396.

Diehl, J. and T. Barrett, 1988. *The Conservation Easement Handbook: Managing Land Conservation and Historic Preservation Easement Programs*, Trust for Public Land and Land Trust Exchange, San Francisco, CA.

Downs, A. and A.K. Stanley, 1995. D.C. should learn from Pittsburgh, *Washington Post*, September 24.

Egan, A.F. and A.E. Luloff, 2000. The exurbanization of America's forests: research in rural social science, *Journal of Forestry* 98: 26–30.

Florida Department of Environmental Protection, 2002. Preservation 2000/Florida Forever, http://p2000.dep.state.fl.us/vbwmemo.htm. [Date accessed: June 30, 2002.]

Florida Exotic Pest Plant Council, 2003. List of Florida's invasive species, http://www.fleppc.org/03list.htm. [Date accessed: August 8, 2003.]

Garkovich, L., 2000. Land Use at the Edge: The Challenges of Urban Growth for the South, Southern Rural Development Center Report No. 13, Mississippi State University.

Glück, P., 2000. Policy means for ensuring the full value of forests to society, *Land Use Policy* 17: 177–185.

Green, G.P., D. Marcouiller, S. Deller, D. Erkkila, and N.R. Sumathi, 1996. Local dependency, land use attitude, and economic development: comparisons between seasonal and permanent residents, *Rural Sociology* 61: 427–445.

Greene, J.L., T. Cushing, S. Bullard, and T. Beauvais, 2000. Effect of Federal Estate Tax on Non-industrial Private Forest Holdings in the U.S., in Fragmentation 2000: A Conference on Sustaining Private Forests in the 21st Century, September 17–20, 2000, Annapolis, MD, L.A. DeCoster and R.N. Sampson, Eds., Sampson Group, Alexandria, VA, pp. 142–144.

Gustanski, J.A. and R.H. Squires, Eds., 2000. *Protecting the Land: Conservation Easements Past, Present, and Future*, Island Press, Washington, DC.

Heimlich, R.E. and W.D. Anderson, 2001. Development at the Urban Fringe and Beyond: Impacts on Agriculture and Rural Land, Agricultural Economic Report 803, Economic Service, U.S. Department of Agriculture.

Hesseln, H., 2000. The economics of prescribed burning: a research review, *Forest Science* 46: 322–334.

Hesseln, H., 2001. Refinancing and restructuring federal fire management, *Journal of Forestry* 99: 4–8.

Hirschorn, J.S., 2000. Smart Growth and Enlibra, in Fragmentation 2000: A Conference on Sustaining Private Forests in the 21st Century, September 17–20, 2000, Annapolis, MD, L.A. DeCoster and R.N. Sampson, Eds., Sampson Group, Alexandria, VA, pp. 25–33.

Libby, L.W., 2001. Policy issues and options at the rural–urban interface: a national perspective, *Southern Perspectives* 5: 12–13 and 15.

Mazza, P. and E. Fodor, 2000. Taking its Toll: The Hidden Costs of Sprawl in Washington State, Climate Solutions, http://www.climatesolutions.org. [Date accessed: June 30, 2002.]

Meacham, S., 2000. Draft Horses, Bird Boxes and Other Keys to an Urban Interface Success Story, in Fragmentation 2000: A Conference on Sustaining Private Forests in the

21st Century, September 17–20, 2000, Annapolis, MD, L.A. DeCoster and R.N. Sampson, Eds., Sampson Group, Alexandria, VA, pp. 215–222.

Mississippi Society of American Foresters, 2000. MS-SAF Position Statement on Estate Taxes, http://www.cfr.msstate.edu/mssaf/taxes.html. [Date accessed: June 30, 2002.]

Opio, C., 1999. Forest management issues in a wildland–urban interface: the case of West Bragg Creek timber license in Alberta, *The Forestry Chronicle* 75: 129–139.

Phillips, J. and E. Goodstein, 2000. Growth management and housing prices: the case of Portland, Oregon, *Contemporary Economic Policy* 18: 334–344.

Poe, G.L., 1999. Maximizing the environmental benefits per dollar expended: an economic interpretation and review of agricultural environmental benefits and costs, *Society and Natural Resources* 12: 571–598.

Rochelle, J.A., 1999. Conference summary statement, in *Forest Fragmentation: Wildlife and Management Implications*, J.A. Rochelle, L.A. Lehman, and J. Wisnieski, Eds., Brill, Leiden, Netherlands, p. 3.

Sampson, N. and L. DeCoster, 2000. Forest fragmentation: implications for sustainable private forests, *Journal of Forestry* 98: 4–8.

Sierra Club, 1998. Sprawl: The Dark Side of the American Dream, http://www.sierraclub.org/sprawl/report98/. [Date accessed: August 8, 2003.]

Sierra Club, 2000. Sprawl Costs Us All, http://www.sierraclub.org/sprawl/report00/. [Date accessed: August 8, 2003.]

Squires, R.H., 2000. Introduction to legal analysis, in *Protecting the Land: Conservation Easements Past, Present, and Future*, J.A. Gustanski and R.H. Squires, Eds., Island Press, Washington, DC, pp. 69–77.

Staley, S.R., 2000. The "Vanishing Farmland" Myth and the Smart-Growth Agenda, Policy Brief No. 12, Reason Public Policy Institute, Los Angeles, CA.

Stevens, T.H., D. Dennis, D. Kittredge, and M. Rickenbach, 1999. Attitudes and preferences toward co-operative agreements for management of private forestlands in the Northeastern U.S., *Journal of Environmental Management* 55: 81–90.

U.S. Department of the Interior/U.S. Department of Agriculture (USDI/USDA), 1995. Federal Wildland Fire Management Policy and Program Review Report, U.S. Department of the Interior, U.S. Department of Agriculture, Washington, DC.

Wear, D.N., R. Liu, J.M. Foreman, and R.M. Sheffield, 1999. The effects of population growth on timber management and inventories in Virginia, *Forest Ecology and Management* 118: 107–115.

Winter, G. and J.S. Fried, 2000. Homeowner perspectives on fire hazard, responsibility, and management strategies at the wildland–urban interface, *Society and Natural Resources* 13: 33–49.

6 Land-Use Planning and Zoning at the Wildland–Urban Interface

Margaret Myszewski and James E. Kundell
Carl Vinson Institute of Government, University of Georgia

CONTENTS

1-56670-602-5/05/$0.00+$1.50
© 2005 by CRC Press LLC

6.1 INTRODUCTION

Land-use regulation is among the most contentious areas of concern faced by any community. For communities within the wildland–urban interface, conflicts between newcomers and long-term residents, as well as differing private and public land management needs, can make reaching a consensus on land-use decisions even more difficult. These conflicts are bound to increase as land development within the interface increases. Between 1982 and 1997, most of the land developed for residential or commercial purposes in the U.S. had previously been forest or agricultural land (U.S. General Accounting Office [GAO] 2000) (Figure 6.1). There are several reasons why this rapid land consumption within the interface has often resulted in poorly planned development, with relatively little attention paid to natural resource protection or growth management. Federal and state policies regarding highway construction, housing, and financing over the past 50 years have acted to promote growth in rural areas by making land within the interface both accessible and affordable. In addition, zoning regulations that separate housing, jobs, and shopping have combined with low-density development that requires the use of a car to generate haphazard, inefficient development. Finally, because of their heavy reliance on property taxes for revenue, most local governments are reluctant to disallow any new development. Therefore, many land-use decisions are driven more by concerns about preserving the local tax base than preserving the character and natural resources of the community itself.

It has become increasingly clear that this type of development, in addition to being economically wasteful, is also environmentally harmful. Land development practices too often result in a loss of natural ecosystem functions, such as flood mitigation and provision of wildlife habitat. One consequence of habitat loss and movement of humans into interface areas is the increased incidence of encounters between wildlife and people. Human run-ins with alligators in Florida, mountain lions in Montana, and black bears in Atlanta suburbs have become increasingly common due to encroachment into what was formerly wildlife habitat (Kreyling 2001). Land development that increases the amount of impervious surface also affects interface lands by preventing stormwater infiltration and allowing sediments, nutrients, and pollutants to flow directly into streams and rivers.

Effective planning and growth management can help mitigate or prevent much of the environmental degradation that too often results from rapid, uncontrolled development. A community's land-use plan acts as an overall set of goals and policies to guide land-use decision making by the local governing body. The plan is implemented through zoning ordinances that traditionally have limited the type of land use by dividing a locality into residential, agricultural, industrial, or commercial districts. Planning and zoning can be coordinated through a combination of growth management techniques that allow a community to determine at what rate and location development will be permitted (Figure 6.2).

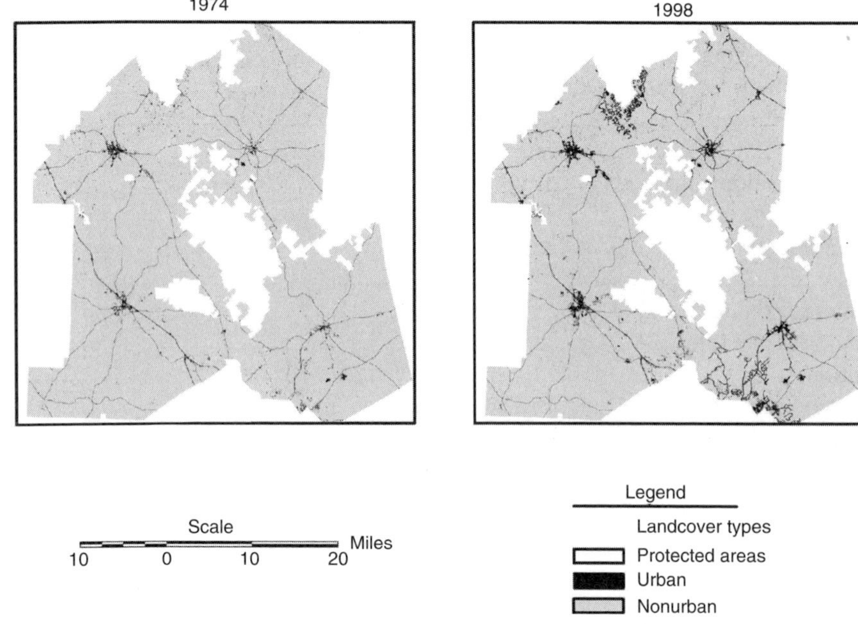

FIGURE 6.1 Agriculture and forestland loss at the wildland–urban interface in four Georgia counties from 1974 to 1998. These GIS maps demonstrate how urban development in four Georgia counties (Butts, Jasper, Jones, and Monroe) has encroached on surrounding forestland and open space. The protected areas in white include the Oconee National Forest and other federal and state lands (Natural Resources Spatial Analysis Laboratory 2002 [NARSAL]).

FIGURE 6.2 Planning, zoning, and growth management techniques allow communities to manage the rate of growth as well as direct development away from environmentally sensitive areas in order to preserve a desirable quality of life for their citizens. (Courtesy of Jim Kundell.)

In this chapter, we will discuss the basic concepts of planning, zoning, and growth management. We will also suggest ways in which these methods and techniques may be used by communities and influenced by natural resource managers to effectuate patterns of land use that are protective of the natural resources within the wildland–urban interface.

6.2 PLANNING

Planning for local communities, including counties, cities, towns, and villages, is the major means by which the natural resources of the locality can be preserved and maintained. Planning is necessary to correct the tendency of unbridled development to follow the line of least resistance. By adhering to a land-use plan, a locality can guide and control growth in a way that is consistent with community goals.

6.2.1 History of Planning

City planning in one form or another has existed in this country since its beginning. William Penn and James Oglethorpe, for example, created extensive plans for the formation of Philadelphia and Savannah, respectively. However, prior to the 20th century, city plans tended to be either blueprints for undeveloped sites or designs for public spaces. As a response to increasing urbanization in the U.S. following World War I, the U.S. Commerce Department issued the Standard City Planning Enabling Act (SCPEA) in 1928. The SCPEA allowed states to delegate planning power to local governments through the creation of local planning commissions (Kaiser and Godschalk 1995). Today, every state has some form of planning legislation, although many states have not updated their enabling statutes to any great extent from that of the original SCPEA (Smith 1993).

6.2.2 Planning Process

The first step in the planning process is for the local governing body to create a planning commission and establish its membership. The planning commission is generally charged with preparing the four traditional tools of the planning process: a land-use plan, zoning ordinances, subdivision controls, and capital improvements programming. These tools provide the legal foundation upon which all other growth management approaches are based.

6.2.2.1 Land-Use Plan

A land-use plan is an official public document used by the local government as a policy guide to decisions about the physical development of the community (Juergensmeyer and Roberts 1998). Usually, the plan itself is compiled by professional planners employed by planning consulting firms or state, regional, or local planning agencies. The planners will research and analyze a wide range of current and future physical and economic conditions of the locality, including population trends, distribution of land uses, transportation networks, and environmentally sensitive areas, and traffic flows on major highways (Keene 1996). This information is presented to the planning commission, which uses the data to examine present conditions and problems, establishes policies to deal with these problems, and recommends specific techniques and priorities for changes and improvements. Usually, the land-use plan includes a list of goals stating where the community is at present and where it wants to be at a set time in the future (Figure 6.3). The proposed plan is then published, and public hearings are set to provide for citizen comment and participation. Finally, the plan is adopted by the planning commission and the local governing body, both of which are responsible for seeing that the plan is implemented and kept current (Smith 1993). For communities that wish to control future growth and development, the importance

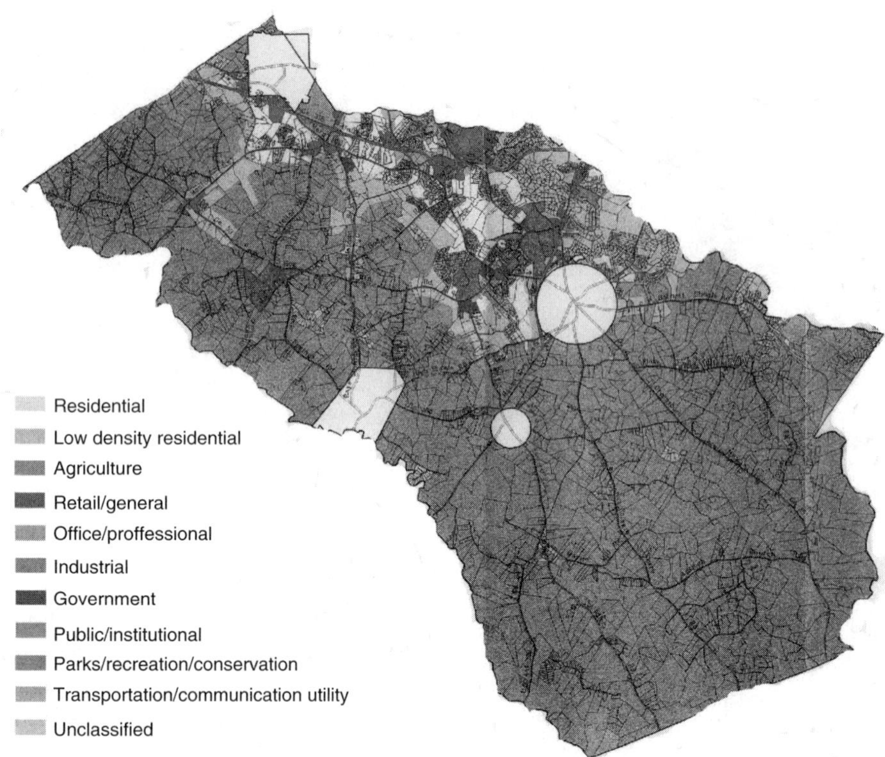

FIGURE 6.3 An important part of the planning process is the development of a land-use map, which depicts the type of land use that will be permitted within each district. (Courtesy of Jim Kundell.)

of having a land-use plan cannot be overstated. A well-researched land-use plan provides the planning committee with a policy basis for granting or denying variances to zoning ordinances that can withstand challenges in court.

6.2.2.2 Zoning Ordinances

Zoning ordinances, which will be discussed in more detail in Section 6.3, are the primary way in which land-use plans are implemented. Zoning ordinances may divide a locality into land-use districts with respect to agriculture, forestry, protective-greenbelt, recreation, residence, industry, trade, conservation, and other activities that will be permitted within the district. Zoning ordinances also set standards within each district for building height, setbacks from lot lines, density of development, and so forth. It is the responsibility of the planning commission to see that a locality's zoning ordinances are consistent with the community's land-use plan in order for the plan to have any meaningful effect.

6.2.2.3 Subdivision Controls

Subdivision controls determine the standards for the subdivision of land for various purposes. Together with zoning ordinances, they can be an important and effective way of implementing the land-use plan. Subdivision controls usually contain standards concerning the installation of streets, curbs, gutters, sidewalks, street signs, and trees and establish minimum standards for each. Their success depends on how carefully ordinances are administered. Each subdivision proposal should be reviewed by the entire planning commission, which should consider the effects that the proposed development would have on the land-use plan and the community. Subdivision standards should also be consistent with the zoning requirements for the area (Smith 1993) (Figure 6.4).

6.2.2.4 Capital Improvements Programming

The capital improvements program is a summary of the needs of the community in terms of public improvements (parks, sewers, water lines, streets), the estimated costs of these improvements, and the development of logical priorities for their provision. The purpose of a capital improvements program is to anticipate the needs of a community for investment in improving infrastructure and other public services. The land-use plan, together with the work of the planning commission in analyzing land-use and population trends, economic pressures, and the financial capacity of the locality to pay for the necessary services, should determine the capital improvement needs of a community (Smith 1993).

6.2.3 Role of Planning at the Interface

Land-use planning can be a powerful tool for communities within the interface that want to be proactive in directing growth that will conserve and protect the area's natural resources. Determining where future development

FIGURE 6.4 Development such as this demonstrates the need for subdivision regulations that require environmental protections such as stormwater retention measures, floodplain protection, and setbacks to buffer streams. Fortunately, such regulations are increasingly being used to ensure that new developments are constructed in an environmentally protective manner. (Courtesy of Jim Kundell.)

will be allowed should begin by identifying environmentally sensitive areas that ought to be preserved. These areas may include surface water, marshes, floodplains, aquifer recharge areas, steep slopes, prime agricultural land, and forests and woodlands (Porter 1997). The state of Georgia, for example, requires local governments to establish regulations for protection of "vital natural areas," defined as water supply watersheds, significant groundwater recharge areas, wetlands and river corridors, and mountains (Georgia Department of Community Affairs [DCA] 2002). Land-use plans that recognize the value of preserving natural resources within the interface are then implemented through zoning, subdivision controls, and/or other measures (e.g., greenspace ordinances, tree ordinances, and building codes). Zoning ordinances can be used to protect sensitive environmental areas by creating conservation, agricultural, and open space districts. Recent trends in subdivision regulations increasingly require environmental protections such as stormwater retention measures for erosion control, setbacks to buffer streams, and floodplain protection. In addition, planned unit development and cluster development provide interface communities with the opportunity to shape development and growth through subdivision review and approval (Smith 1993).

6.2.4 Public Involvement

In order to be successful, land-use planning must reflect the concerns and commitment of the citizens of the community. Most land-use plans are advisory in nature, and even legal tools such as zoning and subdivision ordinances will not succeed unless they have public support. Public participation should be an integral part of the planning process, continuing through the development, administration, and implementation of the land-use plan. Of special importance is the role of the public, in general, and the natural resource manager, in particular, in determining the objectives of the land-use plan through participation in public hearings and meetings. Natural resource managers should work to ensure that those who are interested in and dedicated to effective land-use planning and controlled growth attend such meetings and make their support known. Developing professional relationships with local planners is another way natural resource managers can help to identify environmental concerns about development within the interface and planning options for addressing these concerns. Citizens and natural resource managers can also become involved in the planning process through participation in citizen advisory committees. Planning committees frequently use such committees to study and provide additional public input on specific problems or critical issues (Smith 1993).

6.3 ZONING

Historically, the primary approach to land-use regulation in the U.S. has been zoning. Local governments use zoning ordinances to control the type and density of land uses allowed on a particular piece of property. A zoning ordinance will usually divide an area into residential, commercial, industrial, and other districts. Density is regulated through requirements for minimum lot size, building heights, and building setbacks. By segregating incompatible land uses, zoning is meant to provide local governments with a mechanism to ensure orderly development and protect property values and human health. A zoning ordinance includes a written text and a zoning map, which shows the boundaries of the area, the streets, and the district boundaries, and identifies the district classifications (Smith 1983).

6.3.1 History of Zoning

Traditionally, control over land use has been the prerogative of local governments. Cities began to implement zoning ordinances at the beginning of the 20th century as a way of protecting private investment in homes from commercial and industrial intrusions. New York City passed the first citywide zoning ordinance in 1916, and zoning by local governments gained widespread implementation shortly after World War I. In 1924, the U.S. Commerce Department issued the Standard State Zoning Enabling Act (SSZEA) in reaction to the increase in urbanization then faced by American cities. The SSZEA enabled the states to delegate their police power (such as the power to regulate to protect public health, safety, morals, and welfare) to municipalities in order to remove any question over their authority to enact zoning ordinances. The SSZEA also contained procedures for establishing and amending zoning ordinances. Today, nearly every state has some type of home rule provision enabling municipalities to exercise some degree of self-governance. The degree of zoning authority granted to local governments, however, can vary greatly from state to state. Therefore, one of the first pieces of information a natural resource manager needs to have in order to help bring about change at the wildland–urban interface is what degree of zoning authority the locality possesses. This authority may vary depending on whether the locality is a municipal, township, or county government. For example, a municipality may have authority to zone while the county in which the municipality is located does not. A natural resource manager should realize, however, that while a local government may have the authority to adopt zoning and other land-use controls, many have chosen not to do so. Zoning may simply be politically unacceptable in a particular jurisdiction.

6.3.2 The Zoning Process

The first step of the zoning process involves conducting studies and preparing the initial draft of the zoning text and the zoning district map. A planning commission is usually charged with these tasks if one exists; otherwise, the local government must appoint a zoning commission to begin the zoning process. Either the planning or zoning commission may enlist the services of a professional consultant. The local government's attorney may also be consulted in order to ensure that the zoning ordinance is related to the planning standards (if any) of the community. The planning or zoning commission also acts as a liaison between the local government and the public. This responsibility requires the commission to provide adequate information to the public as well as express opinions obtained from the public to local government officials (Smith 1983).

The second step in the zoning process is to make the findings and recommendations of the commission public. In most states, the agency charged with organizing a zoning ordinance is also charged with submitting the ordinance officially to the public and then allowing the people to express themselves at a formal public hearing. The public hearing is an extremely important part of the zoning process because those responsible for creating and administering zoning ordinances need to hear support for and opposition to the proposed regulations. It is the responsibility of a natural resource manager to be aware of the zoning process within interface communities and to take advantage of the process in order to protect the interface environment. Following the public meeting, the commission meets again to review any comments and suggestions and to develop a revised ordinance and map to submit to the local government (Smith 1983) (Figure 6.5).

Finally, the local government introduces the zoning ordinance and votes on whether it will be given further consideration. Following this first vote, the ordinance must be published in a local newspaper of general circulation. This publication must include the zone district map as well as the text and any other material necessary to present the ordinance completely. After an official public hearing by the governing body, and following a favorable vote, the zoning ordinance becomes law (Smith 1983). Before approval, it is important that the zoning ordinance be reviewed for consistency with the land-use plan. Courts are more likely to uphold zoning ordinances if they are consistent with local policy as stated in the land-use plan.

6.3.3 Role of Zoning at the Interface

Traditional zoning ordinances have played a major role in the development of growth patterns that have fostered interface conditions. This has occurred for a number of reasons. Local governments receive most of their funding from property and sales taxes and, therefore, have little reason to attempt to limit land development in their jurisdictions. In addition, localities often compete with one another to attract new development in pursuit of a larger tax base. This competition, coupled with a desire to maximize property tax revenue, encourages communities to zone more land for development than is likely to be needed in the foreseeable future. In Maryland, for example, zoning by the state's local jurisdictions as of 1996 allowed five times as much development as was required for projected growth to 2020 (Porter 1999). Even when local governments attempt to limit growth, however, the policies they implement can have the effect of increasing development within the interface. For example, when local governments become alarmed about the potential impacts of development on available infrastructure, they often reduce allowable densities to levels supportable by private wells, septic tanks, and roads, thereby spreading out settlement and causing more land to be developed. Again, in Maryland, more than half of the development

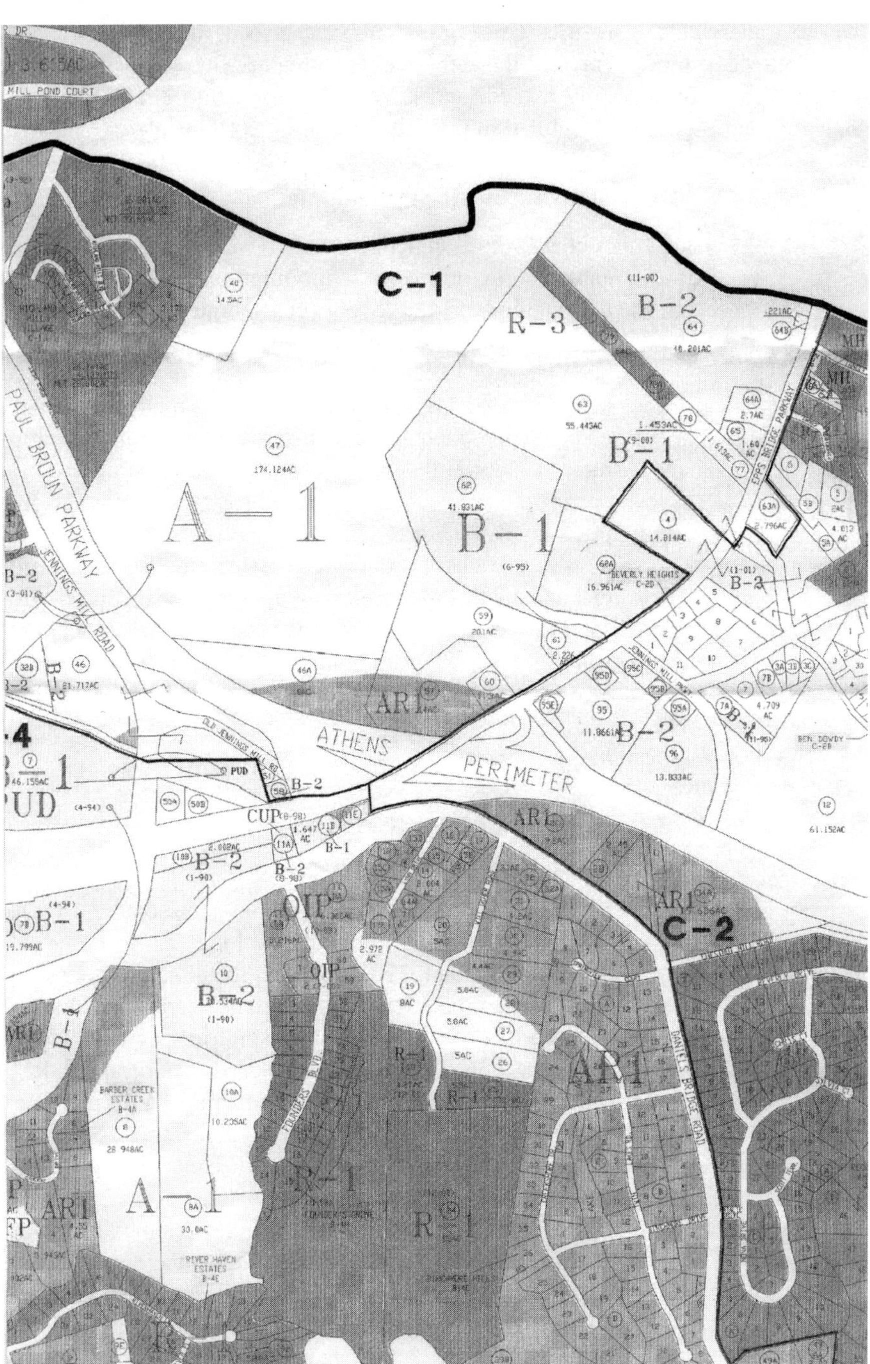

FIGURE 6.5 A zoning map depicts the boundaries of the area included in the zoning ordinance, the streets, and the zoning districts as well as individual parcels of land, natural stream courses, and other natural features. (Courtesy of Jim Kundell.)

capacity allowed by local plans in 1996 was outside current or planned sewer service areas (Porter 1999). In another attempt to control growth, local governments sometimes implement restrictive zoning practices. However, by raising the entry costs for new residents and businesses and limiting undesirable land uses, localities often end up directing new development into interface areas located in other jurisdictions (Lockard 2000).

Despite the problems that can be caused by poorly planned zoning, local governments can also use zoning regulations to manage development successfully within the interface. The following are examples of alternative types of zoning ordinances that communities can use to preserve forests, wetlands, floodplains, or environmentally sensitive land.

6.3.3.1 Limit Development in Environmentally Sensitive Areas

Zoning ordinances can require low-density development or no development on steep slopes, erosion control measures during and after construction, and preservation of stream valleys to reduce erosion that can degrade water quality. Communities that depend on groundwater for their water supply can protect groundwater quality by limiting the amount and/or types of development that might pollute the aquifer. Wellhead protection measures also restrict location of various activities close to wells. Consequently, the zoning ordinance can be used in conjunction with other control measures to protect water resources and environmentally sensitive areas. The town of Dedham, MA, passed a zoning ordinance that provided that lands subject to seasonal or periodic flooding could not be used for residences or other purposes that might endanger the public welfare (*Just v. Marinette*, 56 Wis.2d 7 [1972]). In addition to zoning, subdivision ordinances that require developers to set aside critical areas can be used to protect environmentally sensitive land (Porter 1997). Local governments can also require setbacks or buffer zones to protect watersheds.

6.3.3.2 Floating Zones

The floating zone is a zoning district that is not designated on the municipal zoning map until a landowner or developer applies for rezoning. An owner who requests that the floating zone be applied to a particular parcel must demonstrate that a variety of impacts will be properly handled, such as the project's effect on natural resources and preservation of open space (Juergensmeyer and Roberts 1998). A floating zone scheme was used in approving a 200-acre development on the Patuxent River in the Chesapeake Bay area of Maryland. The project included a central marina, habitat preservation areas, a community beach with restricted access areas, and central sewage. The rezoning allowed developers to increase the housing density from 1.0 units per acre to 2.83 units per acre on part of the site; the total number of units was not increased by the floating zone designation. Additionally, numerous conditions were imposed, covering such areas as street design, parking, and sediment and stormwater control, and placing limitations on pleasure boat mooring and use (Kenney 1985).

6.3.3.3 Overlay Zoning

Overlay zoning is a flexible zoning technique that allows a municipality to encourage or discourage development in certain areas. An overlay zone supplements the underlying zoning standards with additional requirements that can be designed to protect the natural features in an important environmental area. A parcel within the overlay zone is simultaneously subject to the underlying zoning regulation and the overlay zoning requirements (Juergensmeyer and Roberts 1998). In 1989, Portland, OR, applied an environmental overlay zone to 20 percent of the city's land area or 19,000 acres of mostly forested public and private property. The overlay is divided into full protection zones, surrounded by conservation zone buffers that restrict which areas can be disturbed, the trees that can be cut, and the setback from streams. One third of the publicly owned land is set aside for wildlife, and trails are prohibited in this area (Kreyling 2001).

6.3.3.4 Cluster Development

A cluster development is a subdivision in which the applicable zoning ordinance is modified to provide an alternative method for the layout, configuration, and design of lots in order to preserve the natural and scenic qualities of open lands. A locality can allow or require development to be placed on a portion of the parcel and the rest to remain undeveloped open space. Clustering allows for a development density higher than would normally be allowed on part of the property in exchange for retaining other portions in their natural condition. Over the total area, the density remains the same. Clustering can be used to protect parcels with particular natural resource characteristics such as wetlands, valuable viewsheds, prime agricultural soils, or steep slopes (Juergensmeyer and Roberts 1998) (see Box 6.1). For example, Middletown, RI, passed an

Box 6.1

PRAIRIE CROSSING: A CONSERVATION COMMUNITY

In conservation subdivisions, houses are placed closer together than in traditional developments in order to disturb less land. The remaining natural areas are used for recreation, wildlife habitats, and agriculture. Opened in 1994, Prairie Crossing, an exemplar of a conservation subdivision, is located in Grayslake, IL, 20 mi from Chicago. Homes in Prairie Crossing are spaced close together to allow 60 percent, or 350 of its 667 acres, to be preserved as prairies, marshes, gardens, and lakes. Despite the closely clustered homes, there is little sense of congestion because they seamlessly blend into the surrounding prairie as well as the neighboring 2500-acre Liberty Prairie Forest Preserve. Residents of Prairie Crossing are committed to making sure that they hold true to its conservation mission as they assume control of the development. For example, the homeowners' association conducts annual prairie burns and teaches residents proper burn techniques. Prairie Crossing is also home to the first endangered and threatened fish refuge in Illinois. The project, funded by the residents themselves, was sited at Prairie Crossing because of its biotic diversity, high water quality, and commitment to controlled land use and fertilizer use. Prairie Crossing has been held up as a national model for suburban planning and has received accolades from respected land-use organizations, including the Urban Land Institute (Umlauf-Garneau 2000; Corley 2002).

ordinance requiring that cluster developments retain at least 30 percent of the total land area as open space, which may be used for playgrounds, golf courses, beaches, bird sanctuaries, and bridges (Freis and Reyniak 1996).

6.3.3.5 Incentive Zones

Incentive zones allowing significant waivers of zoning requirements are offered to developers as a means of directing larger-scale development into designated growth areas. Land developers can be required to provide public amenities such as transportation, parks, affordable housing, or other infrastructure in exchange for the waivers (Juergensmeyer and Roberts 1998). Many cities and suburbs across the U.S. use incentive zoning. San Francisco, for example, offers zoning bonuses to encourage provision of rooftop observatories; Anchorage, AK, furnishes incentives for climate-controlled courtyards; and Cincinnati, OH, gives incentives for historic preservation (Kayden 1992).

6.3.3.6 Flexible/Mixed-Use Zoning

Local governments are also amending their zoning ordinances to allow for flexible or mixed-use development, which allows denser growth in urban areas by permitting commercial and residential development to take place within the same area. A recent survey of local governments conducted by the U.S. General Accounting Office revealed that 91 percent of the cities and 64 percent of the counties surveyed used their zoning authority to allow mixed-use development (GAO 2000). Arlington County, VA, for example, rezoned land for a high-density complex that combines office, retail, and hotel space. Similarly, Redondo Beach, CA, established a mixed-use ordinance to permit the construction of residential/retail communities (GAO 2000).

6.3.3.7 Performance Zoning

Performance zoning allows or prohibits land uses based on their conformance to preset criteria. While traditional zoning specifies land uses that will or will not be allowed within each district, performance zoning permits a wide variety of land uses as long as each land use is able to meet the performance standards set for that district and use. Communities that use performance zoning usually create different standards to be met for different districts or zones. Performance zoning can be used to protect natural resources within the wildland–urban interface by requiring adherence to standards that explicitly limit development that damages the environment or causes the destruction of important natural resources. Because of its flexibility, performance zoning allows developers to cluster developments in a manner that ensures the preservation of important environmental features such as drainage ways and steep slopes (Figure 6.6). One way in which communities can use performance zoning to protect the environment is through the use of impervious surface ratio (ISR) performance standards. The ISR is the proportion of a site covered by impervious surfaces and is a good indicator of hydrological degradation (see Chapter 12 for a detailed discussion). Many performance zoning systems specify a maximum ISR for a site, which can make them especially effective in managing stormwater runoff and groundwater recharge (Marwedel 1998).

6.3.3.8 Planned Unit Development

Planned unit developments (PUD) are designed to permit the development of entire neighborhoods based on approved plans. The completed development usually includes a variety of residential types, common open space for parks, and in some cases, commercial areas. The PUD principle is that a land area under unified control can be designed and developed in a single operation and according to an officially approved plan. The plan does not have to correspond to the use regulations of the zoning district in which the development is located. Common PUD objectives include encouraging developers to use a more creative approach in their development of land and encouraging a more efficient and more desirable use of open land (Juergensmeyer and Roberts 1998). The Coffee Creek Center, located outside Chicago, is a recent example of a planned unit development. It includes a 160-acre greenspace centered on a variety of home types as well as commercial, retail, and workspace. The Coffee Creek development plans to coexist with the native ecosystems and habitats by creating environmental guidelines and a conservation district deeded to a nonprofit conservation group that will in turn be bound by Coffee Creek's environmental guidelines. The conservation group will be in charge of managing the land but will be financially supported by the property owners' association (Umlauf-Garneau 2000).

6.4 GROWTH MANAGEMENT

Growth management planning involves consolidation of all requirements and control techniques necessary to carry out the intent of a land-use plan. The fundamental principles of growth management are that all growth is not inevitable and that the rate and type of development should be determined by public policy and action (Porter 1997).

6.4.1 History of Growth Management

The interest in protecting the environment that began in the 1960s was the impetus for the idea of using growth management techniques as a means of controlling growth

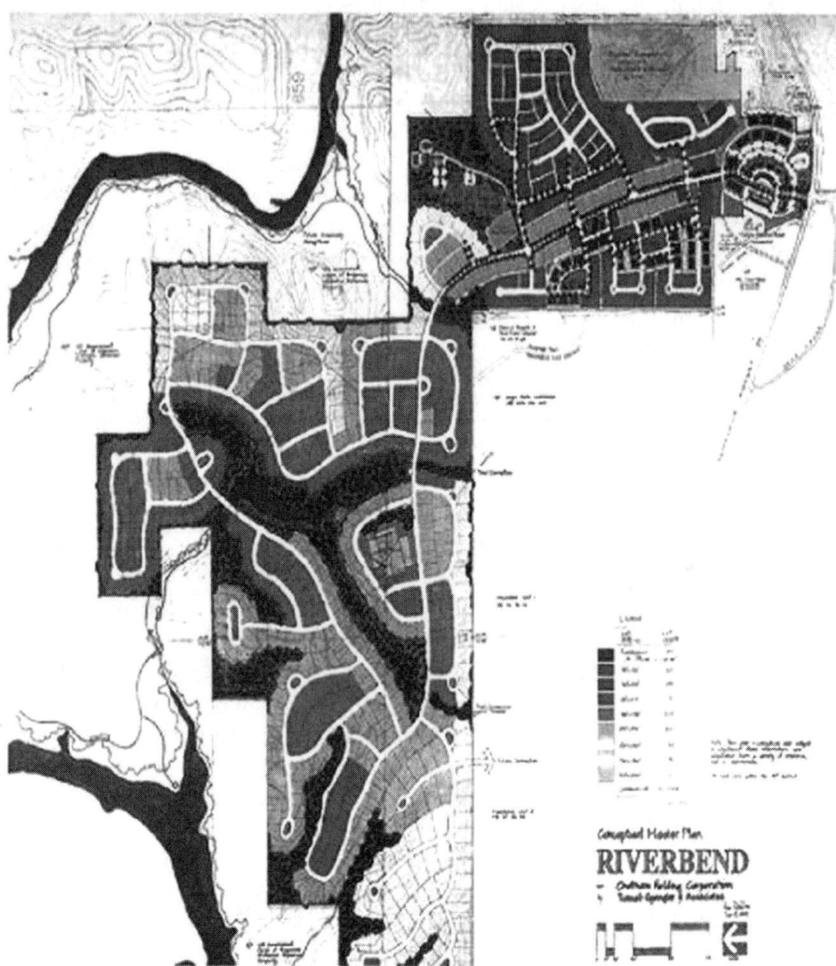

FIGURE 6.6 Alternative zoning ordinances such as performance zoning provide greater flexibility than traditional zoning ordinances and allow planners to design developments that better fit the land and preserve more greenspace. (Courtesy of Jim Kundell.)

to conserve a community's natural resources. Today, growth management is seen as a way of guiding and controlling rather than restricting growth. Most growth management programs are used by local governments as a way of maintaining the quality of life of a community as well as preserving its environmental resources (Porter 1997).

6.4.2 State Role: Comprehensive Growth Management Plans

The authority to control land use lies mainly with the states, which generally choose to delegate this authority to local governments. However, state legislatures can always supersede local zoning where statewide interests are at stake. States are increasingly choosing to manage statewide growth and development through the adoption of comprehensive growth management plans or other statutory requirements. As of 2002, 11 states had enacted comprehensive growth management statutes: Florida, Georgia, Hawaii, Maine, Maryland, New Jersey, Oregon, Rhode Island, Tennessee, Vermont, and Washington. While these approaches differ in the details, they have several commonalities. Generally, they require or encourage local governments to prepare land-use plans that conform to state goals and policies (Porter 1997). For example, the Georgia Planning Act of 1989 does not require local governments to prepare local comprehensive plans, but if a community refuses to plan or if it fails to address state planning goals, it may not apply for state water and sewer grants, community development funds, or other forms of financial assistance. In addition, state growth management statutes usually incorporate or express goals and objectives to guide the statewide planning process. Oregon's land-use law includes 19 goal statements, six of which deal with environmental concerns. Goal Four, for example, requires protection of land for timber production (Beaumont 1996). Likewise, Maryland's statute incorporates seven "visions" as the prime policies to be implemented by local plans. These vision statements serve to define state interests that are to be addressed by the plans and regulations of state agencies, any existing regional agencies, and local governments (Porter 1997). Finally, state comprehensive growth management statutes require some type of review process in an attempt to ensure consistency among the different levels of governments, between jurisdictions, and between planning and regulations within jurisdictions. All the states with comprehensive growth management plans (except Vermont) review local plans for consistency with state goals (Porter 1997).

6.4.3 Role of Growth Management at the Interface

Traditionally, the location of development has been determined through planning, zoning, and subdivision regulations. However, these methods have begun to be supplemented by various growth management techniques that allow communities to exercise a firmer control on growth. In addition, many growth management techniques have evolved to prevent development on lands deemed important for natural resource or environmental purposes (Figure 6.7). These approaches range from complete protection through acquisition of the land itself to partial protection by purchase of development rights or conservation easements to regulatory approaches such as urban growth boundaries, concurrency requirements, impact fees, and conservation planning/zoning.

6.4.3.1 Land Acquisition

The acquisition of land for protection against development is the most certain way of preserving natural resources within the wildland–urban interface. Purchase of environmentally sensitive land can be done by local, state, and federal governments or by private land trusts. Many state conservation programs include open space preservation. For example, through Program Open Space, established in 1969, Maryland has acquired more than 158,000 acres of open space for state parks and natural resource protection (American Planning Association 1999). More recently, Georgia created a Greenspace Trust Fund with the goal of ultimately preserving 20 percent of Georgia's land area as open space (Griffith 2000) (Figure 6.8). In addition to state land conservation programs, there are more than 1000 land trusts currently operating at the local and regional levels in the U.S., protecting more than 4 million acres of land through voluntary land transactions (Wiebe et al. 1997). Examples of land trusts that operate on a national basis include The Nature Conservancy, which protects more than 11 million acres in the U.S., and The Trust for Public Land, which protects more than 1.2 million acres in 45 states (The Nature Conservancy 1999; Trust for Public Land 2000).

6.4.3.2 Conservation Easements

The conservation easement is the most widely used method of protection for private land in the U.S. today. Conservation easements are voluntary legal agreements between a landowner and another party that restrict the development of a tract of land. They are designed to preserve farmland, historical sites, environmentally sensitive lands, and open space. A conservation easement is created when a landowner restricts his or her rights to develop the land in ways that would be incompatible with the purpose of the easement. For example, in Georgia, conservation easements protect the future of longleaf pine habitat on forested land, while allowing the landowners to retain the right to continue selective timber harvesting and quail hunting (Fowler and Neuhauser 1998). Easements can be donated or purchased, usually for a price considerably lower than the market value of the land. In either case, a landowner may be eligible for tax benefits that reflect the donation or the loss of potential income (see Chapter 5). The federal government, along with many state and local governments, implements various programs designed to protect land through the use of conservation easements. For example, the Farmland Protection Program is a federal program that helps protect farmland from conversion to nonagricultural uses by providing matching funds to state, tribal, and local governments with existing farmland protection programs to purchase conservation easements. Since 1996, more than 108,000 acres of farm and ranch

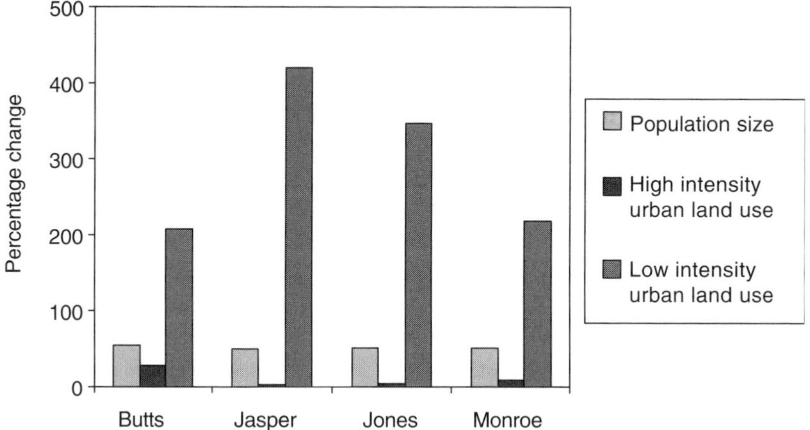

FIGURE 6.7 Percentage of land-use and population change in four Georgia counties from 1974 to 1988. This graph demonstrates how population growth in four Georgia counties has resulted in large increases in low-intensity urban development. The effect of this type of development is frequently the loss of agricultural land, forests, and other forms of open space. Growth management programs offer communities a way to manage growth so that loss of environmentally sensitive areas can be minimized (Natural Resources Spatial Analysis Laboratory 2002).

FIGURE 6.8 In addition to state programs aimed at protecting greenspace, many local governments are also acquiring greenspace as part of their conservation programs. (Courtesy of Jim Kundell.)

land have been protected in 28 states. The 2002 Farm Bill expanded the program to include nongovernmental organizations as eligible entities (U.S. Department of Agriculture, National Resources Conservation Service [NRCS] 2002). Maryland's Rural Legacy Program is an example of a state program designed to protect forest and farmland. Maryland provided funding for the preservation of nearly 92,000 acres of forest and farmland as of 1999, and the state has authorized $71.3 million in funding for this program through 2002 (American Planning Association 1999).

6.4.3.3 Transfer/Purchase of Development Rights

A transfer of development rights (TDR) program divides a jurisdiction into areas where development will be discouraged (sending areas) and areas where development is appropriate (receiving areas). Owners in sending areas are allowed to sell their rights to develop, which they can no longer use, to the owners of land in receiving areas. Montgomery County, MD, is an example of a jurisdiction that has long-term experience with TDR programs. Since 1980, development rights to more than 32,000 acres of farmland in the northern part of the county have been transferred to developing areas in the southern part of the county. More recent examples of TDR programs include those used by the Long Island Central Pine Barrens, New York (1995), Bucks County, PA (1994), and Dade County, FL, where TDRs are helping to preserve more than 100,000 acres of everglades ecosystems outside the Everglades National Park (Porter 1997) (also see Box 6.2). The advantage of TDR programs is that they provide a way to economically benefit owners of environmentally sensitive land without loss of the land to development or a payout of public money (Juergensmeyer and Roberts 1998).

Box 6.2

OLD MISSION PENINSULA

Old Mission Peninsula, located in the northwest corner of Michigan's lower peninsula, is known for its unusual topography and microclimate, which makes the area one of the premier agricultural locations in the U.S. for growing cherries and other fruits. When the peninsula began to be threatened by development, concerned citizens looked for ways to prevent the conversion of this land from agricultural use to commercial and residential use. Citizen groups began the process by identifying key issues affecting the quality of life on the peninsula, with focus on farmland preservation, open space and natural area conservancy, business development, and residential needs. The result of these focus groups was the creation of an Agricultural Preservation Plan that would include the funding of the first locally based purchase of development rights (PDR) program in the nation. Landowners who voluntarily entered the PDR program had a portion of their development rights separated from the property, and they were compensated at the difference between the fair market value of the property with and without the development rights. The township holds these development rights in perpetuity on behalf of all the residents. To pay for the development rights, voters passed a referendum that raised property taxes. After a 5-year effort, more than 1100 acres were preserved through the PDR program. Funding by a combination of private, regional, state, and federal sources allowed for the purchase of conservation easements on another 1000 acres of open space, natural habitat, and farmland. The experience of Old Mission Peninsula demonstrates the ability of citizens to build consensus for land-use planning programs and to elicit change in government when such change is seen as necessary to protect valuable natural resources and farmland (Westphal 2001).

6.4.3.4 Urban Growth Boundaries

Urban growth boundaries identify and separate land that will be available for future growth from land that is designated to remain rural. Boundaries are typically drawn by the locality to take into account the amount of land that will be necessary to accommodate new housing, economic development, and open space for a given amount of time. Oregon has required local governments to implement urban growth boundaries since 1973, while states such as Washington and Tennessee have more recently included boundary requirements in their state comprehensive plans (Beaumont 1996; Porter 1997). Minnesota requires a regional urban growth boundary for a five-county area around metropolitan Minneapolis and St. Paul (Minn.Stat.Ann. '473.861). Chapters 4 and 5 provide more discussion about urban growth boundaries from economic and social policy perspectives, noting that questions remain concerning the effectiveness and equity of these boundaries.

6.4.3.5 Concurrency Requirements

Concurrency requirements are used by local governments to avoid, deter, or overcome many problems associated with growth by conditioning the issuance of building permits or subdivision plat approval on the existence of public infrastructure and capacity. Sometimes, developers are required to pay fees that are to be used by the local government to provide the roads, schools, parks, sewer, and water facilities needed for the new development (Juergensmeyer and Roberts 1998). The state of Washington requires local governments to link development approval and provision of facilities (Wash. Rev. Code Ann., ch. 36.70A). Florida also imposes concurrency requirements as part of its mandatory state planning law (Juergensmeyer and Roberts 1998).

6.4.3.6 Impact Fees

Impact fees are charges levied by local governments against new development to generate revenue for capital funding necessitated by the new development. Local governments usually use impact fees as a way of controlling growth by forcing developers to pay the community for the costs of the new parks, roads, schools, sewer, and water treatment facilities that new development will require (Juergensmeyer and Roberts 1998). However, some communities are using impact fees to acquire conservation lands. An example of such a community is Riverside, CA, which charges an impact fee on all new development to provide funds for the purchase of wildlife habitat (Porter 1997).

6.4.3.7 Conservation Planning/Zoning

State and local governments can use land-use planning and zoning maps to identify environmentally critical areas in which development should be prohibited or curtailed. The first step in conservation planning is the creation of a comprehensive map and inventory of critical natural and scenic resources. This natural resource inventory includes information about the location of high-quality natural resources such as wetlands, prairies, woodlands, streams and stream corridors, shorelines, and significant viewsheds. This inventory also identifies habitat of endangered and threatened species and prime farmland. The natural resource inventory and map then serves as a guide to conservation planning. Conservation goals are stated in the land-use plan and implemented through subdivision and zoning ordinances. Conservation zoning is highly flexible and allows greater unit density in exchange for greater preserved land area. Conservation subdivisions delineate conservation areas and identify undevelopable land as well as sites of unique ecological and environmental importance. Only after these critical areas are determined are housing sites, streets, and lot lines created (Porter 1997).

6.5 WAYS TO IMPROVE PLANNING AND ZONING AT THE INTERFACE

6.5.1 Increase Regional Cooperation

Regional cooperation between local jurisdictions is necessary because environmental problems resulting from unrestrained development often spill over from one local jurisdiction to another, usually from urban areas toward the wildland–urban interface. As long as cities and counties differ in their vision of how development should proceed, developers will be able to shop for lenient jurisdictions and make decisions that yield the highest profits. Local governments need to cooperate with each other in order to promote development patterns that will conform to planning policies that protect natural resources at the interface. One reason why local governments often find such cooperation difficult is their heavy reliance on local property taxes for revenue, which causes them to compete with each other for development that will expand the tax base. To address this problem, Minnesota has instituted an innovative regional tax-sharing program in the greater Minneapolis-St. Paul area in which 187 communities pool some of their property tax revenues. Forty-three communities are net payers, while 144 receive more revenue than they contribute (Smith 1983).

In addition to cooperation among local governments, there is a need to encourage cooperation and collaboration among federal and state governments, regional agencies, and local governments when dealing with multijurisdictional natural resource issues. Growth-related issues are often best addressed at a regional level, especially in the case of large metropolitan areas containing multiple local governments. Across the country, there are numerous examples of local governments working with other jurisdictions to create regional solutions to land-use issues. For example, in the Albuquerque metropolitan area, an organization called the Extraterritorial Land-use Authority is responsible for making land-use decisions in the 5-mile belt surrounding the city limits. Comprising city and Bernalillo County officials, the authority gives the county a role in approving annexations and controlling development at the edge of Albuquerque (GAO 2000). Regional environmental conservation agencies such as the Tahoe Regional Planning Agency and the Chesapeake Bay Commission have also proven to be successful in managing growth in an environmentally protective manner (Porter 1997). Chapter 7 provides more examples of innovative collaborations that have formed across jurisdictional boundaries in efforts to conserve wildlands amid development (see Box 6.3).

6.5.2 Modernize and Strengthen State Land-Use Role

It is important for states to recognize their responsibility to provide a sense of direction to local planning, zoning, and growth management efforts. While local governments traditionally have exercised control over land-use regulation, states have a central role to play in land-use planning, and there are several steps that states can take to support local growth management decisions. Foremost among the contributions that state government can make to the planning process is the creation of a clearly stated, well-implemented state land-use policy. Such a policy should establish statewide growth management principles that can serve as the basis for state agency planning as well as provide guidance and support to local governments crafting their own growth management and land-use planning programs. Ideally, a state land-use policy should incorporate principles that emphasize protection of the state's natural resources, including preservation of air and water quality and wildlife habitat, and provide for rural residential development that preserves the rural character of the area. States also need to revise and improve the tools they currently provide to local governments to allow communities to exercise further flexibility and choice when making effective growth management decisions. States should require communities to meet consistency requirements in order to ensure that local regulations are consistent with the state land-use plan and that projects receiving state and federal funds are also consistent with the plan. States also need to ensure that local zoning codes and subdivision regulations support implementation of the local land-use plan and that various local land-use regulations are consistent within a jurisdiction and do not work at cross-purposes with each other. Finally, states must provide local governments with growth-related education and training to help local officials make responsible development decisions.

Beyond providing local governments with the tools and guidance they need to make growth management decisions, it is important that states prevail upon local governments to develop and implement land-use plans. States with comprehensive growth management plans have mandated or provided incentives for local governments to plan. However, even when states require land-use planning, they may not enforce, and local governments may not follow, requirements for updating the plans or making them consistent with statewide goals. A recent California study found that the general plans of many localities were no longer current, as required by state law. For example, 301 cities and 37 counties, 65 percent of the localities within the state, had not updated the housing element of their general plan in the last 5 years, and 227 cities and 38 counties (51 percent) had not updated at least one other element of their general plan (GAO 2000). Similarly, Georgia's current standards for local comprehensive planning do not require local governments to implement any portion of their plans. This has resulted in a situation where, despite the fact that 99 percent of Georgia's 693 local governments have prepared and adopted comprehensive plans, many have failed to implement these plans successfully through

Box 6.3

The Adirondack Forest Preserve

The Adirondack Park is an example of the implementation of a land-use and development plan that recognizes matters of local concern, as well as those of regional and state concern, while providing appropriate regulatory responsibilities for the administering agency and the local governments of the Park. Management of the public and private lands within the Adirondack Park is the responsibility of the Adirondack Park Agency. The more than 2.6 million acres of public land within the park are managed according to the State Land Master Plan, which classifies public lands in the park into five major categories: wilderness, primitive, canoe, wild forest, and intensive use. Private land within the park is subject to the Adirondack Park Land-Use and Development Plan. This plan is designed to channel much of the future growth in the park around existing communities where roads, utilities, and other services are already present. Under the plan, the agency has the authority to issue land-use permits if the use is supportive of the plan and is compatible with the policies and objectives of the land-use area. Violations of permit conditions are subject to fines of up to $500 per day for each day the violation continues (Adirondack Park Interpretive Centers 2001; N.Y. COMP. CODES R. & REGS. Adirondack Park Agency Act. 27, §§807,813 [1999]).

the use of zoning or subdivision ordinances (Georgia Department of Community Affairs [DCA] 1998). Some states achieve implementation through the creation of a specific agency whose purpose is to oversee land-use planning in the state. For example, Oregon created a bipartisan Joint Legislative Committee on Land Use, which monitors how state and local governments carry out the legislative mandates. In addition, judicial procedures affecting planning are routed through the state circuit court, with appeals being heard by the Land-use Board of Appeals (American Planning Association 1999).

6.5.3 Increase Public Involvement in the Planning and Zoning Process

Land-use planning and zoning efforts and growth management programs within the interface cannot succeed without the strong and active support of citizens. However, many landowners do not support strong land-use controls or else do not understand the necessity of controlling development within the interface. Many landowners still believe that owning property gives them an absolute right to do whatever they want with their land — even if it is detrimental to the community at large (Figure 6.9). Other property owners simply do not recognize the consequences their land-use decisions may have on their neighbors or realize how their individual actions, while seemingly innocuous in themselves, cause serious problems when numerous people act in the same way. Overcoming such negative attitudes and lack of understanding about planning and growth management requires public education, consensus building, and public involvement in all phases of the planning and zoning process.

Educating the public about natural resource and conservation issues within the interface is one of the most important roles that natural resource managers can play in the planning process. Natural resource managers can encourage those who live within the interface to become aware of their connection to the forest and of their responsibility to assist with its stewardship. Many people do not understand, for example, the importance that watersheds have in supplying clean water to communities. Because of this lack of understanding, interface residents may not actively assist or support managers in ensuring that watersheds are sustainably managed. Educational programs could also improve the perceived legitimacy of specific land-use planning and zoning measures. The distribution of information over the Internet through the use of web sites aimed at the general public could be an effective way of conveying such messages.

A strong growth management program formalized by land-use planning and zoning also requires citizens to develop a consensus about what they want the interface community to look like in the future. Consensus-building processes can range from dispute resolution techniques to open discussion and negotiation procedures (Porter 1997). Such consensus-building exercises should, in turn, be reflected in local planning and land-use ordinances. A recent example of consensus building is the Minneapolis/St. Paul Blueprint 2030 project. As part of the Twin Cities' effort to plan for future anticipated growth, the Metropolitan Council, a regional planning agency that coordinates development in a seven-county area, sponsored several Community Dialogues. Public participants, including interested citizens, public officials, business people, environmentalists, and developers, were

FIGURE 6.9 House in Clayton, GA. Many landowners believe that they are entitled to do whatever they please with their property, regardless of the consequences their land-use decisions may have on their neighbors. (From Kundell et al. 1989. With permission.)

asked questions about the kind of metropolitan area they preferred and voted via electronic keypad technology after discussing the issues. The questionnaire could also be answered online, and discussion via the web site was possible. Feedback from these Community Dialogues is being used to develop a draft version of the Blueprint 2030 planning document that will be made available to the public in the fall of 2002 (Metropolitan Council 2002).

In order to protect natural resources within the interface, all stakeholders, including the public, must be engaged in the land-use planning and zoning process. Mechanisms for broad-based and ongoing opportunities for open dialogue must be provided throughout the land-use planning and zoning process. These dialogues should be open to any person, conducted in nontechnical terms, readily understandable to the general public, and structured in a manner that recognizes and accommodates differing schedules, capabilities, and interests. The participation of citizens should be encouraged from the beginning and maintained throughout the planning and zoning process. It is imperative for natural resource managers to remember that without broad-based public understanding and support, land-use planning and zoning designed to conserve and protect natural resources within the interface by controlling growth and development cannot succeed.

6.5.4 Increase Natural Resource Manager Involvement in the Planning and Zoning Process

Natural resource managers are in a unique position to protect the natural resources within the wildland–urban interface by working with communities to implement concepts of environmental preservation and conservation in their current and long-range planning and zoning practices. In order to do this, however, natural resource managers need to understand the role that local planning and elected officials play in managing growth and development within the interface. In particular, natural resource managers need to better understand and influence land-use planning policy as it relates to natural resource management. One way natural resource managers can support and help address land-use planning and zoning issues in their states and local communities is to actively participate in the local land-use planning process. Natural resource managers can do this by initiating communication with planners, elected officials, landowners, and developers, by responding to requests for comments or participation by local communities, and by paying closer attention to the goals and objectives of the local planning process. For example, one USDA Forest Service official stationed in the San Bernardino National Forest works directly with municipal planners and private developers to minimize or mitigate the potential impacts of development on the forest, which is situated 50 mile from Los Angeles (USDA Forest Service 1999). Involvement in the land-use planning process presents an ideal opportunity for natural resource managers to call attention to important natural resource issues within the interface that may not have been recognized by planners and developers. By helping private landowners, local community officials, and planners understand ecological systems, natural resource managers can help ensure that planning and development decisions are made in an informed, science-based manner (see Box 6.4).

6.6 CONCLUSION

The conflict of values and the risk to resources that result when urban development is mingled with wildland areas

Box 6.4

Limits of Acceptable Change

One way in which natural resource managers can influence land-use planning decisions within the wildland–urban interface is by working with planners using such methods as the limits of acceptable change (LAC) process (see Chapter 10). The LAC process works by identifying a community's desired future condition and specifying issues and concerns of future development. LAC then defines desired goals, establishes a standard of acceptable land-use patterns, and makes recommendations regarding planning policies to guide decision making when conditions approach unacceptable land-use patterns (Zipperer et al. 1992). The Mount Rogers National Recreation Area provides an example of how the LAC process is implemented. In July 1999, the USDA Forest Service began the LAC process to formulate a long-term management plan for the Mount Rogers High Country, nearly 20,000 acres located in Smyth and Grayson counties, Virginia. Recent increases in the use of this scenic area caused USDA Forest Service officials to become concerned about the possibility of adverse impacts to the land. The LAC process is being used in the High Country to create a plan that will contain standards and goals to manage future recreational use and impacts in the area. To achieve this end, the USDA Forest Service is conducting the LAC process as a series of public meetings that will capture ideas and help formulate this management plan. As of 2002, the LAC process had moved through step six of a nine-step process. The entire LAC process is expected to take 2–3 years to complete (USDA Forest Service 2002). The Mount Rogers National Recreation Area experience demonstrates how the LAC process can be used by natural resource managers, land-use planners, and the public to effectively detect and prevent the degradation of natural resources within the interface.

FIGURE 6.10 Planning, zoning, and growth management programs can be used to mitigate the conflicts that may occur when urban growth encroaches into undeveloped areas. (Courtesy of Jim Kundell.)

is one of the most difficult issues being faced by natural resource managers (Figure 6.10). These are largely policy issues that need to be addressed by the public as well as elected and appointed officials if any progress toward preserving natural resources within the wildland–urban interface is to be made. Citizens can protect the natural resources of their interface communities through responsible land-use planning implemented by zoning ordinances, subdivision regulations, and other measures that incorporate new strategies for protecting natural resources. Many flexible growth management techniques can be added to this foundation to construct a strong community land-use policy. Natural resource managers can play an important part in this process by raising public awareness of the natural resource and conservation issues within the interface, as well as becoming directly involved in the planning and zoning process in their local communities. By providing planners with current and accurate environmental information, natural resource managers can help a community develop a solid land-use plan that, when implemented through consistent zoning ordinances, is capable of withstanding legal challenges. Ultimately though, it is up to the citizens of communities within the interface to determine whether growth will be managed in an environmentally sensitive manner so that the natural aspects that make these communities such desirable places to live are protected and preserved for future generations.

REFERENCES

Adirondack Park Interpretive Centers, 2001. About the Adirondack Park, http://www.northnet.org/adirondack-vic/about.html. [Date accessed: February 5, 2001.]

American Planning Association, 1999. Planning Communities for the 21st Century, American Planning Association, n.p., http://www.planning.org/plnginfo/GROWSMAR/images/APA_complete1.pdf. [Date accessed: September 25, 2000.]

Beaumont, C.E., 1996. *Smart States, Better Communities: How State Governments Can Help Citizens Preserve Their Communities*, National Trust for Historic Preservation, Washington, DC.

Corley, C., 2002. Conservation Community of Prairie Crossing near Chicago, July 18, 2002, http://www.npr.org/programs/atc/index.html. [Date accessed: August 22, 2002.]

Fowler, L. and H. Neuhauser, 1998. A Landowner's Guide to Conservation Easements for Natural Resource Protection, Georgia Department of Natural Resources, http://www.dnr.state.ga.us/dnr/wild/heritage/consease. [Date accessed: October 24, 2001.]

Freis, J.H., Jr. and S.V. Reyniak, 1996. Putting takings back into the Fifth Amendment: land use planning after Dolan v. City of Tigard, *Columbia Journal of Environmental Law* 21: 103.

Georgia Department of Community Affairs (DCA), 1998. Georgia's Future: Beyond Growth Strategies: Recommendations of the Growth Strategies Reassessment Task Force, http://www.dca.state.ga.us/planning/gafutureintro.html. [Date accessed: November 9, 2000.]

Georgia Department of Community Affairs (DCA), 2002. Environmental Planning Criteria Q & A, http://www.dca.state.ga.us/planning/gafutureintro.html. [Date accessed: August 9, 2002.]

Griffith, J.C., 2000. The preservation of community green space: is Georgia ready to combat sprawl with smart growth? *Wake Forest Law Review* 35: 563–607.

Juergensmeyer, J.C. and T.E. Roberts, 1998. *Land Use Planning and Control Law*, West Group, St. Paul, MN.

Kaiser, E.J. and D.R. Godschalk, 1995. Twentieth century land use planning, *Journal of the American Planning Association* 61: 365.

Kayden, J.S., 1992. Market-based regulatory approaches: a comparative discussion of environmental and land use techniques in the U.S., *Boston College Environmental Affairs Law Review* 19: 565.

Keene, D.B., 1996. Transportation conformity and land-use planning: understanding the inconsistencies, *University of Richmond Law Review* 30:1135.

Kenney, J.A., III, 1985. Problem of people: critical areas and floating zones in the Chesapeake, *Virginia Journal of Natural Resources Law* 4: 209–218.

Kreyling, C., 2001. Not-so-wild things, *Planning* 67: 18.

Kundell, J., R.W. Campbell, J.M. Heikoff, L.R. Hepburn, R. Klant, and S.W. Woolf, 1989. *Land-Use Policy and the Protection of Georgia's Environment*, Carl Vinson Institute of Government, University of Georgia, Athens.

Lockard, O.O., 2000. Solving the tragedy: Transportation, pollution and regionalism in Atlanta, *Virginia Environmental Law Journal* 19: 161–195.

Marwedel, J., 1998. Opting for performance: an alternative to conventional zoning for land use regulation, *Journal of Planning Literature* 13: 220.

Metropolitan Council, 2002. Blueprint 2030, http://www.metrocouncil.org/planning/blueprint2030/overview.html. [Date accessed: July 24, 2002.]

Natural Resources Spatial Analysis Laboratory (NARSAL), 2002. Georgia Land Use Trends Program Database, Institute of Ecology, University of Georgia, Athens.

The Nature Conservancy, 1999. Home Page of the Nature Conservancy, http://www.tnc.org [Date accessed: November 20, 2000.]

Porter, D.R., 1997. *Managing Growth in America's Communities,* Island Press, Washington, DC.

Porter, D.R., 1999. Reinventing growth management for the 21st century, *William and Mary Environmental Law and Policy Review* 23: 705.

Smith, H.H., 1983. *The Citizen's Guide to Zoning,* American Planning Association, Washington, DC.

Smith, H.H., 1993. *The Citizen's Guide to Planning,* American Planning Association, Washington, DC.

Trust for Public Land, 2000. About TPL, http://www.tpl.org. [Date accessed: January 4, 2001.]

Umlauf-Garneau, E., 2000. Sustainable Sites, http://www.housingzone.com/topics/pb/legislation/pb00ia012.asp. [Date accessed: July 19, 2002.]

U.S. Department of Agriculture, Forest Service (USFS), Committee of Scientists, 1999. Sustaining the People's Lands: Recommendations for Stewardship of the National Forests and Grasslands into the Next Century, http://www.fs.fed.us/nes/science. [Date accessed: October 24, 2000.]

U.S. Department of Agriculture (USDA), Forest Service, 2002. Limits of Acceptable Change Study, http://www.southernregion.fs.fed.us/gwj/mr/lac/index.htm. [Date accessed: August 12, 2002.]

U.S. Department of Agriculture, National Resources Conservation Service (NRCS), 2002. Farm Bill 2002: Farmland Protection Program, http://www.ncrs.usda.gov/programs/farmbill/2002/pdf. [Date accessed: August 14, 2002.]

U.S. General Accounting Office (GAO), 2000. Community Development: Local Growth Issues — Federal Opportunities and Challenges, GAO/RCED-00-178, http://www.access.gpo.gov/su_docs/aces160.shtml?/gao/index.html. [Date accessed: August 14, 2002.]

Westphal, J.M., 2001. Managing agricultural resources at the urban–rural interface: a case study of the Old Mission Peninsula, *Landscape and Urban Planning* 57: 13–24.

Wiebe, K.D., A. Tegene, and B. Kuhn, 1997. Finding common ground on public and private land, *Journal of Soil and Water Conservation* 52: 162–165.

Zipperer, W.C., R.L. Neville, and G.L. Stokes, 1992. Managing urban sprawl at the fringe, in *Proceedings of the 5th National Urban Forest Conference*, November 15–19, 1992, Los Angeles, CA, P. D. Rodbell, Ed., American Forestry Association, Washington, DC, pp. 30–32.

7 Developing Land while Retaining Environmental Values: A Modern Search for the Grail

Douglas R. Porter
The Growth Management Institute

Lindell L. Marsh
Siemon, Larsen & Marsh

CONTENTS

7.1 INTRODUCTION

The wildland–urban interface occurs at the urbanizing edges of cities and towns and along surviving natural corridors within urban areas. At these junctures, urban development and environmental protection appear to work at cross-purposes. Community development interests see urgent needs to accommodate a growing population and economy in pleasant neighborhoods and business centers, while conservation advocates are troubled tby the continued spread of American cities and towns into the countryside. Problems at the interface are exacerbated by our conventional low-density forms of suburban development, which consume great swaths of forests and farmlands and threaten to degrade air and water quality. But our approaches to protecting natural qualities and features at the interface are often uncoordinated and a case of too little, too late. Unfortunately, the debate over balancing needs for urban growth with environmental protection traditionally has been framed in terms of growth vs. nature instead of growth with nature or even, as Ian McHarg argued, growth within nature (McHarg 1992).

The man/nature conflict has historic roots, of course. The *Book of Genesis* articulates man's dominion over nature, for example. James Fenimore Cooper's writings reflect the Romantic fixation on bringing order out of chaos. Casper David Friedrich painted darkly threatening landscapes overpowering man's presence. Perhaps, we are

1-56670-602-5/05/$0.00+$1.50
© 2005 by CRC Press LLC

hoping to heal the deep, age-old schism by initiating a modern search for the Arthurian grail.

Our increasing appreciation of the concepts of sustainable development and smart growth may show us a new pathway to reconciliation of human and natural systems. Developers, planners, and public officials are paying increasing attention to forms of urban expansion that weave townscapes and landscapes together — that retain significant natural qualities within urbanized areas as valued "green infrastructure." In a sense, we seek to merge human and natural ecosystems. Although we have much to learn about the "what" and "how" of these concepts, the outlook for environmentally sensitive urban development (a term once presumed an oxymoron) promises some fruitful approaches to balancing development and natural preservation. In this chapter, we argue that we have at hand many of the necessary techniques to accomplish that aim.

7.2 THE DEVELOPMENT/ENVIRONMENT CONUNDRUM

For most of this nation's history, Americans have focused on promoting the development of the seemingly inexhaustible supply of land across the nation. Federal domestic policy has emphasized settlement and economic development — taming the wilderness, building cities, and promoting economic growth. Federal programs from early times encouraged human use of land for farming, resource recovery, habitation, and production at the expense of forests, swamps and wetlands, natural waterways, and wildlife habitats. Highway construction has been a major priority for a century, weaving a network of interstate and intrastate roads that continue to open new territories to development. Federal programs also have invested heavily in services such as water supply and electric power that enable and spur development. Such long-term, major investments have helped to propel the spread of population across the continent and, over the last century, from central cities to suburban jurisdictions and from northeastern and midwestern states to southern and western ones.

In the rush to settle the beckoning landscape, the importance of the environmental qualities of the nation's land often got short shrift. Cities and towns set aside parks for recreational purposes, and national and state preserves were established to protect highly valued natural features, but there was little recognition of the impacts of development on fundamental environmental resources. Perhaps most visible were the degradations during the rise of the industrial age, when factories discharged noxious fumes into the air, dumped wastes into rivers and wetlands, and infected their sites with toxic materials. Noticeable, too, were cities that turned rivers into sewers, captured stream flows to produce electricity, covered stream valleys and wetlands with development, wiped out woodlands and wildlife habitats to provide building sites, and allowed construction in floodplains. Less obvious were the environmental effects, especially since the mid-20th century, of decreasing densities of development that use land extravagantly and promote dependence on highways and automobile travel. Also, only over the past one or two decades have we begun to acknowledge the impacts of these forms of development on natural hydrologic and landscape systems and air quality.

During the late 1960s and early 1970s, however, the nation began to confront the costs of unrestricted settlement patterns on the environment. In those years, a wave of federal laws and regulations enacted to protect wetlands, water and air quality, and threatened species signaled greater interest in checking unbridled development, particularly at the wildland–urban interface. Many states followed the federal lead with environmental regulations, including acts modeled on the National Environmental Policy Act of 1969, which required evaluations of environmental impacts caused by state actions and, in some cases, by major private development projects.

The laws, in general, required federal or state agencies to determine that proposed projects will have little or no impact on environmental qualities or that actions will be taken to avoid, reduce, or compensate for impacts before permits can be issued. In addition, the National Environmental Protection Act provided an evaluation process that offers many opportunities for public review and comment on proposals and for litigation on both the substance and procedures of environmental impact reviews.

But both federal and state initiatives tended to establish "command-and-control" regulatory regimes based on the faulty assumption that regulations alone could manage the development process. Two major problems emerged in administering the laws. First, all three tiers of regulation (local, state, and federal) relied upon permitting procedures that generally deal with individual project applications rather than areawide environmental concerns. The possible cumulative effects of many projects over time are not evaluated. Relationships among permits for different purposes are rarely established. In many cases, opportunities are missed for reconciling competing objectives or making trade-offs for securing better conservation.

Second, federal permitting procedures operate virtually independently of state, regional, and local planning processes. Federal agencies view their role as regulators, administering laws and rules in a top-down manner. They make little attempt to ensure that permits make sense in the context of local planning and development policies. A report by the Maryland Office of Planning makes the point: "The regulations may treat matters in isolation but nature does not." It goes on to observe that "a more ecological approach to regulation, one that looks at the relationship of all the pieces, would prevent some of the defects in the present system" (Maryland Office of Planning 1995). Local governments, therefore, often have a difficult time applying federal environmental mandates in the real world of community development.

Regulatory issues are further complicated by mounting movements in some communities to slow or stop development. These initiatives may be motivated by concerns for conserving especially sensitive natural environments or to preserve farmlands; but frequently, their primary aim is to shield newly built areas from the effects of continuing development. The "not-in-my-back-yard" (NIMBY) approach to guiding community development typically disregards the need to accommodate increases in population and economic activity and sets up acrimonious debates over community expansion policies.

7.3 SOME CURRENT SOLUTIONS

To deal with some of the difficulties encountered in reconciling environmental and development objectives, public agencies have devised a number of approaches that provide more collaborative, predictable, and equitable outcomes.

7.3.1 Federal Approaches

Several federal programs now allow use of *ad hoc* planning processes to craft multiparty negotiated settlements on interface issues. In accordance with the 1980 amendments to the Coastal Zone Management Act, the U.S. Army Corps of Engineers can develop Special Area Management Plans (SAMPs) in conjunction with federal, state, and local resource agencies. The use of SAMPs has been extended to all "waters of the U.S." regulated under the Clean Water Act. The plans provide both natural resource protection and reasonable water-dependent development for designated areas. The EPA and the Corps can also carry out studies to provide advanced designation of wetlands. The studies, by designating wetlands as suitable or unsuitable for disposal of dredged or fill material, give advance notice of conservation needs and thus reduce conflicts between landowners and the agencies in securing Section 404 permits. The Endangered Species Act allows landowners, governmental agencies, and environmental groups to participate in preparing Habitat Conservation Plans (HCPs) to identify habitats to be conserved as well as areas that may be developed. Approval of such a plan allows minor incidental "taking" of low-priority habitat in turn for assurances of protection, restoration, and maintenance of high-priority habitats. In rapidly developing areas such as Southern California and Austin, TX, HCPs have proven an invaluable, although as yet imperfect, tool for reconciling needs for development and conservation at the wildland–urban interface.

All of these federally supported planning efforts provide stakeholders an opportunity to participate in planning, improve the predictability of the federal and state permitting process, and provide more flexibility to address specific environment vs. development conflicts in the regulatory process. Commonly, they focus on a specific concern, such as habitat conservation within a specific geographic area. The collaborative approach allows planning for environmental resources to be considered together with local and regional objectives for such areas, thereby strengthening the subsequent implementation process. The planning processes are time-consuming, however, and depend on voluntary participation of agencies and other organizations whose budgets frequently limit staff travel and time spent on lengthy projects. Also, except for HCPs, they carry no federal assurances that the plans, once completed, will be fully supported by later agency actions.

7.3.2 State Approaches

Many states have established processes for designation of critical areas for conserving important natural features and resources threatened by development. Such designations assure continuing attention to conservation within the areas but, more important, usually establish criteria and standards to guide decisions on development. The Pinelands in New Jersey, which conserves a pine barrens and wetlands ecosystem in the central third of the state, is one notable example (New Jersey Pinelands Commission 1994). Another is the Maryland Chesapeake Bay critical area, established by law in 1984 (Meyers et al. 1995) (see Box 7.1).

7.3.3 Local Approaches

Less dramatic but widely applied are the efforts of many local governments to recognize needs for environmental protection in developing areas. Many local planning offices expend a considerable amount of time and effort to determine the location and characteristics of specific resources in developing areas. In many cases, they employ the technique advocated by Ian McHarg of overlaying resource maps to define areas of greatest value for natural conservation (McHarg 1992). In addition, local planners often undertake some form of "carrying capacity" analysis to determine areas (e.g., floodplains, steep slopes, prime farmlands) that should not be developed, areas that may be developed with special care to avoid adverse impacts, and areas suitable for development. Such an analysis does not absolutely define lands to be conserved. Conservation needs must be balanced by needs to accommodate development, and adverse environmental impacts

Box 7.1

The Chesapeake Bay Critical Area Law

The largest estuary in the U.S., the Chesapeake Bay drains an area of 64000 square mile, including 1.2 million acres of wetlands. Population increases in the bay region during the 1960s and 1970s began to increase nutrient and sediment loading of bay waters, causing dramatic declines in fisheries, waterfowl population, and general water quality, as well as submerged aquatic vegetation. In response, the states of Maryland, Pennsylvania, and Virginia, plus the District of Columbia and EPA, signed the Chesapeake Bay Agreement in 1983 that committed each party to take immediate, substantial measures to restore and protect the bay. One of the most dramatic commitments was Maryland's Chesapeake Bay Critical Area Law of 1984 (Maryland Natural Resources Code Annotated, Section 1801, et seq.).

Maryland's new law defined a critical area consisting of the water of the bay, tidal wetlands and tributaries, lands under these waters, and 1000 feet of upland adjoining the water boundary. A commission developed criteria to minimize adverse impacts of development and farming on water quality, fish and wildlife, and plant habitat, which were enacted into law in May 1996. The criteria placed strict limits on growth, and the 16 counties and 45 municipal governments within the critical area were required to prepare plans adhering to the approved criteria and submit them for commission approval. Eventually, all local plans were approved, the commission succeeded in prompting local jurisdictions to account for environmental factors in planning for development, and substantial preservation efforts are now in place on the shores and waters of the bay (Meyers et al. 1995) (see Figure 7.1).

by development may be avoided or mitigated by various design and other techniques. But the capacity evaluation can establish a better understanding of potential natural constraints on development. Local planning offices use these techniques in formulating comprehensive plans to guide the locations and character of future development (see Box 7.2).

To implement such long-range plans, local governments have transformed their land development regulations over several decades to recognize needs for environmental protection in developing areas. Zoning requirements now routinely promote conservation goals by establishing conservation, agricultural, and open space districts. Subdivision regulations often incorporate requirements such as on-site stormwater retention to promote groundwater recharge and control erosion, setbacks to buffer stream valleys, tree conservation requirements,

Box 7.2

The Desert Spaces Management Plan for Maricopa County, Arizona

Most local comprehensive or general plans designate areas and policies for protection of critical environmental resources. An example is the "Desert Spaces Management Plan" prepared by the Maricopa Association of Governments for the Phoenix, AZ, region (Maricopa Association of Governments 1996). The plan identifies (1) public and private lands with outstanding open space values that should be protected from development; (2) public and private lands with high open space values whose environmental features should be retained through sensitive development; and (3) existing or designated parks, wilderness, and wildlife areas. The plan recommends acquisition of the first category of lands, some 1.5 million acres, through a combination of regulatory restrictions, reservations through the subdivision exaction process, donations by individuals and conservation groups, purchase of easements, or acquisition. The second category, about 2.2 million acres, should be managed through zoning and subdivision approval processes that would secure reservations of important natural features in development plans. The plan was adopted by the Association of Governments in 1995. The plan's primary significance lies in its identification of major natural areas worthy of preservation, which can then guide the future actions of governments, landowners, and conservation organizations in the region.

and wetlands protection and restoration. Also, many communities are adopting growth management measures intended to curb development in rural areas, such as urban limit lines, urban growth boundaries, farmland preservation programs, incentives for clustering development, and programs to promote recycling of land and buildings in existing urbanized areas (see Boxes 7.3 and 7.4). (Many of these measures are described in Chapter 6.)

7.4 FAILINGS AND FLAWS

The approaches described above are not meant to imply that conservation needs are always recognized by growing communities or that local planning goals are always achieved. Many communities simply fail to invest the effort necessary to anticipate needs for protecting environmental resources, and are frequently more excited about welcoming development than assuring its quality and sustainability. Often, local governments managing growth in the urbanizing edge of development are still too low on the learning curve to recognize the long-term

FIGURE 7.1 The Chesapeake Bay Commission aims to protect the water quality of the bay, a task that requires regional cooperation and public education. (Photo by Larry Korhnak.)

downsides of development, and property rights advocates may threaten legal action to avoid restrictions on development, especially curbs on developing sensitive natural areas. Planning and management of environmental conservation in the face of pressure for development have tended to generate costly and continuing conflicts between development and environmental interests, with little appreciation among the parties for meshing objectives rather than creating opposition to specific proposals. Nowhere is this problem more evident than in the wildland–urban interface.

Consequently, in many respects interface areas resemble battle zones, littered with disconnected bits and pieces of development that may or may not mature into desirable built environments and lying amid dysfunctional remnants of open land and scarred landscapes. Interface areas are battle zones as well, because of conflicts between landowners and builders eager to move ahead with development and others concerned with protecting natural features and/or keeping urban growth at bay. In many communities, occasional skirmishes between environmental and development interests have evolved into open warfare on practically every development proposal that comes down the pike.

These circumstances stem from a number of factors at work in most growing urban areas:

- Long-established preferences for low-density living and travel by automobile that in combination generate large-scale demands for urban expansion
- Expectations of property owners, including land speculators, to benefit from value increases occurring at the urbanizing fringe, coupled with public policies, including zoning practices, that support those expectations
- Ingrained practices of the development industry, including financial backers, that reward short-term value capture and virtually ignore long-term externality effects
- Fragmentation of public decision making on development issues that generates competition rather than coordination among jurisdictions and generally prevents large-scale, long-term considerations for balancing urban growth and environmental protection
- Uncoordinated and mostly reactive efforts by environmental interests to identify and prioritize key natural features and qualities in advance of development pressures.

Among the consequences of these trends are the capital and process costs incurred by both public and private interests in the face of rapidly escalating property values, and the frequently piecemeal and disconnected results of conservation campaigns. For example, the HCP process to preserve two dozen species, nine listed as endangered, in 75000 acres of the Balcones Canyonlands near Austin, TX, evolved over 11 years. During this period, environmental and development interests fought over such issues as the necessity of preserving the identified species, the scope of preserved lands, and responsibilities for the

Box 7.3

Conservation Planning and Implementation in Lincoln, MA

The town of Lincoln, MA, a community of 7710 residents covering 15 square mile near Boston, established a conservation-centered plan years ago, reflecting residents' concerns for careful stewardship of the land (Figure 7.2). Steady growth from the 1960s onward, which threatened to erode the town's rural character, led Lincoln's leaders to plan, prioritize, and manage open space acquisition while allowing carefully planned development. When Lincoln adopted its first master plan in 1958, local citizens also established the Lincoln Land Conservation Trust to manage open space tracts already in town ownership and to solicit new donations. After rezoning 1200 acres in 1973 to preclude development of wetlands, the town adopted an open space plan in 1976 that designated 1450 acres as "land of conservation interest," and issued a prioritized list of lands it hoped to acquire or preserve in some manner.

Over the years, the town and land trust together have acquired about 1700 acres of fields, woods, trails, and wetlands. Another 375 acres in private hands have been protected by conservation restrictions. The open space plan, updated in 1983 and 1989, remains a core working document in guiding town actions for conservation. Among other attributes, it demonstrates how Lincoln's open spaces combine into a network of connected lands that provide for ecosystem and recreational needs (Porter 1996).

estimated $134 million in costs of land acquisition. The complex and lengthy negotiations resulted in decisions to acquire half the land as part of a federal wildlife refuge, to raise $40 million from City of Austin revenues, and to tap developers for the remaining land requirements. But the costs of extended conflict — not counted in the acquisition costs — were immense in terms of dollars and time for all involved (The Growth Management Institute 1996).

In Florida's Sarasota County, the establishment of a growth boundary in 1975 to control the spread of development into open lands failed to curb low-density development. Resulting needs to expand the boundary to accommodate continued development caused sharp opposition, including challenges to a series of county comprehensive plans and a batch of litigation stretching over years (Porter 1996). Moreover, Maryland, despite aggressive state programs to fund purchase of farmland and sensitive areas, was pressured by environmental groups to spend $25 million — considerably more than its annual budget for open space acquisition — to acquire a tract on

Box 7.4

Voter Support for Open Space Protection

Advocates of open space protection have been heartened in recent years by the willingness of local voters to tax themselves to pay for acquisition of open space. A survey in 1998 found that voters from California to New Jersey approved, often with large majorities, over 170 state and local ballot measures to protect, conserve, and improve parks, farmland, historic resources, watersheds, greenways, and other environmental enhancements. The measures authorized more than $7.5 billion in state and local spending for conservation (Myers 1999). Another survey in 2000 revealed that voters approved more than three fourths of over 100 state and local ballot measures that committed $3 billion for protection of open space and enhancing recreational opportunities (*Common Ground* 2000). One major program was Florida's Preservation 2000 program, enacted in 1990 to provide $300 million annually for 10 years to protect wildlife habitat, water resources, forests, beaches, and parks. The program has been succeeded by a similar one labeled Florida Forever, and beginning in 2001, the Florida Communities Trust will dedicate $70 million per year to implement open space, recreation, and coastal management elements of local plans.

the Potomac River, paying top dollar to prevent development already sanctioned by the county (Governor's Press Release 1998).

These inadequate and conflict-ridden approaches to resolving development/environment issues at the wildland–urban interface should be transformed into more constructive and effective processes. Future efforts should focus on providing broader, more certain protection by improving the delineation and implementation of development and conservation areas, by incorporation of environmental values into development policies and project designs, and by wider use of collaborative processes for making these decisions.

7.5 SOME HOPEFUL DIRECTIONS

A host of innovative approaches for achieving an equitable and sustainable relationship between urban growth and conservation in interface areas are being pursued throughout the U.S. Some arise from ideas incorporated in the concepts of sustainable development and smart growth, which are gaining increasing political and citizen support. Others involve regional and local initiatives to achieve more effective management of environmental resources in areas affected by rising development pressures.

FIGURE 7.2 Lincoln's conservation plan preserves clusters of houses amid protected open spaces. (Photo by Alex S. MacLean/Landslides.)

7.5.1 Sustainable, Smart Development

Popular in some circles, the concept of sustainable development builds on McHarg's assertion that human and natural systems are intricately bound together (McHarg 1992). Cities, he believed, exist in natural settings whose qualities are essential to maintaining life. Surface water, marshes, floodplains, aquifers, aquifer recharge areas, steep lands, prime agricultural land, and forests and woodlands form natural systems that are an essential component of our living environment. Land, in other words, is not simply a commodity awaiting development and stream valleys are not just handy places to dump trash. Old-growth forests are a finite resource that should not be indiscriminately covered with concrete. Marshes and floodplains perform valuable functions for man's benefit as well as for the natural order.

Making development sustainable calls for maintaining the integrity of complex ecological systems while promoting economic viability and social equity. Although development is necessary to further economic and social ends, it should be undertaken in ways that minimize impacts on the natural functions of landscapes. Development should be designed to maintain sensitive lands and habitats, to minimize its footprint on the land in order to retain natural features, and to retain natural land and water functions rather than using engineered facilities (Beatley and Manning 1997). These high-minded principles may seem difficult to apply in the everyday world of growing communities, but planners and designers are finding ways to blend development with nature, as illustrated below.

Smart growth principles are focused especially on guiding the character of community development in ways that will improve the quality of life. The concept of smart growth packages broad statements that promise to widen choices of living environments and conserve natural and economic resources through inclusive, consensual processes of decision making. Many organizations have drawn up lists of smart growth principles that vary according to their particular interests (see, e.g., American Planning Association 1998; National Association of Home Builders 1999; Hirschhorn 2000). Generally, however, the lists include:

- Promoting compact, mixed-use development
- Conserving open space and natural features and qualities
- Efficiently maintaining and expanding infrastructure systems
- Broadening choices of community and neighborhood living environments
- Encouraging infill, redevelopment, and adaptive reuse in existing built environments that will reduce sprawl into the countryside
- Expanding mobility through multimodal transportation
- Applying these principles by involving all interests and considering regional needs.

Like the ideals of sustainable development, these principles are quite general and capable of many interpretations. But they express goals and hopes, or targets and directions, which communities can use as guides to planning for future development at the wildland–urban interface. Some states and many communities are incorporating smart growth principles within their planning and implementation processes for community development (Porter 2002) (see Box 7.5).

"Conservation subdivisions" offer one model for accomplishing some of the aims of sustainable development and smart growth. Championed by Randall Arendt as an alternative to conventional low-density suburban projects, conservation subdivisions are designed to cluster development in one part of a site to allow retention of substantial open spaces in the remainder of the site (Arendt 1996). Development may be clustered in the grid patterns propounded by advocates of new urbanist or traditional neighborhood designs or simply grouped together on small lots (Katz 1994). The retained open spaces are protected and enhanced to preserve in working order farmlands, woodlands, wetlands, or other environmental features and qualities (Figure 7.3).

Many development designers are taking this concept a step further toward sustainability by using on-site wetlands, for example, to manage stormwater runoff, or maintaining native vegetation and/or woodlands as wildlife habitats. Their plans for new projects pay attention to maintaining the natural systems on the site and even enhancing them to provide amenities and support functions for development (Porter 2000). Advocates of this approach offer the following guidelines for developing in greenfields areas:

- Respect the existing natural landforms and landscape features, and understand the historic and current ecosystem, its natural processes, and the stresses that adjacent development may place on it.
- Take advantage of a site's natural assets as much as possible by preserving the existing landforms and vegetation that define its natural structure and character.
- Create landscapes that can be sustained as a permanent, ongoing natural environment, and plan to restore the site's landscape character and vegetative palette, using native plant species in undeveloped or open space parts of the site.
- Refrain from breaking up or promoting intrusion into contiguous expanses of sensitive habitats and wildlife corridors, especially those of threatened or endangered species.
- Avoid construction in stream valleys and floodplains.
- Endeavor to create landscapes that define spaces and create places for a variety of activities and that reinforce relationships between buildings and landforms (Porter 2000) (see Box 7.6).

These designs that respect and adapt existing landscapes and hydrologic functions are clearly preferable to conventional "land-scraper" subdivisions, especially if the preserved open spaces remain linked to offsite natural

Box 7.5

A Tangle of Terminology

Many terms are now in use to describe contemporary concepts of forms of development that respect environmental qualities and values. The subtle distinctions among some of the most popular are briefly described below.

Sustainable development: Development that, by integrating and respecting environmental, economic, and social concerns, meets the needs of the present without compromising the ability of future generations to meet their own needs.

Green development: Development that promotes conservation and restoration of land and water qualities on building sites and the use and recycling of renewable resources for building designs, materials, and functions.

Green infrastructure: Emphasizes the integration of natural resources within urban development, including creation of multiuse greenways, retention and restoration of existing landforms, vegetation, and hydrologic systems within developed areas.

Smart growth: Describes the physical qualities of development that will create more livable communities, preserve open space, and provide more efficient facilities and services in cities, towns, and villages.

Growth management: The variety of development policies and regulatory techniques, programs, and incentives that can assist in implementing community development plans.

Box 7.6

Coffee Creek Center: A Model of Green Development

Coffee Creek Center, a project now in development near Chesterton, IN, illustrates the ecological approach of designer William McDonough. Located in an urbanizing area near the commuter train line to downtown Chicago, the 640-acre development combines a modified grid street system with neighborhood greens and a mixed-use town center. The centerpiece of the development is an existing creek (Figure 7.4) bordered by 240 acres of parklands, including existing and constructed wetlands, restored prairie lands, and cycling and walking trails. The riparian corridor will handle stormwater runoff and provide wastewater treatment as well as a central recreation space.

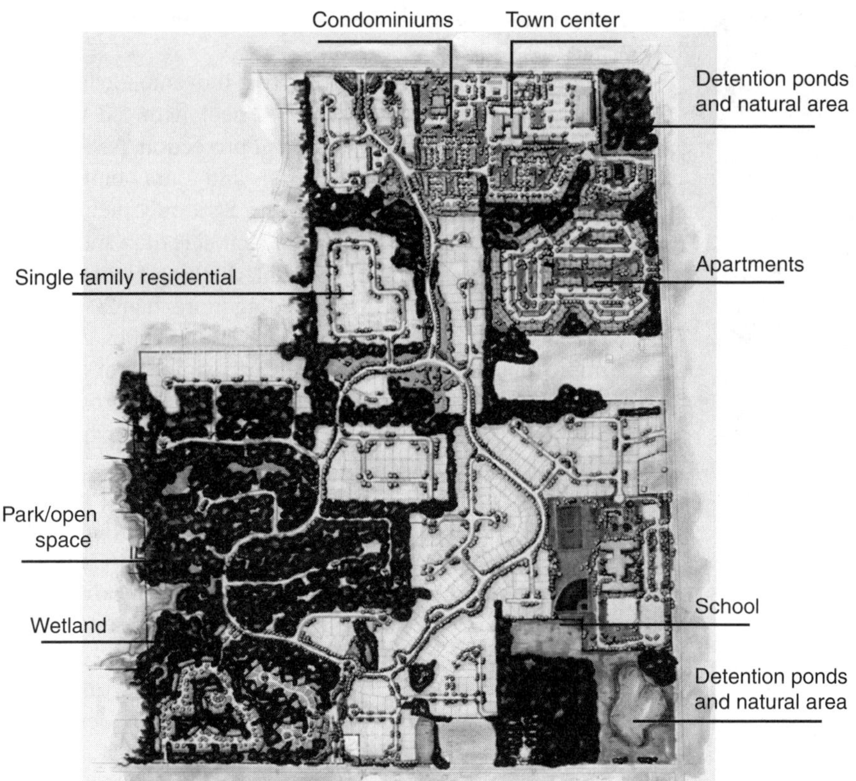

FIGURE 7.3 The plan for Bailey's Grove illustrates how greenways and conserved open spaces can be woven into a clustered development. (Courtesy of David Jensen Associates.)

systems. But 1000 conservation subdivisions do not make a city. Although they respect existing natural conditions, they maintain the low-density aspect of much suburban development. Developing truly urban communities, as proposed by smart growth principles, demands more careful attention to weaving nature and development together. Here, the concept of "green infrastructure" offers a useful direction.

7.5.2 Green Infrastructure: Integrating Development and Conservation

People usually think of infrastructure (when they think of it at all) as the systems of roads, fire stations, sewer and water lines, and other facilities that provide services necessary to support development. The concept of green infrastructure advocates a systems approach to open space as an essential element of development (Benedict 2000). Natural systems are integrated within the framework of other infrastructure systems and constructed buildings to produce a satisfying living and working environment. Essential elements of the natural ecosystem are retained and enhanced to provide a natural setting for urban development. The concept of green infrastructure parallels the concern of green development for conserving land and water resources, but places greater emphasis on strengthening the interrelationships between natural and developed areas. Using the term "infrastructure" also underscores its essential role as an important and supportive component of developing areas.

A green infrastructure system is composed of rivers and stream valleys, wetlands and floodplains, woodlands and wildlife habitats, hills and ridges, plus green spaces meant specifically for human use, such as parks, recreation lands, reservoirs, and retention ponds. These hydrologic systems and natural landscapes can be linked by greenways — strips of land often conserved along rivers, streams, lakes, abandoned canals, and railroad rights-of-way that function as public open spaces. Walking and cycling pathways along greenways allow them to serve as the connective tissue of communities, tying together a series of open spaces, linking one neighborhood to another, and connecting people to the natural world.

Green infrastructure systems therefore keep nature alive and appreciated within urbanized areas, drawing the natural systems existing at the wildland–urban interface into the built environment. The examples of Bailey's Grove and Coffee Creek Center described earlier demonstrate how this can be designed in new developments. Within urbanized and urbanizing areas, the riverfronts being renewed as valued visual and recreational assets in hundreds of communities (as varied as old industrial cities and growing Sunbelt cities) provide another kind of example. Riverfronts provide central green spaces and pathways that often connect regionally important facilities and institutions through urban neighborhoods to developing

FIGURE 7.4 Coffee Creek is being restored as the functional, visual, and environmental centerpiece of the development. (Photo by Jerry Mobley.)

suburbs. At the interface, development can be designed to preserve these natural systems as logical extensions of regional networks of green spaces.

7.5.3 Regional Initiatives

Guiding development to respect natural systems at the wildland–urban interface is most effective if conceived and carried out in a regional context that recognizes the commonalities of built and natural systems across the artificial boundaries of local governments. Environmental qualities, transportation networks, water and sewer systems, and urban economies, for example, all function as regional systems. Determinations about the direction and character of urban growth and environmental conservation benefit from consideration of regionwide patterns of development and natural features. However, except for a few outstanding examples such as the New Jersey Pinelands and Chesapeake Bay areas described earlier, regional initiatives to define appropriate interface relationships have been largely ineffective, victims of community aversion to nonlocal control of development. Although many surveys have documented Americans' applause for regional thinking about development, ultimately, citizens cast their vote for local decision making.

Nevertheless, some productive movement can be discerned on the regional scene to deal with the interface between urban spread and natural protection. As an example, voters in the City of St. Louis and four counties in the Illinois and Missouri parts of the St. Louis metropolitan area approved a one tenth of a cent sales tax increase in 2000 to create and manage an interlocking greenway system along river corridors, including more than 200 mile of hiking and biking trails. Half the tax revenues will fund improvements in metropolitan parks and recreation areas, and the other half will finance local park improvements. The Metropolitan Park and Recreation District was formed to administer the program on the Missouri side (Metropolitan Park and Recreation District [St. Louis Region] 2002). Other examples from Maryland, New York, Santa Clara, and Riverside counties in California, and Portland, Oregon, illustrate some of the possibilities for intergovernmental management of interface issues (see Boxes 7.7 and 7.8).

Urban growth boundaries provide another means for establishing an effective interface between developing areas and conserved lands. Adopted by ordinance, growth boundaries draw a line between permitted development areas and protected conservation areas. Boundaries are implemented by zoning, which encourages development within the boundary and discourages development outside it. However, to succeed in establishing a sustainable integration of development and conservation at the interface, boundaries must be supplemented by policies, programs, and regulations that promote compact development, create green infrastructure systems, and maintain conserved open space.

Portland, OR, is perhaps the premier example of long-term, effective use of an urban growth boundary coupled with additional supportive programs and policies. The Metro regional organization, the only elected regional growth management body in the U.S., oversees the regional growth boundary established in 1980 and provides or coordinates a number of regional services (Porter and Lassar 1990; Abbott and Abbott 1991; Porter 1995). In 1992, Metro adopted the Metropolitan Greenspaces Master Plan to supplement the growth boundary as a means of consolidating urban development and retaining significant natural spaces outside it (Portland Metro 2002). The plan led to a proposal to protect 14 key natural resource areas, which was approved by voters in 1995. Voters also approved a number of funding measures to support land acquisition and capital improvements, including a bond issue of $135.6 million for purchasing open spaces, parks, and stream valleys. As of August 2001, Metro has acquired more than 7118 acres of land in 214 separate property transactions and is working on protection plans for streams and floodplains, parks and

Box 7.7

MARYLAND'S RURAL LEGACY PROGRAM

Maryland's Rural Legacy program was part of the state's landmark smart growth legislation enacted in 1997 (Chapter 759, Sections 5-9A Annotated Code of Maryland). The program strengthened existing state programs for purchasing development rights to protect open space by offering state grants to local governments to acquire land and development rights. For the first 5 years of the program, the state committed $128 million to preserve up to 75000 acres of land in Maryland's 23 counties. Criteria for proposed acquisitions favor a multijurisdictional regional approach, including the degree to which purchases would protect contiguous tracts of lands, greenways, or corridors; the size of the protected area and the quality and value of the protected resources; and the degree of threat to the resources. Proposals also gain favor if submitted by two or more local jurisdictions and/or nonprofit partners. Many initial grants involved protection of large blocks of land in several jurisdictions that formed significant natural corridors in areas threatened with development (Maryland Rural Legacy Program 2002).

natural areas, and fish and wildlife habitats. In addition to protecting significant scenic, recreational, and habitat areas, Metro is creating a network of trails and greenways connecting important natural sites.

California cities are increasingly using urban growth boundaries to guide urban development. Santa Clara County, home of Silicon Valley, initiated a policy in 1973 to contain development within the 15 cities and limit development outside cities to nonurban uses (Association of Bay Area Governments 2000). All 15 cities adopted urban service areas, which acted as short-term urban growth boundaries, but beginning in 1995 the county urged cities to establish long-term urban growth boundaries to further discourage land speculation in rural areas and promote more compact urban development. As of 2000, seven have done so. County land-use policies for rural areas encourage retention of open space, resource conservation, agricultural preservation, and limited amounts of very low-density development. Thus, the county and its cities, working in partnership, have created a regulatory framework for managing growth and environmental protection in the rural–urban interface.

Another example of regional efforts to manage the wildland–urban interface is the Multiple Species Habitat Conservation Plan being prepared for western Riverside County, CA. Although modeled to some extent on the multispecies plan for San Diego County prepared by the county and several utility and other organizations, the Riverside County effort is part of an integrated planning process that

Box 7.8

NEW YORK'S REGIONAL "GREENSWARD" CAMPAIGN

New York's Regional Plan Association was organized as a private nonprofit organization by New York's civic and business leaders in the 1920s to promote regional planning in the New York metropolitan area. Its initiative to consolidate preservation efforts in the New York-New Jersey Highlands west of New York City demonstrates a regional approach writ large. A major element of the association's Third Regional Plan published in 1996 was preservation of an extensive regional "greensward" of parks, recreation lands, and environmentally sensitive lands and a network of greenways (Regional Plan Association of New York 1996). To further this strategy, the association backed acquisition of the 5700-acre Sterling Forest property to protect the largest unfragmented forest in the region and ratification of a historic agreement to protect drinking water supplies in the Croton, Catskills, and Delaware basins for nine million New Yorkers, as well as other important initiatives. With these extensive areas secured as part of the regional "greensward" wrapping around the northern and western edges of the metropolitan area, the association has mounted a campaign to protect the two-million-acre New York-New Jersey Highlands, an area rich in wildlife and water resources and offering recreational opportunities that draw two million visitors annually. Association staff are working with a coalition of conservation organizations and with county and state planners to raise the awareness of this area as a significant regional resource. New Jersey's state plan and intergovernmental planning process and the state's two-billion-dollar commitment to acquire open lands are providing a considerable amount of leverage for implementing the Highlands protection strategy (Yaro 1997; Pirani 2001).

also includes preparation of an updated general plan for the entire county and a transportation corridor plan that will achieve community and environmental acceptability for the county's western section (County of Riverside, California, Planning Department 2001). The rapid rate of land conversion to urban development in the 2000-square mile western area threatens the loss of natural habitat and associated species. The multispecies plan will extend existing reserves established by the previous habitat conservation plan for the Stephen's kangaroo rat and create new reserves, habitat linkages, and wildlife corridors. The coordination of the multispecies planning process with regional growth management and transportation planning will ensure the retention of the economic viability of the region while conserving Riverside County's biodiversity.

These examples demonstrate that regional organizations can play an immensely important role in dealing with conflicts and change at the urban edge.

7.6 ELEMENTS OF A CONSTRUCTIVE APPROACH

Experience with determining an equitable and workable balance between development and environmental protection at the wildland–urban interface indicates that several factors can make all the difference between success and failure. Commitments by federal, state, and local agencies to take constructive steps toward reconciling competing needs and objectives are essential. Professionals working in the fields of natural resource protection and urban development can be instrumental in pursuing agency commitments for:

- Early notification (in advance of development) of intentions to conserve (preserve, enhance, restore) significant natural resources
- Conservation of green infrastructure concurrent with urban development
- Promotion of local and regional plans that encourage infill and redevelopment in existing urbanized areas and designs for newly developing areas that balance development and protection of natural resources
- Long-term management of resources, including retrofitting, restoring, and enhancement as appropriate
- Establishment of dedicated funding sources for accomplishing goals
- Adoption of effective processes for reaching agreement and resolving conflicts
- Creation of enforceable agreements and assurances for implementing plans
- Recognition of regional needs and responsibilities for implementing a shared vision.

7.6.1 An Unresolved Issue

Affecting these commitments is an area of unresolved tension between advocates of environmental conservation and proponents of compact, clustered development that reduces needs for urban land. Land planners and designers who preach the virtues of resource-sensitive development as a preferred alternative to conventional subdivisions finesse the issue of appropriate development densities at the community and regional scales. They propose clusters of buildings to conserve substantial parts of development sites in some form of open space, but the resulting overall site density generally remains rather low — frequently in the range of one or two housing units per acre or floor-area ratios considerably less than one. (This consequence does not bother some advocates of back-to-nature development, who are, of course, convinced that cities are too large and already dense and that the good life is found in rural self-sufficiency.)

By contrast, the compact densities of development desired by supporters of smart growth usually aim for an average of over four units per acre and may range up to scores or hundreds of units per acre, depending on local markets and practices. Development at these densities allows preservation of relatively limited amounts of open space and must rely to a great extent on "engineered" facilities (e.g., pipes) to provide water supply, drainage, and wastewater treatment. The walkable, highly interactive centers of civilization such as central Paris, San Francisco, or downtown Portland, OR, cannot be created without generating concentrations of impervious surfaces, some degree of river channelization, and considerable dependence on piped hydrologic systems. Inspired designers and innovators may conserve some features of natural landscapes and hydrologic systems within built-up sites, weaving in greenways, and installing permeable pavements, for example. But the concept of compact development depends on building more densely than propounded by the models of conservation subdivisions and green development. Ideally, natural resource losses associated with compact development can be offset or mitigated by a combination of improved on-site technologies and a provision of a higher level of resource protection and restoration outside the periphery of metropolitan development. Unfortunately, our ability to arrange that trade-off at the regional scale is haphazard and unpredictable, falling between the cracks of myriad local jurisdictions, ineffective regional institutions, and state and federal commitments.

The conflict between these ideas of desirable forms of development has not been well recognized or explored, perhaps because a great deal of development still occurs in conventional forms generally insensitive to preserving qualities of the natural environment. In these circumstances, the occasional green development project is welcomed. Still, interface issues cannot be fully reconciled until environmental and development interests manage to find common ground for optimizing the balance between the interests.

7.6.2 One Model for Achieving Agreement

In the contentious climate found in many growing communities, reaching agreements that resolve interface issues requires an unusual degree of collaboration between the parties at interest. An example of such an approach is the ongoing experience of the Santa Ana River Watershed Group in determining the future course of development and conservation in the Santa Ana River Valley east of the Los Angeles metropolitan area—an experiment in "shared governance."

The Santa Ana River Watershed Group was formed in 1998 to work out solutions to the environmental impacts of the dairy industry in the 50-square mile Chino Basin (Figure 7.5). The 270 dairy operations and 350000 dairy animals constitute a $1 billion industry that provides one quarter of California's milk production. It also produces close to one million tons of manure annually. At one time, manure was spread on fields that grew feed for the cattle, but the increasing concentration of dairy operations and rising land prices gradually reduced this synergistic practice. Now, about two million tons of manure are stockpiled on the ground and increasing surface water runoff from up-slope urban development has been carrying manure salts into the groundwater and river system.

Addressing these problems is complicated by the imminent move of many dairy operations out of the basin and their replacement by urban development spreading out from nearby cities. This trend stimulated the working group to broaden its mission — to craft collaborative initiatives relating to urban and economic development as well as the groundwater pollution issues.

This is not a small task. The Santa Ana River Valley watershed comprises 2650 square mile, including parts of three counties and numerous local jurisdictions, as well as the domains of several water and wastewater utilities. The watershed's population is expected to grow from five to seven million people in less than 20 years, increasing pressure on a broad range of ecosystems and wildlands. The impacts are not only the direct loss of habitat and open space to urbanization, but indirect impacts such as increased competition for limited water supplies, changes in stream and river flow regimes, and impairment of recreational experiences.

7.6.2.1 The Process: Shared Governance

The working group brought together representatives of the local governments, the dairy and utility interests, environmental organizations, and state and federal agencies to explore issues in the watershed and collaboratively determine potential solutions. The collaborative process, which the group called "shared governance," evolved from a need to involve a wide and shifting constituency of interests that allows a variety of individual but connected efforts to meet challenges for future development and conservation. Funding for the group's efforts has been provided by key participants such as the dairies, local agencies, and the U.S. Environmental Protection Agency, as well as grants from utility agencies and foundations.

The objective of shared governance is to promote strategic collaborations — *ad hoc* arrangements that transcend the typical fragmentation of interests and goals often encountered in multiparty processes. Because the need to think systemically may involve strange mixes of agencies and interests, the Watershed Group is an open forum rather than a fixed organization; amoeba-like, it changes shape depending on the issues at hand. From the hundreds of potential participants notified about the project, over 100 have elected to join in facilitated discussions at various times, with a core of about a dozen attending most meetings. Voting (even by consensus) is not sought, since adopting "official" positions and recommendations is difficult to manage among so many interests.

However, NEPA's provision for "scoping" of concerns, issues, and opportunities, which results in reports describing possible approaches to their reconciliation, offered a workable approach that allows the group to evolve in constituency and focus to consider a great

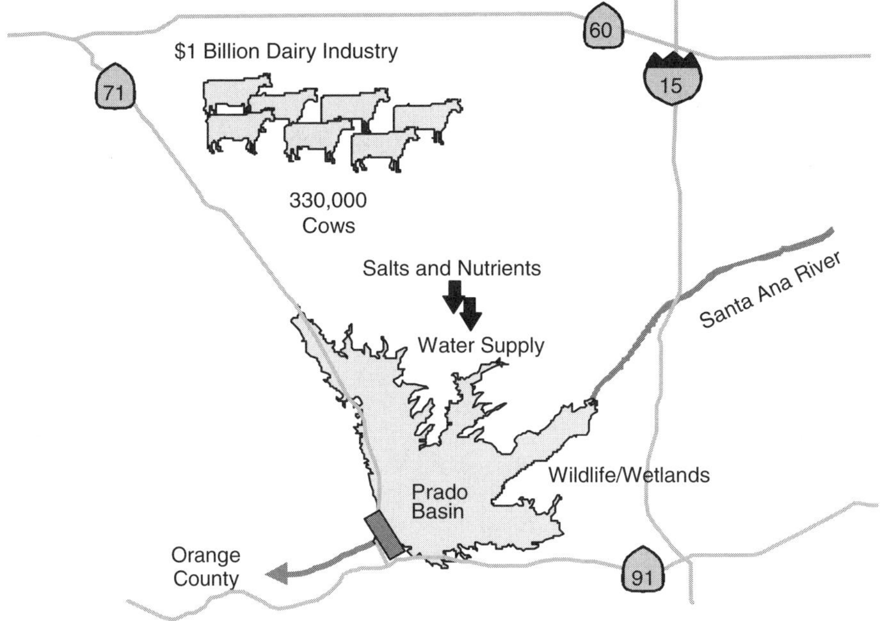

FIGURE 7.5 Chino Basin concerns include environmental impacts of a large dairy industry and spreading urban development. (Courtesy of Lindell Marsh.)

variety of issues. The group employs a facilitation team that, in consultation with participating members, prepares draft scoping reports that identify issues and lay out potential strategies to resolve them. As specific tasks are identified, affected decision makers (those who will implement the tasks) take on assignments to move them forward, in many cases through collaborative processes. This working methodology allows group members to articulate views and positions in a context that fosters cooperation and constructive actions.

7.6.2.2 Some Outcomes

The working group has been involved in a dozen or more initiatives over the past 3 years. They range from preparing a plan for manure management in the Chino Basin and arranging for $10 million in funding to build a drainage retention basin to intercept urban stormwater flows through the dairy area to preparation of a scoping report for a conservation program for 10000 acres upstream of the Prado Dam on the Santa Ana River. The program uses a Corps of Engineers Watershed Plan as an umbrella for a three-part strategy: (1) a habitat conservation/special area management plan, (2) a strategy for managing surface water flows (for water supply, flood control, and water quality), and (3) a recreational program (including a river center and a section of the coast-to-crest trail system). It is anticipated that the program will provide for wildlife habitat, trails, and riparian corridors reaching north from the regional open space into urbanizing areas — the very interlacing of wildland and urban areas espoused by the concept of green infrastructure.

Recently, the group realized the usefulness of formalizing its process through two Memoranda of Understanding. Signatories to one were the Boards of Supervisors of San Bernardino, Riverside, and Orange counties, the Santa Ana Watershed Project Authority (constituting the five major water districts in the watershed), and the Orange County Sanitation District. To add state and federal representatives to this framework, a second one was signed by the above organizations and key state and federal agencies. Both memoranda committed the signatories to continuing participation in the Watershed Group and in specified initiatives.

Perhaps the most important outcome of the process to date has been an unprecedented degree of collaboration between a wide variety of interests. The structure of facilitated discussions centered on participant-determined issues and focused on defining specific action strategies has proven effective in engaging and committing many interests to constructive solutions.

7.7 CONCLUSION

Reconciling what seem to be conflicting aims of conservation and urbanization at the edges of growing communities is gaining increased attention in political and professional circles. Given the continuing expansion of urban populations and economies and the heightened awareness of the potential impacts of that expansion on fundamental qualities of the environment, the importance of reaching a sustainable compromise that satisfies both objectives cannot be overstated. Our layered and fragmented system of governance erects major obstacles to attaining a constructive balance, but a variety of fresh approaches show promise for creating environmentally sensitive urban development. Among them are the revival of smart, sustainable development designs that protect and restore environmental qualities within developing areas and innovative collaborations that form networks across established jurisdictional, institutional, and conceptual boundaries to plan and implement interface conservation amid development.

Professionals engaged in resource protection and urban development can be instrumental players in promoting these approaches to reconciliation of conservation and development objectives at the interface. They can urge participation and enlist in the forming of multi-interest and multijurisdictional groups that can collaboratively explore solutions in specific areas. They can inform themselves in order to be able to inform others about innovative techniques and programs that have been successful in other communities and regions. They can reach across agency boundaries to secure advice and support.

The wildland–urban interface in most communities and regions is a moving target. Waiting for conflicts and issues to be somehow resolved in the future will ensure the needless damage of natural resources and the wasting of natural assets that could provide amenities for developing areas. The time to grapple with emerging concerns at the interface is before they become crises. For every growing community, that time is now.

REFERENCES

Abbott, C. and M.P. Abbott, 1991. Historical Development of the Metropolitan Service District, prepared for the Metro Home Rule Charter Committee, Portland, OR.

American Planning Association, 1998. The Principles of Smart Development, Planning Advisory Service Report No. 479, American Planning Association, Chicago.

Arendt, R., 1996. *Conservation Design for Subdivisions*, Island Press, Washington, DC.

Association of Bay Area Governments, 2000. *Theory in Action: Smart Growth Case Studies*, Association of Bay Area Governments, San Francisco.

Beatley, T. and K. Manning, 1997. *The Ecology of Place: Planning for Environment, Economy, and Community*, Island Press, Washington, DC.

Benedict, M.A., 2000. Green Infrastructure: A Strategic Approach to Land Conservation, Planning Advisory Service Memo, American Planning Association, Chicago.

Common Ground (Newsletter of the Conservation Fund), November/December 2000. Open space wins across country.

County of Riverside, California, Planning Department, 2001. Draft Multipurpose Open Space Element for the Riverside County General Plan.

Governor's Press Release, 1998. Governor Announces "Chapman's Landing" Agreement, Press Office of Governor Parris Glendening, Annapolis, MD.

Hirschhorn, J.S., 2000. *Growing Pains: Quality of Life in the New Economy*, National Governors Association, Washington, DC.

Katz, P., 1994. *The New Urbanism: Toward an Architecture of Community*, McGraw-Hill, New York.

Maricopa Association of Governments, 1996. Desert Spaces Management Plan for Maricopa County, Phoenix, AZ.

Maryland Office of Planning (now Maryland Department of Planning), 1995. Achieving Environmentally Sensitive Design, No. 11, Flexible and Innovative Zoning Series, Baltimore.

Maryland Rural Legacy Program, 2002. Website, www.dnr.state.md.us/rurallegacy. [Date accessed: April 2002.]

McHarg, I.L., 1992. *Design with Nature*, John Wiley and Sons, New York.

Metropolitan Park and Recreation District (St. Louis Region), 2002. The Clean Water, Safe Parks, Community Trails Initiative (Fact Sheet).

Meyers, E., R. Fischman, and A. Marsh, 1995. Maryland Chesapeake Bay critical areas program: wetlands protection and future growth, in *Collaborative Planning for Wetlands and Wildlife*, D.R. Porter and D. Salvesen, Eds., Island Press, Washington, DC, pp. 181–201.

Myers, P., 1999. Livability at the Ballot Box: State and Local Referenda on Parks, Conservation, and Smarter Growth, Election Day, 1998, Discussion Paper Prepared for the Brookings Institution, Washington, DC.

National Association of Home Builders, 1999. *Smart Growth: Building Better Places to Live, Work and Play*, National Association of Home Builders, Washington, DC.

New Jersey Pinelands Commission, 1994. *A Brief History of the New Jersey Pinelands and the Pinelands Comprehensive Management Plan*, New Jersey Pinelands Commission, New Lisbon, NJ.

Pirani, R., 2001. Presentation at the Forum on "Ad Hoc Regionalism", sponsored by the Lincoln Institute of Land Policy, Cambridge, MA., April 14, 2001.

Porter, D.R., 1995. Reinventing Portland: Metro's 2040 plan, *Urban Land*, July, 1995, pp. 37–42.

Porter, D.R., 1996. *Managing Growth in America's Communities*, Island Press, Washington, DC.

Porter, D.R., Ed., 2000. *The Practice of Sustainable Development*, Urban Land Institute, Washington, DC.

Porter, D.R., 2002. *Making Smart Growth Work,* Urban Land Institute, Washington, DC.

Porter, D.R. and T. Lassar, 1990. Urban–rural boundaries: the limits of limits, *Urban Land*, December.

Portland Metro, 2002. Website, www.metro-region.org/parks/openspaces. [Date accessed: April 2002.]

Regional Plan Association of New York, 1996. *The Third Regional Plan*, Regional Plan Association of New York, New York.

The Growth Management Institute, 1996. Final Report of the National Wildlife Conservation/Economic Development Dialogue, The Growth Management Institute, Washington, DC.

Yaro, R.D., 1997. Implementing RPA's third regional plan for the New York Metropolitan Region, *Environmental and Urban Issues* (Newsletter of the FAU/FIU Joint Center for Environmental and Urban Problems), pp. 9–16.

8 Landscape Assessment

Elizabeth Kramer
Institute of Ecology, University of Georgia

CONTENTS

8.1 INTRODUCTION

The 1997 USDA National Resources Inventory estimated that developed land in the contiguous U.S. increased by 25 million acres or 34 percent between 1982 and 1997 (USDA Natural Resources Conservation Service 2000). The increase in this 15-year period represents a quarter of all development in the U.S. that has occurred since the original European settlement. In addition, the U.S. Census Bureau estimated that the U.S. population increased by 15 percent during this same period (U.S. Census Bureau 2000). These two statistics indicate that land consumption rates are occurring at twice the rate of population growth. In fact, between 1982 and 1992, land was developed at the rate of 1.8 times the rate of population increase and between 1992 and 1997 at the rate of 2.5 times the rate of population increase (Beach 2002). The Census Bureau estimates that the population of the U.S. will increase by an additional 110,000,000 people by the year 2050. If land consumption continues at the same rate, this population increase will result in an additional increase of 275 million acres of developed land, an area twice the size of the state of Texas. This new growth in developing areas will greatly expand the wildland–urban interface.

Most of this newly developed land will be low-density residential or suburban areas and to a lesser extent higher-density or urban areas, which are typically composed of commercial and industrial activities. These newly developed areas are most likely to result from the conversion of forested and agricultural lands, leading to a loss of natural and seminatural systems that provide ecosystem services to communities.

Other human activities such as forest cutting, agricultural practices, road development, and the diversion of waterways also alter the landscape (Dale et al. 1998). Most of these produce a number of changes in the spatial patterning of natural vegetation, including fragmentation, patch shrinkage, bisection of patches (by road building), and deterioration within patch edges (Collinge 1996). These modifications to the landscape make it more vulnerable to the invasion of nonnative species and have an impact on forest function and the maintenance of native biodiversity. In addition, these changes alter the ability of the landscape to protect water and air quality.

Understanding the impacts of these human activities requires an understanding of the effects of changes at both the local and regional levels (O'Neill et al. 1999). Because humans alter landscapes at multiple scales, the ecological consequences of these alterations must be identified and predicted at multiple scales (Fausch et al. 2002). To do this, tools that can provide spatially explicit information over a large area are needed. Remotely sensed imagery and geographic information systems (GIS), combined with the theory of landscape ecology, provide the ability to monitor and assess large-scale ecological systems. Landscape assessment identifies the trends and status of resources and answers questions regarding "what is there" and "what is its condition." The assessment goal or question will provide the scale at which an assessment needs

1-56670-602-5/05/$0.00+$1.50
© 2005 by CRC Press LLC

to occur. For example, a manager concerned with the effect of forest management practices on red cockaded woodpeckers may need to conduct a landscape scale assessment that considers the extent and condition of habitat throughout the range of the woodpecker as well as potential management activities that may occur on other lands managed by different organizations. Often, private landowners must undertake landscape planning to avoid potential regulatory activity for wildlife or clean water protection (Loehle et al. 2002).

A landscape assessment can be as simple as an initial inventory to determine what is there, or it may be more complex because it needs to evaluate an impact of some type of management on a particular resource. With either approach, the assessment involves a definition phase, where the goals and questions of the assessment must be identified. These include when to measure baseline conditions, the types of practices that will be used, the length of time for which the assessment will be used to predict impacts, and the boundaries of the area to be assessed. The second phase of a more complex landscape assessment is the solution phase, in which the effects of the modifications brought about by the identified activities are evaluated (Hunsaker et al. 1990).

In this chapter, I present an overview of the geospatial technologies currently utilized for landscape assessments and a number of case studies representing landscape assessments performed at various scales from the state to national levels. The technologies include GIS, remote sensing, and global positioning systems (GPS). I also briefly describe spatial pattern analysis. Finally, I provide an example of how these data can be used for greenspace planning in areas that are changing from resource-based management to residential communities.

8.2 TOOLS FOR LANDSCAPE ASSESSMENTS

Many of the common issues being faced by resource managers — forest fragmentation, loss of biodiversity, edge effects, integrated resource management, cumulative impacts, endangered species protection, endangered ecosystems, and so forth — are spatial in nature (Crow and Gustafson 1997). Tools such as GIS improve our ability to organize and understand large quantities of spatially structured data, providing us with an insight into how patterns affect processes, and ultimately shape new ways to manage landscapes (Franklin 1994). Simply speaking, a GIS is the hardware and software that allows for the development, analysis, modeling, and maintenance of spatially referenced data. In some cases, the output is a map; however, the power of GIS is its ability to bring together spatially referenced data and provide a new way of exploring and visualizing these data. GIS is a tool that allows the user to organize data from small plots to large areas such as regions and watersheds. GIS can also support inventory and monitoring programs, provide support for planning, assist in policy-making, and can be used for consensual decision making. See Sample (1994) for a good overview of the use of GIS and remote sensing for natural resource management applications.

Another spatial technology, remote sensing, provides a source of data for inventory and monitoring programs. Remote-sensing technologies such as aerial photography and satellite imagery have been available for decades; however, widespread use has been limited in the past because of high costs and technological limits. Remote sensing can be defined as the science of gathering data about the earth's surface or near-surface environment through the use of a variety of sensor systems that are usually borne by aircraft or spacecraft, followed by the processing of these data into information that can be used for understanding and/or managing our environment (Hoffer 1994). In the simplest context, remote sensing is the science of observation from a distance (Barrett and Curtis 1992).

Remote sensors for the most part measure electromagnetic energy using either passive or active sensors. Passive systems include satellites such as Landsat, SPOT, and IKONOS, as well as aerial photographic systems. These sensors collect reflected or emitted energy from the surface of the earth. A typical passive sensor is designed to collect data at various wavelengths of the electromagnetic spectrum. For example, Landsat ETM+, a multispectral scanner, collects electromagnetic data from nine different bands of wavelengths. In fact, today there are hyperspectral scanners, which measure electromagnetic radiation in hundreds of wavelengths.

Active sensors include radar and laser systems, such as Lidar (Lefsky et al. 2002). These systems supply their own source of energy and measure characteristics from the return of this energy to the sensor. They essentially employ artificial sources of radiation as a probe and have the unique ability to penetrate cloud cover and smoke.

Remote-sensing devices not only collect data at multiple wavelengths but also collect data at different spatial resolutions. The picture element or "pixel" represents the smallest spatial unit on the ground for which data are collected via digital remote-sensing systems. Spectral resolution can range from centimeters for aerial photographic data to meters and even kilometers for satellite imagery (Figure 8.1). Every object on the surface of the earth has a characteristic spectral response referred to as a "signature," which is a function of the way objects or phenomena reflect, emit, or transmit electromagnetic energy. Identification and separation of objects can be accomplished in part through analysis of spectral signatures. Landsat ETM+ collects data at 15, 30, and 60 m, SPOT data are available at 5, 10, and 20 m, and IKONOS data

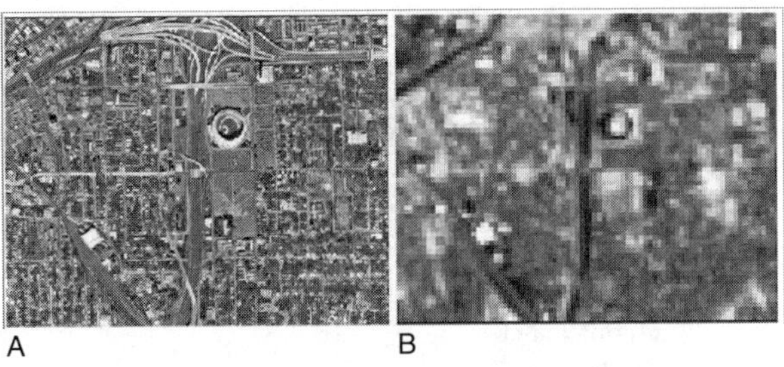

FIGURE 8.1 Images of the same area of downtown Atlanta in the early 1990s. Image A represents a 1-m resolution aerial photograph converted into a Digital Ortho Quarter Quad. Image B represents a 30-m resolution Landsat TM image.

are available at 1 and 4 m (Table 8.1). Also, submeter satellite imagery is now commercially available.

Because of its multispectral capabilities, remote sensing presents a unique perspective for observation and measurement of biophysical characteristics. In addition, remotely sensed data can be collected at multiple scales and at multiple times, thereby offering the opportunity for analysis of various phenomena synoptically from local to global scales through time (Quattrochi and Pelletier 1991). With the advent of Landsat ETM+, the imminent availability of new commercial satellites, and recent changes in licensing policies by commercial satellite vendors, remote-sensing data have become economical for many types of applications.

Data collected from remotely sensed technologies must be converted into useful information that facilitates the assessment of natural resources. This is done through image-processing techniques. Table 8.2 lists many types of natural resource information and applications that can be derived from remotely sensed data. Image-processing techniques vary from photo interpretation to automatic feature extraction. Although image processing is typically divided into two categories — computer interpretation, which is called unsupervised, and human interpretation, which is called supervised — these are seldom performed independently of each other. For the most part, we are still not able to completely automate the image interpretation process. A good technician who is familiar with the area being mapped is invaluable to the process, as well as good ground control and training information.

Another tool available for the collection of geospatial data for landscape assessment is the GPS. GPS consists of a constellation of satellites orbiting the earth that broadcast radio signals allowing users with receivers to determine their spatial position (Kennedy 1996). Each of the 24 satellites in the constellation sends out a unique radio signal and a time stamp for that signal. The receiver sends a signal back to the satellite with a time stamp, and the lag time between the satellite and receiver signal is measured to determine the distance between the two units. The receiver picks up multiple satellite signals and, using triangulation, calculates the position of the receiver on earth. The more satellite signals the receiver is able to pick up, the more accurately it will determine location. Then, by linking position with an attribute of the site, the user can incorporate the spatial information into a GIS. For example, GPS facilitates identifying the location of field plots for long-term monitoring. Also, the user can locate field points generated with a GIS system to ground-truth processed imagery or collect training data to assist in image processing. Hand-held systems are available, which facilitate the integration of GIS and GPS in the field. These allow the user to collect field data and GPS points in a digital format that can be uploaded directly into a desktop GIS system. Eventually, these hand-held systems will effectively integrate wireless technology and allow the user to edit an agency's GIS database while in the field, thereby allowing for real-time updates.

GPS units are now being linked with digital photographic technologies to allow for low-cost access to highly accurate remotely sensed images. Examples of the integration of these technologies are the digital videography systems and airborne digital photography systems used by the Gap Analysis Program (Slaymaker et al.

TABLE 8.1
General Comparison of the Specifications for Commonly Available Satellite Imagery

	Imagery			
Specifications	**SPOT**	**IRS-C**	**IKONOS**	**Landsat 7**
Panchromatic resolution (m)	10	5	1	15, 60
Multispectral resolution (m)	20	30	4	30
Number of bands	4	4	4	9
Availability[a]	1998+	1998–2000	1999+	1999
Scene coverage (square mile)	1400	60	User defines	12,075
Pricing per square mile	\$0.25–\$0.50	\$8–\$18	\$57–\$196	\$0.04–\$0.05

[a]Year launched and length of service.

[b]These prices are representative of the cost in 2002. Pricing varies based upon the licensing agreement, quantity purchased, and the amount of preprocessing performed by the vendor.

TABLE 8.2
Examples of Applications of Remotely Sensed Data

Resource Inventory	Physical Properties	Planning and Resource Management Applications
Soil classification	Landforms	Land cover/land use
Soil moisture	Geologic structures	Land cover/use change
Soil erosion	Topography	Coastal zone monitoring
Crop inventory	Cloud cover	Environmental impact assessment
Crop condition	Drainage characteristics	Transportation planning
Crop irrigation	Elevation	Site evaluation
Forest inventory		Wildlife habitat assessment
Forest condition		Flood prediction
Disturbance effects		Mitigation planning
Forest yields		Urban heat islands and climate modeling
Wetland area		Impervious surface estimation and water quality monitoring
Natural areas identification		

1996). These systems allow users to collect spatially explicit high-resolution digital photography for training image-processing procedures and for ground-truthing vegetation maps. Newer technologies are incorporating laser altimeters to develop forest canopy models for forest inventory and biomass assessments.

8.3 SPATIAL PATTERN ANALYSIS

Spatial pattern metrics are frequently used as surrogates for measures of ecosystem function at the landscape scale (O'Neill et al. 1999). Literally hundreds of measures have been developed to quantify various aspects of spatial heterogeneity (Gustafson 1998), driven by the increased need for measurement and monitoring of landscape-level patterns and processes. The premise is that ecological processes are linked to and can be predicted from patterns at large spatial scales. For example, some landscape metrics have been successfully related to habitat integrity and water quality measures. The availability of GIS technologies and new software that measures patterns make it easy for those not trained in landscape ecology to calculate hundreds of metrics from maps. However, it is important to make a distinction between heterogeneity that can be mapped and measured and heterogeneity that is ecologically significant (Turner 1989).

Landscape metrics are measures that are used with categorical maps to quantify the characteristics of landscape patches, classes of patches, or landscape mosaics. These metrics typically fall into two categories: those that quantify the landscape without regard to its spatial configuration and those that quantify the landscape's spatial configuration.

Compositional measures are an example of landscape indices that do not take into account spatial configuration. They include the proportional abundance of a particular class, richness, evenness, and diversity. Richness is simply the number of classes in a particular landscape, whereas evenness provides a measure of dominance of the various class types. Diversity is a composite measure of richness and evenness (McGarigal and Marks 1995).

Spatial configuration metrics provide information on the arrangement and position of a particular patch type in the landscape, including patch isolation, placement relative to other patch types, and connectivity of the landscape. Some examples of spatial configuration metrics are patch size, patch density, patch shape complexity, patch isolation or connectivity, and contrast. Patch size is a fundamental metric within a landscape and is often used to derive other metrics such as ratios of the largest patch of single classes to the total area of the landscape. Patch density provides information regarding the degree to which fragmentation has occurred in the landscape, and patch shape can also be used as an indicator of human influence on a landscape. The most common shape complexity measure is a perimeter to area ratio. The lower the number the less the complexity observed, which usually indicates a human-dominated landscape. For example, rectangular and circular features characterize agricultural systems, due to the requirements of farming equipment, whereas natural forested areas have more irregular edges and show a higher shape complexity.

Pattern analysis is often used to determine the landscape impacts of land-use activities. For example, numerous indices provide information on the effects of activities such as harvesting or urbanization on forests (Vogelmann 1995; Dale et al. 1998), and they may be used to predict the consequences for biodiversity. Many neotropical migrant birds require large areas of intact forest for breeding. A landscape with a high patch density and small patch areas or limited core areas (a measure of shape complexity) may indicate a highly fragmented region that would not support bird populations. Spatial pattern metrics can also be used to compare one landscape with another to assess priority areas for greenspace protection or potential areas for restoring landscape connectivity and expanding habitat.

Numerous software products are now available that allow users to calculate spatial pattern metrics from digital maps (McGarigal and Marks 1995). Because of the difficulties associated with predicting ecological processes from patterns, users of these products should exercise caution when interpreting data outputs. A number of review papers discuss the types of metrics available and the application of these metrics to forest management (see Haines-Young and Chopping 1996; Gustafson 1998; O'Neill et al. 1999).

8.4 CASE STUDIES

8.4.1 National Scale Assessment Programs

A number of federal agencies maintain monitoring and assessment programs that regularly provide information concerning the status and trends of natural resources in the U.S. These include the U.S. Geological Survey's (USGS) breeding bird surveys, USDA Forest Service's Forest Inventory and Assessment (FIA) program, USDA Natural Resources Conservation Service's (NRCS) National Resources Inventory (NRI), and USGS's National Water Quality Assessment (NAWQA) program.

The FIA program and the NRI are currently using remote-sensing technologies (aerial photography and satellite imagery) as primary data sources. Figure 8.2 shows a forest cover map of the southeastern U.S. derived from Advanced Very High Resolution Radiometry (AVHRR) imagery by the FIA program. These surveys also incorporate data collection from a series of fixed field plots to provide trends in natural resource and forest status for national-, state-, and county-level analyses. Natural resource managers can use these data as a first cut in assessing the current status of forest resources in their area of interest (available at http://fia.fs.fed.us/). Recently, the FIA program shifted its field data collection to an annual cycle by working in partnership with state forestry agencies. These annual surveys include information on area, species composition, mortality, health, harvesting, and utilization of forests. In addition, data collection has expanded to include more ecological information such as understory composition and community types. The information is published on a 5-year cycle for each state. The USDA Forest Service often combines these data with data from other programs to produce special reports. These have included an assessment of urban forests (Dwyer et al. 2000), which integrated FIA data with demographic information, and the Southern Forest Resource Assessment (Wear and Greis 2002), which was done in collaboration with southeastern state forestry and wildlife resource agencies.

The NRI collects data every 5 years from 800,000 sample sites, including data on land cover and use, soil erosion, prime farmland soils, wetlands, habitat diversity, selected conservation practices, and other resource attributes. The program compiles the field data and information from imagery, field office records, and historical records and produces reports on national, regional, and statewide trends (USDA Natural Resources Conservation Service 2000). These data are used to help develop agricultural and natural resource policies at all these levels. They also provide useful contextual information for local landscape assessments by giving the user a big-picture view of nat-

FIGURE 8.2 Distribution of forest cover in the southeastern U.S. The data were produced by the FIA program using AVHRR satellite imagery from the 1999 growing season. The resolution of the imagery is 1 km and the various shades of gray represent different forest cover types. (From National Atlas of the U.S. 2003.)

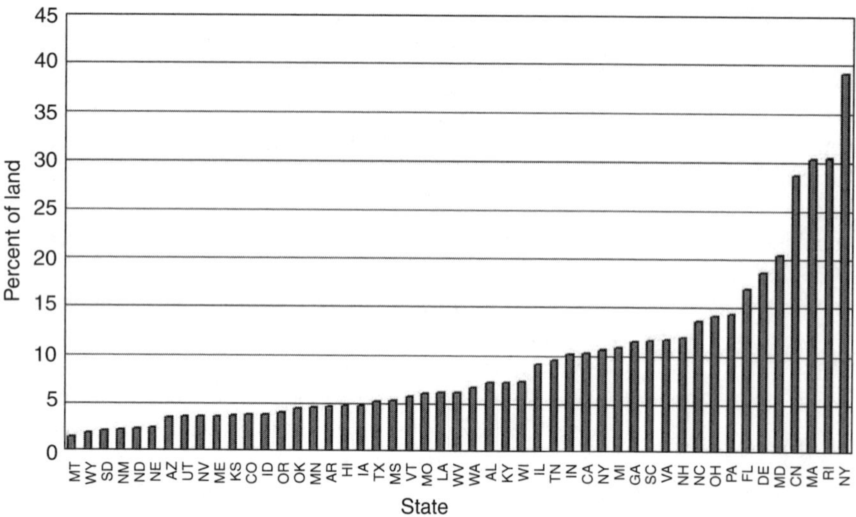

FIGURE 8.3 The percentage of each state that was developed as of 1997. (Adapted from USDA Natural Resources Conservation Service 2000.)

ural resource status and trends. Figure 8.3 represents the 1997 estimates of the percentage of developed land in each of the lower 48 states.

8.4.2 Statewide Assessment Programs

8.4.2.1 GAPs

The loss of biodiversity is a major concern for natural resource managers. The driver for the current loss of biodiversity is habitat destruction and modification due to changes in land-use practices. This is especially true at the wildland–urban interface, where urban sprawl consumes more and more of our natural resources. The GAP was established by the U.S. Fish and Wildlife Service to provide an initial approach to identify areas of high biodiversity that are not protected by the current distribution of nature reserves (now GAP is part of the USGS). A Gap analysis is a scientific method for determining the degree to which native animal species and natural communities are represented in our present-day mix of conservation lands. Those species and communities absent from the existing network constitute conservation gaps. GAP's focus is on ordinary species and their habitats, as well as threatened and endangered species. The program provides information to land managers, planners, scientists, and policy makers to allow them to make better-informed decisions at the local, state, regional, and national levels. The program does this by developing a series of data sets that include land cover and natural community maps, modeled distributions of terrestrial vertebrate species, and maps of the distribution of lands managed for some type of conservation and biodiversity protection. An analysis is performed by overlaying the three types of information and identifying where they do and do not overlap. Areas with high native biodiversity or unique communities that are not protected by the current mix of state, federal, and private conservation lands represent the gaps in protection. Natural resource managers, who must determine and prioritize where scarce financial resources should be applied, can then use this information.

Figure 8.4 shows a map of the existing network of lands that are managed for natural resource and biodiversity protection in the state of Georgia. The lands represent less than 9 percent of the total area of the state. The larger landholdings are federally owned and are mainly military bases.

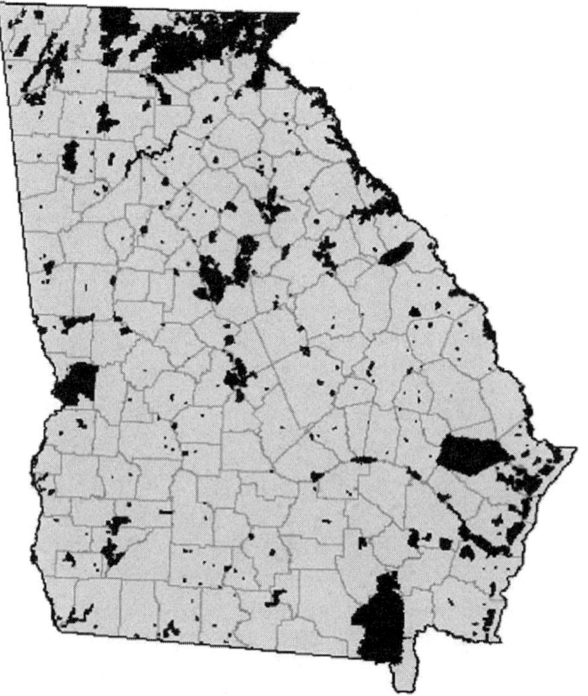

FIGURE 8.4 The distribution of lands that had some level of management for natural resource and biodiversity protection in the state of Georgia in 1998. (From NARSAL 2003a.)

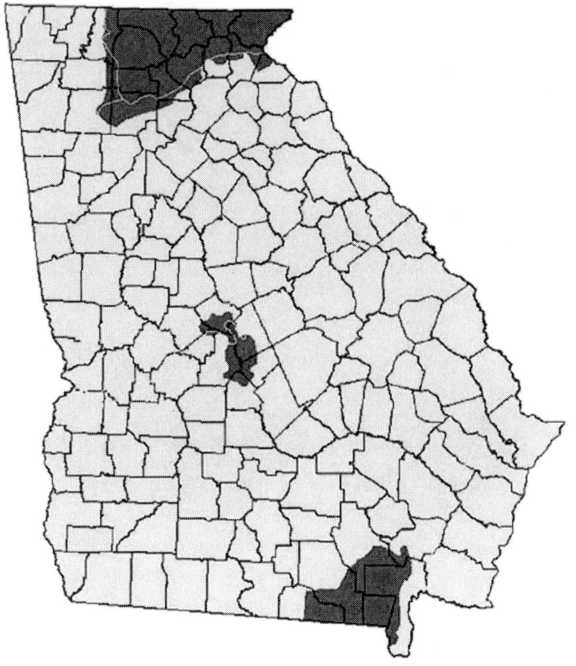

FIGURE 8.5 Potential range of black bear, *Ursus americanus*, in Georgia. (From NARSAL 2003a.)

FIGURE 8.6 Distribution of natural longleaf pine stands in Georgia, as mapped by the Gap Analysis Program. The data were derived from Landsat TM satellite imagery from 1998. (From NARSAL 2003a.)

Figure 8.5, a map of the potential distribution of black bear in Georgia, represents a typical data set developed for the vertebrate distribution maps (NARSAL 2003a). These distribution maps are created in a two-step process: the first is to create a range map for the species, and the second is to identify the potential habitat within the range. Range maps are produced for each species using various sources of information such as museum collections and literature reviews and then verified by taxonomic experts. Habitat models are derived from vegetation maps, such as the map of the distribution of longleaf pine forests in Georgia (Figure 8.6), and other types of landscape information that may affect the species distribution. Spatial pattern data as well as other information derived from topography (e.g., aspect) are used to identify suitable habitat. In the case of the black bear model, large patches of forest and wetlands were classified as habitat and areas with high road density were considered nonhabitat typically avoided by bears.

Once all of the vertebrate distribution maps are completed, they are overlaid to produce species richness maps. These results are then compared with the conservation lands database to determine how well areas of high biodiversity and critical natural communities are protected.

The value of the GAP goes beyond the identification of conservation needs. In many cases, the land cover maps developed for the program provide valuable information for resource planning, and for the first time biodiversity information is available to the local planning community. State Natural Heritage programs often only track and provide information on threatened and endangered species, and these data are maintained as points of occurrence. By contrast, GAP provides information to local planners and natural resource managers on the potential distribution of all vertebrate species found within the state. These data can be used in a comprehensive or greenspace planning program to identify critical areas where natural resource protection may be needed.

A GAP is under way or complete in each of the lower 48 states and Hawaii. The state-by-state model was first used to facilitate data sharing among state agencies. Attempts to merge state project data into regional and national maps have proven to be difficult because each state project used different map legends, data, and methodologies. These problems have led to the limited use of GAP data for multistate and national planning. As part of its next iteration, GAP is taking a regional approach, with programs in place in the southwestern and southeastern U.S. The data products are now developed by ecoregion rather than by state, which should facilitate planning for biodiversity protection within animal and natural community ranges.

8.4.2.2 Georgia Land Use Trends Program

The GAP presents an assessment at a single snapshot in time, but assessments need to answer additional questions, such as how we got here and where we are going. We know that the current mix of land uses is the result of decisions made 50 years ago and that the pattern of land use 50 years from now will be the result of decisions we make today. In Georgia, as in other states, previous mapping efforts have produced land-use and land-cover maps. However, these maps were created using various methods, data sources, and classification schemes, resulting in maps of differing accuracies and land-use categories. The Georgia Land Use Trends (GLUT) program was established to create a common set of historic land-cover maps from satellite imagery that can be used to study the impacts of past land-use decisions on the current mix of land use in the state (NARSAL 2003b).

The GLUT project has developed land-cover maps beginning with 1974 Landsat MSS imagery and continuing with newer Landsat TM imagery, using the same methodologies and classification system for all maps. Because we used the coarser resolution Landsat MSS data, the classification system is more limited than if we had used Landsat TM imagery alone. However, we were able to capture land-cover activity from the early 1970s to the present, and to date we have completed land-cover maps for 1974, 1985, 1992, and 1998.

Data from the land-cover maps are being used to understand the impacts of land-cover change on natural resources and environmental quality in Georgia. In addition, the data are being used to project future patterns of land-cover change due to urbanization. These results will be used to create visualizations of the effects of current land-use decisions on future growth patterns. Figure 8.7 represents a map of urbanization trends for Gwinnett County, GA, derived from data produced by the GLUT program. Areas representing low-intensity urban use, composed mostly of single-family residences, increased by 242 percent from 1974 to 1998. High-intensity urban areas composed mostly of commercial and industrial lands increased by 176 percent. In fact, more than 21,000 acres were converted to urban uses during this 24-year period. Gwinnett County is representative of the rapidly suburbanizing areas within metropolitan Atlanta.

Overall, statewide land-cover change is not as extreme as in Gwinnett County (Figure 8.8). From 1974 to 1998, Georgia showed a steady increase in urban land uses as well as an increase in evergreen forest, mostly plantation pine. Both deciduous forests and agricultural land uses showed a steady decline. Georgia lost almost 1 million acres of agricultural land and another 1 million acres of deciduous forest during the 24-year period.

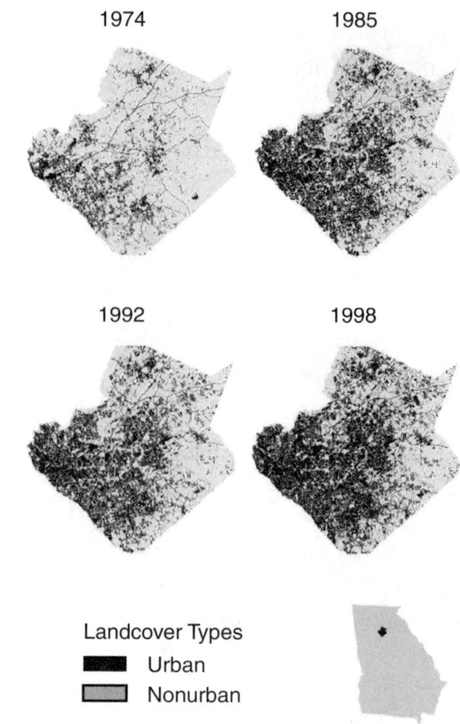

FIGURE 8.7 Urbanization trends for Gwinnett County, GA. Urban land uses include both single-family residential areas and commercial and industrial land-use activities. More than 21,000 acres were converted to urban land uses during this 24-year period. The maps were produced from 30 m Landsat imagery. (From NARSAL 2003b.)

8.4.3 Local-Level Assessment and Planning

In response to the rapid urbanization occurring in Georgia, in 2000, Governor Roy Barnes announced the Georgia Community Greenspace program. The fastest growing counties in the state were offered access to a $30 million fund to assist in purchasing lands for community greenspace. To qualify, the county commission must develop a plan to set aside 20 percent of the county's land base as permanently protected greenspace. In the first year of the program, 40 of Georgia's 159 counties were eligible and 39 elected to participate. One was Jackson County, a predominantly rural county that lies on the outer edge of metropolitan Atlanta.

The county commission began by bringing together a citizens' committee to work with the county planning staff in the development of the plan. Committee members reflected the geographic and cultural diversity of the community. For example, they represented all the participating municipalities and unincorporated areas of the community and various interest groups such as farmers, cattle ranchers, developers, environmentalists, and industry.

The group initiated the process by deciding which goals they wanted the greenspace plan to meet. These included water quality, agricultural protection, maintenance of the county's rural character, and economic development. The next step was to perform a natural resource assessment. No new data were collected; instead, data of the types described in the previous case studies were used. These included agricultural and forest lands statistics,

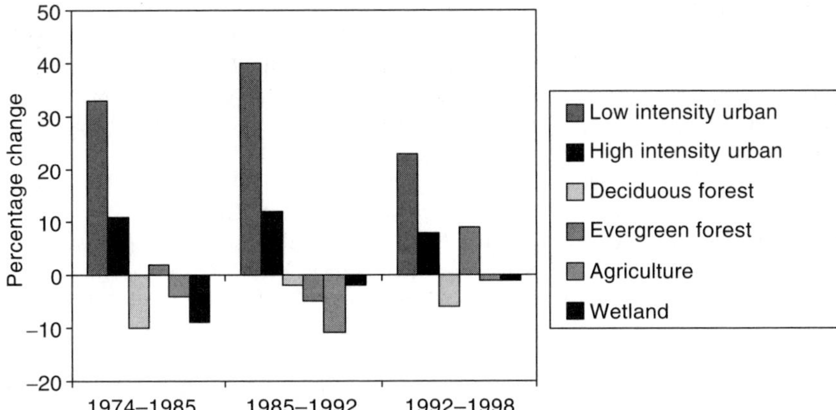

FIGURE 8.8 Land-cover changes in Georgia for the periods of 1974–1985, 1985–1992, and 1992–1998. (From NARSAL 2003b.)

land-use change data, biodiversity data, floodplain maps, and other data sets available from federal agencies. These data were brought together in a GIS that was queried to produce information for the assessment.

The assessment identified key environmental areas for protection of water quality, including water bodies, floodplains, and wetlands. Land-use information was used to locate areas where growth was occurring and areas where there was substantial agriculture for protection. On the basis of this assessment, the committee decided where in the county it was appropriate to develop, as well as which areas needed to be protected.

Once the committee had identified potential greenspace lands, land-use tools and mechanisms were proposed to place these lands under permanent protection. These tools included conservation easements, conservation subdivisions, floodplain and buffer regulations, and funding mechanisms such as a special option sales tax. The plan was then presented to the citizens of the county through a number of public hearings. Finally, the county commissioners adopted the plan and submitted it to the state greenspace program.

8.5 CONCLUSION

As communities grapple with rapid land-use change due to population growth on a limited land base, conflicts from land-use decisions become inevitable. These conflicts are likely to be most challenging at the wildland–urban interface, where there is an ever-expanding need to balance resource uses with ecosystem function and aesthetics. Addressing these issues requires an understanding of ecological phenomena at multiple scales. Managers must look beyond individual landholdings to the surrounding landscape to understand the context in which their decisions must be implemented. Moreover, because these decisions affect large areas and multiple landowners, they must be made in a process that involves the public.

Landscape assessments provide the information needed to manage these complex systems and to help reduce the conflicts. National landscape assessments, including the FIA and NRI programs, provide local managers with the context for making management decisions. Statewide assessments such as those of the GAP and GLUT identify spatial patterns of key resources and how land areas are being modified. Finally, local planning efforts such as the Jackson County Greenspace program bring this information together and allow communities to participate in the management of their natural resources.

GIS, GPS, and remote sensing provide the tools necessary to assist managers in numerous ways, including inventory and monitoring, planning on both a local and regional scale, analyzing alternatives, and communicating with stakeholders. The key to the success of using these technologies is accurate and timely information. During the past decade, data have become more readily available via state and federal clearinghouses, prices for imagery have been reduced due to advances in technologies and increased competition, hardware costs have been reduced while capability has increased, and software for analysis has become more readily available. All of these factors have come together to provide the tools for better planning and resource management.

REFERENCES

Barrett, E.C. and L.F. Curtis, 1992. *Introduction to Environmental Remote Sensing*, Chapman & Hall, London.

Beach, D., 2002. Coastal Sprawl: The Effects of Urban Design on Aquatic Ecosystems in the U.S., Pew Oceans Commission (http://www.pewoceans.org/reports/water_pollution_sprawl.pdf). [Date accessed: September 2, 2003.]

Collinge, S.K., 1996. Ecological consequences of habitat fragmentation: implications for landscape architecture and planning, *Landscape and Urban Planning* 36: 59–77.

Crow, T.R. and E.J. Gustafson, 1997. Concepts and methods of ecosytem management: lessons from landscape ecology, in *Ecosystem Management: Applications for*

Sustainable Forest and Wildlife Resources, M.S. Boyce and A. Haney, Eds., Yale University Press, New Haven, pp. 54–67.

Dale, V.H., A.W. King, L.K. Mann, R.A. Washington-Allen, and R.A. McCord, 1998. Assessing land-use impacts on natural resources, *Environmental Management* 22: 203–211.

Dwyer, J.R., D.J. Nowak, M.H. Noble, and S.M. Sisinni, 2000. Connecting People with Ecosystems in the 21st Century: An Assessment of Our Nation's Urban Forests, General Technical Report PNW-GTR-490, U.S. Department of Agriculture, Forest Service, Pacific Northwest Research Station, Portland, OR.

Fausch, K.D., C.E. Torersen, C.V. Baxter, and H.W. Li, 2002. Landscape to riverscapes: bridging the gap between research and conservation of stream fishes, *BioScience* 53: 483–498.

Franklin, J.F., 1994. Developing information essential to policy, planning, and management decision making: the promise of GIS, in *Remote Sensing and GIS in Ecosystem Management*, V.A. Sample, Ed., Island Press, Washington, DC, pp. 18–24.

Gustafson, E.J., 1998. Quantifying landscape spatial pattern: what is the state of the art? *Ecosystems* 1: 143–156.

Haines-Young, R. and M. Chopping, 1996. Quantifying landscape structure: a review of landscape indices and their application to forested landscapes, *Progress in Physical Geography* 20: 418–445.

Hoffer, R.M., 1994. Challenges in developing and applying remote sensing to ecosystem management, in *Remote Sensing and GIS in Ecosystem Management*, V.A. Sample, Ed., Island Press, Washington, DC, pp. 25–40.

Hunsaker, C.T., R.L. Graham, G.W. Sutter, R.V. O'Neill, L.W. Barnthouse, and R.H. Gardner, 1990. Assessing ecological risk on a regional scale, *Environmental Management* 14: 325–332.

Kennedy, M., 1996. *The Global Positioning System and GIS: An Introduction*, Ann Arbor Press, Chelsea, MI.

Lefsky, M.A., W.B. Cohen, G.G. Parker, and D.J. Harding, 2002. Lidar remote sensing for ecosystem studies, *BioScience* 52: 19–30.

Loehle, C., J.G. MacCracken, D. Runde, and L. Hicks, 2002. Forest management at landscape scales: solving the problems, *Journal of Forestry* 100: 25–33.

McGarigal, K. and B.J. Marks, 1995. FRAGSTATS: Spatial Pattern Analysis Program for Quantifying Landscape Structure, General Technical Report PNW-GTR-351, U.S. Department of Agriculture, Forest Service, Pacific Northwest Research Station, Portland, OR.

National Atlas of the U.S., 2003. http// nationalatlas.gov. [Date accessed: September 2, 2003.]

Natural Resources Spatial Analysis Laboratory (NARSAL), 2003a. Georgia Gap Analysis Program, http:// narsal.ecology.uga.edu/gap.html. [Date accessed: September 2, 2003.]

Natural Resources Spatial Analysis Laboratory (NARSAL), 2003b. Georgia Land Use Trends Program, http:// narsal.ecology.uga.edu/glut.html. [Date accessed: September 2, 2003.]

O'Neill, R.V., K.H. Riitters, J.D. Wickham, and K.B. Jones, 1999. Landscape pattern metrics and regional assessment, *Ecosystem Health* 5: 225–233.

Quattrochi, D.A. and R.E. Pelletier, 1991. Remote sensing for analysis of landscapes: an introduction, in *Quantitative Methods in Landscape Ecology: The Analysis and Interpretation of Landscape Heterogeneity*, M.G. Turner and R.H. Gardner, Eds., Springer-Verlag, New York, pp. 51–76.

Sample, V.A., ed., 1994. *Remote Sensing and GIS in Ecosystem Management*, Island Press, Washington, DC.

Slaymaker, D.M., K.M.L. Jones, C.R. Griffin, and J.T. Finn, 1996. Mapping deciduous forests in southern New England using videography and hyperclustered multi-temporal Landsat TM imagery, in *Gap Analysis: A Landscape Approach to Biodiversity Planning*, J.M. Scott, T.H. Tear, and F.W. Davis, Eds., American Society for Photogrammetry and Remote Sensing, Bethesda, MD.

Turner, M.G., 1989. Landscape ecology: the effect of pattern on process, *Annual Review of Ecology and Systematics* 20: 171–197.

U.S. Census Bureau, 2000. U.S. census, http://www.census.gov/. [Date accessed: September 2, 2003.]

U.S. Department of Agriculture (USDA), Natural Resources Conservation Service, 2000. Summary Report: 1997 National Resources Inventory (revised December 2000), USDA Natural Resources Conservation Service, Washington, DC, http://www.nrcs.usda.gov/technical/ NRI/1997. [Date accessed: September 2, 2003.]

Vogelmann, J.E., 1995. Assessment of forest fragmentation in southern New England using remote sensing and geographic information systems technology, *Conservation Biology* 9: 439–449.

Wear, D.N. and J.G. Greis, 2002. Southern Forest Resource Assessment: Summary Report, General Technical Report SRS-54, U.S. Department of Agriculture, Forest Service, Southern Research Station, Asheville, NC.

Part III

Human Dimensions and the Interface

9 Tools to Reach, Educate, and Involve Citizens

Martha C. Monroe
School of Forest Resources and Conservation, University of Florida

CONTENTS

9.1 INTRODUCTION

Wildland–urban interface issues, by proximity and definition, always involve people. The people may be nearby rural residents, activists in a wise-use or environmental organization, planners and developers, townspeople, or urban visitors. Whether these people are knowledgeable, helpful, disinterested, or antagonistic is often a function of the outreach activities that communicate, educate, and involve citizens in natural resource management (see Box 9.1).

Natural resource agencies and organizations are not newcomers to education and communication activities. They typically use a host of opportunities to conduct outreach with different audiences, from schoolchildren and teachers to garden club members and hunters. Many of these techniques are appropriate for managing interface issues as well, with one small caveat: in the interface, communication, education, and public involvement are more critical because the interface is close enough to some people to be part of "their territory." Nearby residents may feel responsible for it, even if they know they do not own all the land they enjoy. Planners may wish to control it, so they can better manage the developed regions. Those who drive by it may believe their view of landscape should never change. This chapter will explore some techniques and strategies for engaging citizens in the interface in natural resource management issues and describe several examples of successful projects.

9.2 A CONTINUUM OF OUTREACH OPPORTUNITIES

Most resource agencies and environmental organizations conduct a variety of education and outreach programs. At first glance, it might appear to be an unorganized combination of brochures, public meetings, curricula, and workshops. When sorted out by objective and audience, however, a pattern emerges.

Scott and Fien (1999) evaluated the global programs of World Wide Fund for Nature (WWF) and developed the following definitions to organize its many diverse programs. Each category of programs meets a specific need, and each subsequent category should include all previous ones. Thus, all aspects of an outreach program are interrelated and synergistic, but each has identifiably separate goals, objectives, strategies, and outcomes.

1. *Information* includes the distribution of conservation messages that raise awareness. Advertisements, news releases, TV programs, media campaigns, posters, brochures, and stickers are some of the tools used to prompt and increase awareness.
2. *Communication* establishes a two-way dialogue between communities and the organization to create improved understanding and collaborative partnerships. Interpretive programs,

1-56670-602-5/05/$0.00+$1.50
© 2005 by CRC Press LLC

Box 9.1

The Outreach Challenge

Resource managers often complain about the problems of educating and involving elements of the public. You might have heard these comments:

1. The people do not know enough to realize how our management activities help the forest — they are ecologically illiterate.
2. The "public" is not of one mind — different factions want competing things and everyone can't be happy.
3. The people are apathetic, don't read our materials, and don't participate in our planning process, but then they think they have a right to block projects.

These are real problems and often generally accurate perceptions. Most people do not know a great deal about resource or ecosystem management, but they usually have an opinion. A variety of ideas, interests, and values can be found in any community. And most people are busy doing other things and may not have the time or energy to adequately participate in resource decisions. Natural resource education programs must overcome these challenges.

participatory campaigns, newsletters, town meetings, and interactive exhibits are among the diverse tools that can support communication. Communication programs also include information activities.

3. *Education* programs increase knowledge, support attitudes, and develop motivation, commitment, and skills to work cooperatively to achieve conservation goals. Such programs are usually offered in the context of an institution or setting that is dedicated to enhancing human capacity rather than managing natural resources, such as youth groups, schools, technical colleges, extension services, and professional development opportunities. Education programs often involve communication and information strategies.
4. *Capacity-building* activities improve the ability of communities to work for conservation through training, policy development, and institutional strengthening. While education is a key element of capacity building, activities in this category build on educational programs to create social capital for effective and sustainable participation. Partnerships, collaborative decision making, and strategic planning are elements of capacity-building programs. Communication and information strategies are also important to use in activities to build capacity.

All four of these categories are needed to adequately portray WWF's complex set of activities around the world. In developing countries, for example, they work to enhance institutional capacity of local nongovernmental organizations (NGOs) and train community leaders in participatory appraisal techniques. In many nations, they provide curriculum materials for teachers, and use stickers or posters to bring attention to critical conservation issues.

In the U.S., natural resource agencies also use this full array of outreach activities to inform, educate, listen, plan, partner, train, and decide how to manage wildland–urban interface issues. Some agencies have a strong tradition in just a few categories; others have programs that would fall into each of the categories. Any individual district, however, may not have the need or ability to conduct programs in all of these categories. With regard to capacity building, some natural resource agencies may prefer to partner with rural development programs or other social agencies that specialize in leadership development and civic participation. Perhaps a helpful modification in this category would be to emphasize the variety of ways in which agencies work to change behaviors to reduce resource management challenges. Thus, these four categories create a framework for organizing outreach in the interface. Table 9.1 helps demonstrate that different types of programs meet different needs and have the possibility of building on each other to achieve a broader goal.

9.3 STRATEGIES FOR INTERFACE OUTREACH

Scott and Fien's modified categories can be used by agencies and organizations to create strategies for outreach programs in the wildland–urban interface. These strategies may clarify the variety of outreach efforts that an agency already conducts and provide a framework to explore additional programs. These additions would complement and support existing outreach efforts. The wildland–urban interface creates a variety of opportunities for outreach programs because of the increased concentration of people and ready source of timely issues (Table 9.1).

These four strategies form the backbone of the remainder of this chapter. Taking each in turn, the theoretical foundations of education and communication activity in support of each strategy will be discussed, along with a variety of examples and procedural tips. The chapter concludes with a brief review of the program development process.

TABLE 9.1
Outreach Opportunities in the Interface

Category	Strategy	Activities
Information	Provide facts and knowledge about interface ecosystems and issues	Brochures, posters, flyers, fact sheets, radio and TV programs or public service announcements, billboards, magazine articles
Communication	Develop opportunities to engage the community in dialogue about interface issues	Workshops, field trips, town meetings, newsletters with feedback forms, surveys, focus groups, advisory councils, forums
Education	Support programs that enhance problem-solving skills through existing schools and certificate programs	Curriculum development, service learning programs, in-service workshops, programs that meet institutional benchmarks and goals
Behavior change	Encourage changes in behavior to reduce interface resource management problems	Focused communication activities on behavior with broad community support through demonstration areas, neighborhood gatherings, prompts, social commitment, testimonials, role models

9.3.1 Provide Information about Interface Ecosystems, Issues, and Opportunities

Information is critical to improve public understanding of wildland–urban interface issues. It is a challenge, however, because providing information does not necessarily mean the public is more aware. Churning out more brochures, broadcasting nature specials, and hosting a community festival may not be the answer to increasing understanding. Several different problems may be at work, perhaps in concert (see Box 9.2).

1. Sometimes the audience just does not know enough about the problem, or does not have accurate information. A Roper-Starch poll commissioned by the National Environmental Education Training Foundation (NEETF 1998) revealed that only 23 percent of the public correctly identified runoff as the leading cause of water pollution. Agencies working in water quality need to provide basic information about nonpoint source pollution in a compelling and interesting way so that the audience reads it, believes it, and remembers it. In any case, knowing what the audience already knows and believes is the first step to producing effective informational materials.
2. Sometimes the audience knows about the issue, but does not really believe it is a problem. It is much harder to provide convincing information that changes someone's mind. The number of people who believe factories are the leading cause of water pollution is about twice the number who know that nonpoint source pollution tops the list of water pollution problems (NEETF 1998). Previously held information, whether right or wrong, affects our perception of additional information. In addition, our perception of the messenger affects the believability of the message. Some agencies have a difficult time making headway on changing public opinion because people do not respect or believe anything they say. In such cases it is important to partner with other messengers, understand the source of the perception problem, and provide information that the public can understand about those topics.
3. Sometimes the public has the information but does not understand the consequences. While experts easily make connections from one familiar topic to another (e.g., invasive exotic, waterborne seeds can be swept through the stormwater system and grow into plants where the storm sewer empties into the marsh), most people cannot follow the same leaps of assumptions. They may have discrete bits of information without the fabric to tie them together. Agency materials should present both the "what" and the "so what" — including information about the consequences and making connections to meaningful elements.
4. Sometimes the audience understands the consequences, but lacks specific procedural information. In a classic study of recycling in

Box 9.2

Attention!

Cognitive psychologists tell us that a few important things always attract attention, such as loud noises, wild animals, and traffic accidents. We involuntarily notice these events, perhaps because they play a role in basic survival. We also learn to pay attention to the things we care about: children and family, personal health, or economic benefits. If we do not sell information about the interface with gunfights and sex, we need to use relevant motivators that most people care about, like health, family, and future (Kaplan and Kaplan 1982).

Ann Arbor, MI, De Young (1988–1989) revealed that nonrecyclers had as much background information on the landfill, the value of recycling, and problems of municipal waste as the recyclers. The nonrecyclers, however, were stymied by their lack of knowledge about how to package newspapers, whether to remove labels on cans, and other details. Consider the specific skills of the practice you are promoting; do people have the necessary know-how?

Information can be conveyed in a variety of creative ways; agencies should not limit themselves to a brochure and a poster.

- *Contests* can be used to build public support for a topic; winning entries can continue to share the message. The Population Connection (formerly Zero Population Growth) sponsors a contest where youths develop television scripts on a selected topic, such as sprawl. The Dahlem Environmental Education Center in Michigan sponsors a children's art contest, picks 13 winners, and produces a calendar complete with seasonal events. Youth groups sell the calendars and keep some funds for their own activities.
- *Field trips and tours* also help participants see the problems and solutions up close and personal. A bicycle tour of suburban Athens, GA, was conducted in 2000 to demonstrate the problems of interface development and sprawl. The 25-mile "Tour de Sprawl" was punctuated by rest stops with speakers, drinks, and snacks. An accompanying bus allowed additional people to participate.
- *Demonstration areas* have been frequently used by the Cooperative Extension Service to showcase new agricultural techniques. Extension has developed other types of demonstration areas, such as yards that promote xeriscaping and use of native plants to reduce water use and improve water quality. Signs are posted in certified yards that announce to neighbors that the resident is enhancing environmental quality. Demonstration areas are also useful for illustrating the benefits of prescribed fire (Figure 9.1).
- *Festivals* help agencies promote the good things they do, such as producing fry in a fish hatchery for commercial and sport fisheries or celebrating migrating birds as they pass through the vicinity. Festival-goers cannot help but pick up a little information as they engage in activities, listen to presentations, and enjoy the day.

FIGURE 9.1 This large sign designates land that is managed with prescribed fire. (Photo by Larry Korhnak.)

- *Short-range radio* programs often introduce visitors to the rules and regulations of an area, but can also be used to explain management techniques that are visible as people drive through the forest.

It is imperative for resource agencies and organizations to provide science-based information to help answer questions and enhance understanding on interface issues. There are times, however, when the essence of an issue may involve more than science. These issues are often policy questions that connect values and science, and although science provides a portion of the required information, it cannot provide the answer. In these cases, resource managers should develop an information campaign that recognizes and distills information for the public and explores the consequences of each action that may be taken (Farnum and Dean 2001). The agency's science probably cannot resolve questions of values and determine the best policy.

A recent event in Gainesville, FL, provides an excellent example of a policy question that involves but cannot rely on science. A southern pine beetle population of epidemic proportions (Matus 2001; Lockette 2001) created great confusion and expressions of anger at city council meetings. The issue was whether to declare a local state of emergency that would allow the removal of infected trees without homeowner approval (Lippincott 2001). It would be appropriate for a resource agency to provide information on the life cycle of the southern pine beetle, how far they travel at which seasons of the year, and how much damage they are likely to do to what percentage of trees within the county. The community must wrestle with issues of private property rights, public responsibility, the economic viability of the local timber industry, and the degree of trust they have in the tree removal personnel to take only infected trees. The perception that the community has of the risk and the consequences of each potential solution will frame the debate. Resource agencies should be involved and should provide the best science available to inform the decision makers, but they should also recognize the limits of their information.

9.3.2 Develop Opportunities to Enhance Two-Way Communication and Understanding with Community Groups

While informative materials may form the basis of a resource agency's outreach program, there are probably cases where providing information is not sufficient. Perhaps community groups want more than a brochure; they want a guest speaker to answer their questions. Perhaps a divisive interface issue has created a situation where groups talk at each other, instead of with each other. Before the divisive issue erupts into conflict, perhaps the agency personnel need to better understand how various factions perceive the issue. A broad array of activities can be designed to foster two-way communication and understanding (Jacobson 1999). Where interface issues are creating conflict, these strategies may be critical tools for resource agencies (see Box 9.3).

Conflict in the wildland–urban interface is usually a result of different factions of the community having different needs, opinions, or visions of the future. The first step in resolving such conflicts is improved understanding of these opinions, values, and interests (Jacobson 1999). Because each faction may have slightly different perspectives, it is very difficult to generalize from one situation to another, or from one group to another. Mushroom farmers in Chester County, PA, for example, were fearful that newcomers from the suburbs might complain about the odor from their compost. Neighbors do not complain equally, and the farms that have compost do not receive more complaints than those that do not. Three factors influence the number of complaints: size of farm, number of neighbors, and the rate of increase of new neighbors (Abdalla and Kelsey 1996). Similarly, landowners adjacent to public land in the West probably disagree over the best way to manage invasive weeds. If one wants to try a biocontrolling beetle and abhors herbicide, another will believe the beetle is too slow and wants the immediate benefit of an herbicide. Each group has a slightly different perception of the problem, different degrees of information, and different visions of the future of their land. An agency wishing to serve all residents or to involve them in decisions should launch a multipronged program to best meet all of their needs.

Box 9.3

Research Assistance

Although ideal, an agency or organization does not need to launch its own research program to understand its audience. A growing body of literature may be used to help distill relevant questions and gain insights. An excellent database on wildland–urban interface literature is available on-line at www.interfacesouth.org.

There may be sources to assist an agency with a study of interface opinions. Graduate students at a nearby university may need a project or the local Friends Group may be able to sponsor a survey of the public.

Most agencies are familiar with public participation strategies for resource planning. Wondolleck and Yaffee (2000) suggest that the hearings and newsletters to involve local representatives do not go far enough to achieve collaborative resource management. Good collaboration is based on strong communication and public involvement at all stages of resource management — planning, implementation, and monitoring. Partnerships between agency staff and local communities can help pave the way for collaboration. When a National Wildlife Refuge manager uses the visitor center for a local display of wildlife art, when public festivals at a hatchery bring tourist dollars to local restaurants and hotels, and when agency staff assist the school science curriculum coordinator with field trip locations, partnerships are helping to smooth the pathway to better understanding (see Box 9.4).

After understanding the positions and interests involved, either by partnerships or by research, the next step in resolving interface conflicts is obtaining the participation required to make progress. If one faction decides to boycott a decision making process, the outcome is not likely to be sustainable, simply because one group was not a part of the process. Winning this participation is often a function of basic common courtesy, honest communication, and willingness to go the extra mile to make people feel welcome and comfortable.

The following strategies should support a communication program's goals to improve understanding and participation:

- Invite people to attend an introductory meeting. Hold that meeting on neutral territory — a public library or a school. Be considerate of how locations are perceived; some buildings in the South have an unfortunate history of having a "white only" front door. Encouraging African-American participation may require meetings at a building of their choice (Hale 2000). A series of introductory meetings on each group's home turf might be required to build the trust needed for everyone to attend a broader meeting. Start the meetings as early in the process as possible. No one likes to come to the first meeting to hear that decisions have already been made.

Box 9.4

The Case of Research to Understand the Audience

In response to public sentiment against timber harvesting on public lands, a graduate student at the University of Florida conducted a qualitative study to examine the general attitudes held by locally active environmentalists (Bowers 1999). The study consisted of nine in-depth interviews with purposively selected informants who identified themselves as environmentalists. All belonged to at least one of the following environmental organizations: Sierra Club, Audubon Society, Florida Defenders of the Environment, or the Florida Coalition for Peace and Justice.

The interview covered three broad main topics: definition of a forest, benefits of a forest, and management of public land. Participants responded to questions and used 10 pictures of forest scenes to prompt discussion. Participants labeled each photo as "not a forest," "sort of a forest," or "forest" and commented on how the forests in the pictures should be managed if they were publicly owned (Figure 9.2). During the interviews, the participants confirmed that they were strong proponents of the environment, often writing letters to the editor, actively campaigning for proenvironmental legislature, and working with organizations to further environmental causes.

Although these participants might be initially viewed as radical tree-huggers by land managers, their attitudes and beliefs suggest otherwise. Participants consistently recognized the economic as well as ecological benefits of forests. They understood that many of the everyday products used in society come from trees and forests. Participants uniformly called for a more balanced management of forests, incorporating both economic and ecological benefits of forested lands. Just as many land managers view environmentalists as being one-sided, ignoring practicalities and science, blinded by their desperate need to defend the environment, these environmentalists worry that land managers may be too focused on the utilitarian benefits of forests. The qualitative interviews demonstrated that environmentalists and land managers can share a common goal — balanced management of publicly owned forests. The more land managers can understand what their supposed opposition believes, the easier it will be to create a message that accurately portrays management plans that are more acceptable to environmentalists.

- Consider the literacy rate of all the participants. For those whose native language is not English and for those who dropped out of school to work on the family farm, community meetings that require advanced reading, note-taking, or writing could be a significant barrier.
- Use the existing leadership in each group to win support. By talking to people it is often easy to discover who has respect in the community — a store owner, a church leader, an original family. These opinion leaders are key to obtaining access to the broader group and building a base for good discussions (Rogers 1995).
- Consider bringing on an outside facilitator or mediator, particularly if all parties are strongly committed to a different outcome.
- Be patient. Participants in these multigroup negotiation efforts often describe the importance of venting frustration, explaining positions, and questioning authority. This "storming" period is typical of group work and necessary to prepare for productive activity.
- Maintain flexibility in pursuit of a resolution. Initial explanations of "we aren't allowed to do X because of our legislative mandate" do not express willingness to try something new. Those inflexible positions may be at the root of the conflict.
- Try a charette or workshop, two popular strategies to engage the public in planning decisions (Figure 9.3). By offering several options, an agency may learn what the public believes makes a better solution. A meeting and informational materials may be needed to prepare participants for the workshop.

When a group of willing participants has been established, managers should consider how decisions will be made. A simple majority vote means that the side with the most representatives always wins; a minority perspective will always lose. A consensus strategy may help the group discover a tolerable compromise, one that everyone can live with (Susskind et al. 1999) (see Box 9.5).

The U.S. Fish and Wildlife Service has effectively used communication techniques to improve its resource management strategies. On the Mississippi River, for example, refuge managers found it difficult to protect nesting habitat on linear islands because tow captains operating large barges used the trees to tie up their loaded vessels. Between the waves and the tows, nearby vegetation was heavily damaged. By assigning biologists to ride with the tows, refuge managers obtained a better understanding of the reasons why captains tie up to islands and engaged them in creating appropriate solutions, such as concrete piers for foul weather emergency stops (Ady et al. 1999).

FIGURE 9.2 Florida environmentalists have a special feeling about cypress trees and would prefer that they not be harvested from public land. (Photo by Larry Korhnak.)

FIGURE 9.3 Community workshops can provide input to a local plan that identifies significant environmental areas to be protected. (Photo by Renaissance Planning Group.)

9.3.3 Support Education Programs that Enhance Problem-Solving Skills Regarding Interface Issues

Information and communication efforts help educate the audience. These programs provide information and help people know more about interface issues. But that does not make these "education programs." An education program in this case refers to a long-term effort with a stable population and some degree of testing, certification, or graduation at its conclusion. Public and private schools, technical schools, community colleges and universities, religious institutions, and zoos are examples of institutions that manage educational programs.

A great many materials and efforts by agencies and organizations targeted to schoolchildren and teachers would fall into the "information" category, even though the audience is part of an educational institution. Worksheets, activities, field trips, guest speakers, workshops for teachers, and coloring books provide information about interface issues in an appropriate way to this audience. If produced in collaboration with educators and youth, they may be

Box 9.5

A Case of Negotiating a Wildland Fire Message

The Cooperative Extension Service in Florida reacted to the 1998 wildfires by producing a fact sheet on how homeowners could reduce their risk of wildland fire. In surveying the existing materials on the topic and speaking to local experts, it became clear that a simple message of creating defensible space might not be sufficient or appropriate.

A variety of agencies and organizations have advice for homeowners about designing and maintaining their landscaping; some of these messages conflict with each other. To provide a consistent and meaningful message on landscaping for wildland fire, extension specialists organized a 1-day workshop and invited representatives of various agencies and organizations. Twenty-three different groups attended, including the state transportation, wildlife, and forestry agencies; native plant society; city and county fire services; energy utility; and horticulture, energy, soil and water, wildlife, and forestry extension specialists.

During the meeting, fire officials described their situation, the risks to firefighters, and their view of how homeowners could reduce their risk. Other participants raised concerns about the impacts of these actions: removing shade trees increases utility bills, reducing ground cover deters wildlife, removing mulch increases the need for irrigation, and planting nonnative species or lawn reduces native habitat. Small groups were formed to develop a negotiated message for a target audience: What should we tell small landowners at the interface? What guidelines can we provide to developers? (Monroe and Marynowski 1999).

Organizers reported the day's work back to participants and began to formulate guidelines for the public and developers. Several participants worked diligently to review and respond to each draft until a common ground was established: homeowners must first assess their risk of wildland fire and take precautions if their risk is medium or high; the home should have enough cleared area to allow fire trucks to maneuver, but trees and flower beds can be located near the house; vegetation should have vertical and horizontal separation to prevent fire from spreading; and combustible mulch should not be used near the foundation (Figure 9.4) (Monroe and Long 2000). This message is not simple, but it may be more practical and easier to implement than one that conflicts with homeowner values. It may also be more broadly communicated if several agencies are willing to deliver the same message.

excellent examples of communication programs. To become an effective education program, however, agency messages must be blended into the structure and purpose of the existing educational programs, such that by using the agency program, the educators are accomplishing their own objectives as well. Clearly, the first step in designing these programs is to understand these educational objectives.

Many public schools, for instance, are reforming their educational systems to follow the state curriculum and improve accountability. Teachers and principals are less likely to pick up an agency-produced packet of activities if it does not clearly promote their state curriculum or provide appropriate opportunities for students to practice skills assessed on the state achievement test. Agencies are teaming up with educators to produce new materials that meet these goals. A water management district in Florida, for example, sponsored the production of a resource box on watersheds that helps teachers improve reading skills. Stories, puppets, worksheets, and practice test prompts enable students to build writing skills and reading comprehension while they learn about rivers and the water cycle.

Most existing environmental education curriculum materials have not yet been assessed for their ability to help schools improve achievement in reading, writing, or mathematics, but they might be effective resources because the environment is an interesting topic to many youngsters. New curriculum resources can be designed to improve student achievement by building in lessons that allow students to practice the tested skills. A recent project to make the collection databases from the Florida Museum of Natural History available to the public included the development of five lessons on biodiversity for high school teachers. Each lesson includes a writing assignment, and students grade each other's essays with the same rubric used to score their state writing tests. In a pilot test of the curriculum, students significantly improved their writing when essays were scored with the state rubric (Randall 2001).

By the same token, educational programs with scout groups should help youngsters earn badges and pins, activities with 4-H should help youngsters complete independent project books that build life skills, and collaborations with high schools should foster career exploration and skill building. Natural resource agencies and organizations must help the educational institutions achieve their own objectives if agencies expect their interface-management materials to be used.

Some states have adopted "service learning" as their tool to help youngsters gain skills to become responsible citizens. Service learning is an opportunity for young people to actively participate in thoughtfully organized service experiences that benefit the community (Florida Learn & Serve 1999). Programs include four phases: *preparation*, when youth analyze a problem and learn about solutions; *action*, or the service; *reflection*, when youth think critically about their experience; and *recognition*, to complete

FOR 71

Landscaping in Florida with Fire in Mind

Martha Monroe and Alan Long
School of Forest Resources and Conservation

Determine Your Risk

Two factors contribute to wildfire risk:

- **the land use in your area, and**
- **the kind of vegetation around your home.**

Surrounding Land Use

If you live in a subdivision surrounded by homes and lawns, or in an urban area, it is unlikely that a wildfire would reach your house. Like the majority of Floridians, you are at low risk of wildfire and the rest of this brochure does not apply to the safety of your home.

If you have undeveloped or wooded land near your home, however, you could be at some risk in the event of a wildfire. Use the following criteria to assess your risk.

Vegetation

Walk around outside your home and look carefully at the nearby land. The type, size, and density of the plants determine wildfire risk. Some places may have characteristics of more than one category.

Open this brochure and choose the appropriate guidelines to change the conditions that affect your risk level.

You Are at Low Risk if You See...

- Bare ground, improved pasture, or widely spaced grassy clumps or plants.
- Moist forest mostly leafy trees, or mostly large trees.
- Few plants growing low to the ground.
- Oak leaves or other broad leaves covering the ground.

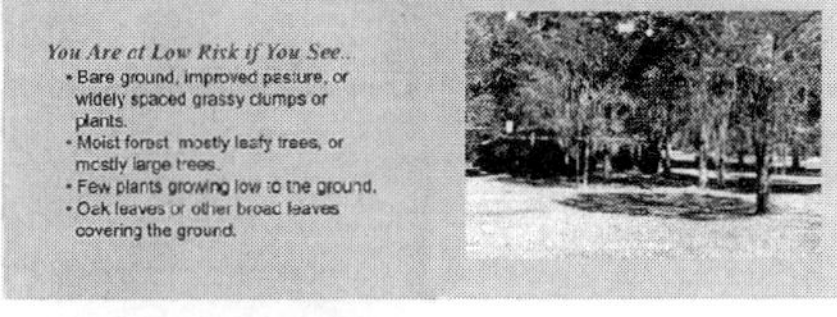

You Are at Medium Risk If You See...

- Thick, continuous grasses, weeds, or shrubs.
- Continuous thin layer of pine needles and scattered pine trees.
- Scattered palmettos or shrubs up to 3 feet tall separated by patches of grass or sand.
- A clear view into or across the undeveloped area.

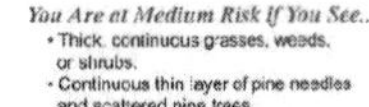

You Live in a High Risk, Fire-Prone Area if You See...

- A thick bed of pine needles and lots of pine trees.
- Continuous palmettos, shrubs, or sawgrass more than 3 feet tall.
- Vines and small-to-medium trees or palms beneath taller pine trees.
- Impenetrable shrubs or young pines.
- No clear view into the undeveloped area because of dense growth.

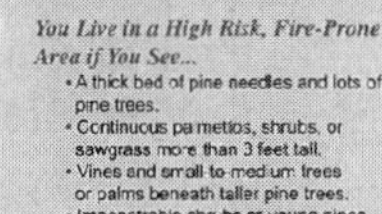

FIGURE 9.4 A negotiated product guides homeowners in assessing their risk of wildland fire before suggesting what they should do to reduce this risk (Monroe and Long 2000).

the program (Duckenfield and Madden 2000). Other programs supporting similar goals with real-world, community-based, problem-solving activities have indicated an impressive ability to improve student motivation and achievement (Lieberman and Hoody 1998). Teachers report that teaming up with their colleagues and approaching a community problem even rejuvenates their practice. These programs can be ideal avenues for agencies to assist educators in meeting school-based objectives while addressing their own environmental priorities.

Several substantial programs have been developed that guide youth through a process of assessing environmental problems, evaluating possible solutions, and taking actions to resolve them (Hammond 1992; Bardwell et al. 1994). Give Water a Hand was developed by the University of Wisconsin-Extension to enable youth to tackle community problems in their own watershed. The Global Rivers Environmental Education Network (GREEN) uses water quality of rivers to connect youth and empower them to resolve environmental and community challenges (Figure 9.5). By beginning with the process of collecting data, youth discover the essence of the problems without being directed by an adult. By exploring all angles of the problem, listening to a variety of experts, and contacting others involved with the issue, teachers avoid the perception of brainwashing or steering youth into an advocacy role. Because these programs also help schools achieve their goals of improving citizenship, problem-solving, group process, communication, and decision making skills, many educators embrace these community activities as exciting supplements to the school day.

Many of these programs can be adopted by interface managers to promote positive community activity (see Box 9.6). If educators become involved independent of an agency sponsor, an appropriate role for a natural resource manager may be as a partner to the school. Agency personnel can provide background information, sponsor field trips, offer scientific data, and support a local action project without the effort and expense of developing their own service learning program.

9.3.4 Encourage People to Practice Certain Behaviors to Support Resource Management Objectives

Information is a necessary ingredient to any outreach activity. It can support the development of new actions,

FIGURE 9.5 Youth test local waters, compare findings, and discuss possible actions in a variety of water quality testing projects. (Photo by Julie Athman.)

Box 9.6

Case of a State Agency Developed Education Program

The Washington Department of Fish and Wildlife (WDFW) recently identified a need to create opportunities for citizen stewardship and to promote its mission of conserving biodiversity. At the same time, the Washington Cooperative Research Unit Gap Analysis Project (WAGAP) was developing a process to collect and display data on terrestrial plant and animal species. A partnership between WDFW and WAGAP created the *NatureMapping* Program to "promote community-based environmental protection by mapping wildlife sightings and habitat" (Tudor and Dvornich 2001). The program is firmly rooted in environmental education goals of community-based fieldwork, which can be integrated into different subject areas and involve higher order thinking skills.

Teachers are interested in the program because it enables them to provide solid, real-world educational opportunities to study biology and watersheds in their own community and opportunities for youth to practice stewardship of community resources. As youngsters learn where to look and what to look for in their hunt for local wildlife, they gain skills in observation and data collection. Schools with long-term data banks have provided valuable information to local cities about wildlife corridors; others have taken their experiences and written a story about floating past different habitats on the Little Spokane River. Several agencies have relied on student and adult observations to build a data set as well as public understanding and involvement in policy decisions.

An evaluation of the program indicates that the program improves communities' impressions of schools and their students, increases communication between resource managers and the public, and builds a broad data set of wildlife and plant community information (Tudor and Dvornich 2001). All of these outcomes are important aspects of an agency-based educational program.

but rarely results in the changes to citizens' behaviors that many resource managers expect from outreach programs (Hungerford and Volk 1990; McKenzie-Mohr and Smith 1999). In-depth communication and education programs can do more than informational campaigns; they can help raise awareness, bridge gaps, begin to resolve conflicts, and support skill building, but these activities take time. In resource management cases where time is critical, as in endangered species recovery or flood and fire danger, the disseminated information need not focus on the background advantages and disadvantages, but on the key steps that citizens must take. Outreach in these cases should borrow from the social marketing literature to create effective behavior change programs (see Box 9.7).

Social marketing uses a set of communication tools in a strategic way to encourage and maintain behavior change (McKenzie-Mohr and Smith 1999). Tools like social commitment, appropriate incentives, reminder prompts, and models help create a social norm that supports the new behavior. It is commonly used in the health field to encourage people to get their blood pressure tested, not to drive while drinking, or to practice safe sex. There is a growing interest in using the same techniques to change inappropriate behaviors to more acceptable ones in order to resolve natural resource problems. Several programs encourage people to exercise care when filling lawn mowers with gasoline, to bundle water heaters with insulating blankets, and to use fluorescent light bulbs to conserve energy. In a demonstration of the power of verbal commitment, for example, home assessors with Pacific Gas and Electric asked their customers, "When do you think that you'll have the weather-stripping completed?" and promised to call back to see if any problems arose. Changes in the assessors' door-to-door

Box 9.7

THEORIES ABOUT BEHAVIOR CHANGE

Educators, psychologists, sociologists, and others have long debated the factors that determine human behavior. Ajzen (1991) suggests that, in addition to knowledge and the associated attitudes, behavioral intention is also a product of the individuals' perceptions of social approval of the behavior and their own ability to adequately perform and achieve results. Intention may not lead to the behavior if alternatives are not available, if barriers of time and convenience are insurmountable, or if external motives are not present. Social marketers develop tools to reduce all of these barriers and enhance any of these factors to change behavior.

conversation tripled the likelihood that homeowners would retrofit their homes (Gonzales et al. (1988) as cited in McKenzie-Mohr and Smith 1999).

There is a danger of orchestrating a program to manipulate human behavior, of course. Some citizens believe government agencies should not undertake a Big Brother role. Natural resource agencies can avoid this trap by:

1. Clearly operating within the legislative mandate that governs their agency's mission to reduce wildland fire risk, conserve water, or protect coastal ecosystems, for example.
2. Using community groups to identify local problems and potential solutions, since organizations and agencies are expected to promote community-accepted behaviors.
3. Sponsoring complementary information programs about the benefits of the new behavior.

Social marketing programs begin with a thorough understanding of the problem and the audience's current behavior and opinions. Sometimes this clarifies the issue such that extensive programs are unnecessary, the solutions are obvious, or the program is reoriented. If potential recyclers express that the collection centers are inconveniently located, a new location may be the only response needed. Florida's rural residents acknowledged that fire plays an important role in maintaining forest ecosystems, but expressed confusion over the definition of prescribed fire (Jacobson et al. 2001). The ensuing educational efforts targeted the differences between wild and prescribed fire.

A variety of tools can be used to market appropriate interface behaviors. Prompts are short messages provided at the site where the behavior occurs. By themselves they do not usually achieve notable change, but in conjunction with background education, these little reminders can help people turn off lights, buy recycled products, or use cloth grocery bags. Prompts are generally used when habitual behaviors can be altered without significant consequences.

When the change is really an alteration of an accepted community norm, more complex social marketing efforts are needed. The U.S. Fish and Wildlife Service is working with the Alliance for Chesapeake Bay and other partners to protect water quality by reducing nutrients from suburban lawns. The BayScapes Program works with targeted communities to establish demonstration gardens to attract attention and interest (Figure 9.6). Workshops, homeowner association meetings, gardens in public spaces, and participatory displays at festivals help build a social norm that supports a new way of landscaping and caring for suburban yards.

Information can be provided in such a way that people are more likely to notice and remember. Community leaders or movie stars can be used to promote the proper disposal of household hazardous materials or the removal

FIGURE 9.6 A demonstration planting in a public place illustrates an alternative to traditional, high maintenance lawns. (Photo by Britt Slattery, U.S. Fish and Wildlife Service.)

of invasive exotic plants. If these leaders are modeling the new behavior, others are more likely to follow suit (McKenzie-Mohr and Smith 1999). Making people aware of "normal people" who are engaged in the appropriate behavior may also help change the perceived social norm. Case studies and success stories are often used to build imagery of the possible to help people think about their choices (Bardwell 1991).

Encouraging interface residents to carpool to work also requires more than prompts and reminders. Kearney and De Young (1994) compared the use of fact sheets and interesting stories to encourage workers to share a ride to work. Both tools resulted in greater knowledge about carpooling, and both treatment groups reported thinking more about carpooling; these are important elements of creating a willingness to engage in a new behavior. The stories appear to be excellent tools for conveying the intangible aspects of carpooling and building complex knowledge structures that generally predict increased problem-solving ability. Unfortunately, the perceived barriers of scheduling problems, pre- and post-work obligations, and arranging work hours to accommodate a carpool prevented a significant increase in carpooling behavior. Success in encouraging carpooling will require a commitment to reduce these barriers in addition to providing information.

Social marketing tools are best used together to maximize effectiveness. Educational programs that help people increase confidence in one's skills may be more successful with the addition of demonstration areas that show the results of a new practice to interested but unsupportive neighbors (Monroe et al. 1999). Prompts may be more effective after obtaining verbal commitments from citizens who are willing to try something new. The key to social marketing is thinking strategically about all the potential barriers to the behavior and all the possible tools that educators and communicators can use to change perceptions (see Box 9.8).

Box 9.8

Social Marketing Helps Reduce the Risk of Wildland Fire

In Texas, several educational programs sponsored by volunteer fire departments, the Texas Forest Service, and homeowner associations are designed to reduce the risk of wildland fire. One program began in Bastrop County, where a partially built subdivision (600 homes on 4000 acres) in a steeply sloped, pine/juniper forest with narrow unimproved roads was at high risk for wildland fire. The Tahitian Village Wildfire Mitigation Project is aggressively promoting the use of defensible space around homes. Several homeowners volunteered their properties to become demonstration areas to show neighbors the feasibility of defensible space (Figure 9.7). At neighborhood gatherings, homeowners complete a hazard assessment form and identify how they might work together to reduce their risk. Although information dissemination is an important goal, they are also achieving awareness of wildland fire issues by creating fun community events that bring neighbors together. To that end, the MulchFest gave residents a chance to learn about wildland fire and defensible space; obtain new street addresses to facilitate emergency access; buy foam; and dispose of brush, leaves, and needles while enjoying the day with friends. One ongoing barrier for the program is that residents have few options to safely dispose of the brush and slash they create from their mitigation efforts. The concepts and experience from Tahitian Village are being shared throughout the 60,000-person county and the state.

9.4 THE PROGRAM DEVELOPMENT PROCESS

Whether producing a brochure to inform, sponsoring a festival to celebrate, collaborating with homeowners to understand better, designing a campaign to take action, or working with teachers to enhance student achievement, programs should be developed according to a fairly standard and intuitive process (Bennett and Rockwell 1995; Jacobson 1999; Day and Monroe 2000). Each step is important in the development of a high-quality program.

1. *The need for the program is assessed.* The need for the program includes both the need to enhance the environment and the need for the audience to participate in the program. The assessment process enables program organizers to understand what the audience already knows and cares about. Various stakeholders are consulted to make sure the program is on target. The capabilities of the sponsoring organization in terms of budget, time, and people are also assessed to ascertain the program possibilities and the need for partners. This step results in a solid foundation that supports the development of the program.
2. *The program or material is designed and developed.* This step includes writing objectives that will guide both the program and the evaluation. Educational objectives target the learner's behavior (e.g., by the end of this program, students will be able to state three environmental changes as a result of spreading suburban development). Principles of learning and communication are used to build a quality product from these objectives. Design parameters

FIGURE 9.7 Volunteer fire fighters and Texas Forest Service staff created several demonstration areas to illustrate landscape and construction modifications that can help reduce the risk of wildfire. The crawl space of this house is enclosed to keep leaves and embers from under the structure. (Photo by Martha Monroe.)

(length of curriculum, number of posters, etc.) are based on the assessment results. Draft copies of the program are produced and implementation activities are designed (such as volunteer training and assessment tools).

3. *The program or material is pilot-tested.* In addition to an expert review to check for accuracy, the program is pilot-tested with the real audience. Feedback from this trial run will help organizers know if the material is performing appropriately, for example, if the reading level is adequate, if the activities make sense, if the design is interesting, or if the presentation is engaging. Revisions are made, and the pilot test is repeated until the results are satisfactory.
4. *The program is implemented.* Materials are produced in quantity and distributed. Leaders are trained. Broadcast time is purchased. Monitoring schemes are put into place to record levels of use, number of contacts, changes in behavior, changes in environment, copies printed, requests reviewed, or whatever is required.
5. *The program is evaluated.* The original objectives may be used to assess whether the program achieved its goals, or other tools may be developed to measure audience contacts, learner growth, leader use, program interest, participant satisfaction, or environmental change. The results of the evaluation can be used to improve the program and the dissemination strategies, to document success, and to report to stakeholders.

Claude Bennett designed a planning process to help guide the development of educational programs that are directly tied to resource management goals (Bennett 1979; Bennett and Rockwell 1995). It may be most useful to use his hierarchy in the program design and development phase. Working for the Cooperative Extension Service, he recognized that a huge variety of activities fall under the rubric of "educational programs" that include such diverse exercises as producing posters and news releases, guiding demonstration tours, and writing curriculum. Rather than immediately launching into materials production, he suggests that managers first consider the environmental problem they wish to solve and then the aspects of that problem that are influenced by people. As managers move through guiding questions, they eventually conceptualize a program that is realistic and targeted to the problem (see Box 9.9).

By starting the program development process with a focus on the issues to be resolved, outreach activities should be in keeping with the agency or organizational mission and goals. The activities may be in the realm of providing infor-

Box 9.9

TARGETING OUTCOMES OF PROGRAMS

1. *Environment*: What environmental condition should be improved? How do actual conditions compare to targets?
2. *Behavior*: What practices must people adopt in order to meet the environmental target? How are these target practices different from current practices?
3. *Knowledge, attitudes, skills*: What knowledge, attitudes, and skills are needed to help the audience adopt the target behaviors? How different are each from the current condition?
4. *Reaction*: How should program participants react to the program, and how will their reaction reflect the degree of change needed in knowledge, attitudes, and skills?
5. *Participation*: Who are the program participants? How many can be engaged in the program? What level of involvement is needed to achieve the required reactions?
6. *Activities*: What are the methods of delivering information and experience to the program participants? What promotional strategies can be used to achieve desired levels of participation?
7. *Resources*: What resources (e.g., staff, volunteers, money, partners) are available to support these activities?

mation, enhancing communication, working with schools, or changing behavior, although the first two categories tend to fit Bennett's model more neatly than the others.

These seven levels also offer an opportunity to track program outcomes at each stage, suggesting ways to evaluate a program that includes resources used, activities conducted, knowledge changed, behaviors practiced, and environment improved.

9.5 Summary

A host of opportunities exist for natural resource managers in agencies and organizations to develop communication and education programs to help solve wildland–urban interface issues. The strategies for developing these programs will vary based on the intent and context. By answering several questions, resource managers can determine what type or category of program is necessary:

- Is information available to prepare people adequately for their role in resolving interface problems? If not, use information strategies that provide the audience with what they need to know.
- Does everyone understand each other? If various groups express different and conflicting perspectives, perhaps some communication strategies designed to encourage two-way discussions would be helpful.
- Is there a conflict? Do not rely on information tools when people are already upset or angry. Take the time to help people listen and understand other views. Develop a strategy to collect data jointly to resolve questions of uncertainty. Use communication strategies.
- Is there an interest in working with an educational institution to build knowledge and skills? Education strategies are useful when working with schools and teachers. In the best case, natural resource programs become the vehicle to achieve educational goals.
- Is there a good match between the educational institution's objectives and those of the resource agency or organization? If there is no match, resort to information and communication tools. If there is a match, move into education strategies.
- Is there a need to change behaviors of a target audience? Information and communication strategies alone may not achieve behavior change, but in conjunction with the behavior change strategies that work in social marketing programs, differences can be made.
- Are there resources to do the job? Effective outreach requires some investment of staff time and resources. If the need overpowers the budget, partner with other agencies to share the load. Seek support from universities for student service learning projects and research opportunities. Funding may be available from a variety of sources for programs that are well designed and meet a need.
- Do I know where to begin? A variety of resources are available to offer guidance and support. A recent compilation of best practices in boating, fishing, and stewardship education provides a host of references and ideas (Fedler 2001). Agencies and environmental groups are often willing to share their expertise and experience, and many have quite a lot to offer.

When the need for a program has been identified, a program development process can be used to develop and implement a solid, well-designed program.

Since a variety of environmental education programs already exist, similar programs in other agencies should be used as models. It is far more effective to adapt an existing tool than to start from scratch. Bennett's model may be used to identify how to adapt a program most efficiently.

Education and communication programs have the power to engage citizens in making substantial changes in policy and personal behavior patterns. As our wildland–urban interface regions expand and grow with new residents, the need for effective outreach programs increases. Agencies and organizations have a variety of tools from which to choose the strategies that will best serve them and the natural resources they manage.

REFERENCES

Abdalla, C.W. and R.W. Kelsey, 1996. Breaking the impasse: helping communities cope with change at the rural—urban interface, *Journal of Soil and Water Conservation* 51: 462–466.

Ady, J., B. Byers, H. Johnson-Schultz, and B. Stieglitz, 1999. Multifaceted Approaches to Outreach in a Natural Resources Agency, paper prepared for the 2nd International Wildlife Management Congress in Gödöllo, Hungary, June 28–July 2, 1999.

Ajzen, I., 1991. The theory of planned behavior, *Organizational and Human Decision Processes* 50: 179–211.

Bardwell, L., 1991. Success stories: imagery by example, *Journal of Environmental Education* 23: 5–10.

Bardwell, L., M. Monroe, and M. Tudor, 1994. *Environmental Problem Solving: Theory, Practice, and Possibilities in Environmental Education*, North American Association for Environmental Education, Troy, OH.

Bennett, C., 1979. Analyzing Impacts of Extension Programs, Report ESC-57, U.S. Department of Agriculture, Science and Education Administration, Washington, DC.

Bennett, C. and K. Rockwell, 1995. Targeting Outcomes of Programs (TOP): An Integrated Approach to Planning and Evaluation, unpublished manuscript, University of Nebraska, Lincoln.

Bowers, A. W., 1999. Environmentalists' Definitions of Forests and Beliefs about Forest Management on Public Lands, Master's Technical Paper, School of Forest Resources and Conservation, University of Florida, Gainesville, FL.

Day, B. and M.C. Monroe, Eds., 2000. *Environmental Education and Communication for a Sustainable World: Handbook for International Practitioners*, Academy for Educational Development, Washington, DC.

De Young, R., 1988–1989. Exploring the differences between recyclers and non-recyclers: the role of information, *Journal of Environmental Systems* 18: 341–351.

Duckenfield, M. and S.J. Madden, 2000. An orientation to service learning, in *Service Learning across the Curriculum*, S.J. Madden, Ed., University Press of America, Lanham, MD.

Farnum, P. and C. Dean, 2001. Science in Service to Society: The Role of Research, John Gray Distinguished Lecture, April 24, 2001, School of Forest Resources and Conservation, University of Florida, Gainesville, FL.

Fedler, A.J., 2001. Defining best practices in boating, fishing, and stewardship education, Recreational Boating and Fishing Foundation, www.rbff.org.

Florida Learn & Serve, 1999. Florida Learn & Serve: K-12 1998-99 Project Descriptions, Florida Learn & Serve, Tallahassee.

Hale, J., 2000. Participation, Power and Racial Representation: Negotiating Ecotourism Development in Hamilton County, Florida, Master's Technical Paper, School of Forest Resources and Conservation, University of Florida, Gainesville, FL.

Hammond, W., 1992. *The Monday Group, Project WILD*, Western Regional Environmental Education Council, Boulder, CO.

Hungerford, H. and T. Volk, 1990. Changing learner behavior through environmental education, *Journal of Environmental Education* 21: 8–22.

Jacobson, S.K., 1999. *Communication Skills for Conservation Professionals*, Island Press, Washington, DC.

Jacobson, S.K., M.C. Monroe, and S. Marynowski, 2001. Fire at the wildland interface: the influence of experience and mass media on public knowledge, attitudes, and behavioral intentions, *Wildlife Society Bulletin* 29: 1–9.

Kaplan, S. and R. Kaplan, 1982. *Cognition and Environment: Functioning in an Uncertain World*, Praeger, New York.

Kearney. A.R. and R. De Young, 1994. Promoting Ride Sharing: The Effect of Information on Knowledge Structure and Behavior, unpublished report from the University of Michigan's School of Natural Resources and Environment to the U.S. Environmental Protection Agency, Ann Arbor, MI.

Lieberman, G.A. and L.L. Hoody, 1998. *Closing the Achievement Gap: Using the Environment as an Integrating Context for Learning*, State Education and Environment Roundtable, San Diego, CA.

Lippincott, C.L., 2001. Native Pine Beetles a Problem, Not an Emergency, *Gainesville Sun*, July 22, 2001, Gainesville, FL.

Lockette, T., 2001. Officials: Pine Beetle Outbreak is Much Worse Than '94, *Gainesville Sun*, July 10, 2001, Gainesville, FL.

Matus, R., 2001. Survey Predicts Area Invasion, *Gainesville Sun*, April 20, 2001, Gainesville, FL.

McKenzie-Mohr, D. and W. Smith, 1999. *Fostering Sustainable Behavior: An Introduction to Community-based Social Marketing*, New Society Publishers, Gabriola Island, BC, Canada.

Monroe, M.C. and A. Long, 2000. Landscaping in Florida with Fire in Mind, FOR 71, Florida Cooperative Extension Service, University of Florida, Gainesville, FL.

Monroe, M.C. and S. Marynowski, 1999. Summary of March 15, 1999, Workshop, Landscaping in Florida with Fire in Mind, unpublished report from the University of Florida to the Florida Fish and Wildlife Conservation Commission Advisory Council for Environmental Education, Tallahassee.

Monroe, M.C., G. Babb, and K.A. Heuberger, 1999. Designing a Prescribed Fire Demonstration Area, FOR 64, Florida Cooperative Extension Service, University of Florida, Gainesville, FL.

NEETF, 1998. 1998 National Report Card on Environmental Knowledge, Attitudes, and Behaviors, National Environmental Education Training Foundation, Washington, DC.

Randall, J., 2001. Enhancing High School Student Writing Skills with Florida Biodiversity Education, Master's thesis, University of Florida, Gainesville, FL.

Rogers, E.M., 1995. *Diffusion of Innovations*, 4th ed., The Free Press, New York.

Scott, W. and J. Fien, 1999. An Evaluation of the Contributions of Educational Programmes to Conservation within the WWW Network: Final Report, unpublished report to the Worldwide Fund for Nature, Gland, Switzerland.

Susskind, L., S. McKearnan, and J. Thomas-Larmer, 1999. *The Consensus Building Handbook: A Comprehensive Guide to Reaching Agreement*, Sage Publications, Thousand Oaks, CA.

Tudor, M.T. and K.M. Dvornich, 2001. The NatureMapping Program: resource agency environmental education reform, *Journal of Environmental Education* 32: 8–14.

Wondolleck, J.M. and S.L. Yaffee, 2000. *Making Collaboration Work: Lessons from Innovation in Natural Resource Management*, Island Press, Washington, DC.

10 Planning and Managing for Recreation in the Wildland–Urban Interface

Taylor V. Stein
School of Forest Resources and Conservation, University of Florida

CONTENTS

10.1 INTRODUCTION

Where else in the world can people escape from work, learn about nature, reduce stress, become physically fit, bond with family members, interact with a diversity of people, and have fun all at the same time? And if these benefits were not enough, sensitive and endangered ecosystems can be protected while still providing jobs and income for local economies. Recreation is one of the most popular activities in the wildland–urban interface, but it is also one of the least understood in terms of who is being served, how to serve those people, and how to serve the needs of local communities and the environment while providing recreation. However, a well-planned and managed recreation area in the wildland–urban interface has the potential to provide visitors, surrounding communities, and ecosystems with numerous benefits.

This chapter provides an overview for managing and planning for recreation in the wildland–urban interface. It will explain the most important issues, concepts, and methods for nature-based recreation management in these areas. Moreover, it will go beyond simply helping managers work with unacceptable environmental impacts and social conflicts and discuss ways in which planners and managers can help provide for the diverse benefits potentially derived from recreation in the wildland–urban interface.

To provide these benefits, planners must identify and clarify them at the initial stages of decision making. Throughout the world, forests, deserts, wetlands, and other natural areas serve as recreation areas that are in high demand. But, what do people really demand from these areas? Although one might write off recreation as simply *fun and games*, society places enormous value on leisure and recreation because these activities help provide for

1-56670-602-5/05/$0.00+$1.50
© 2005 by CRC Press LLC

important psychological (e.g., stress relief), sociological (e.g., family bonding), and economic benefits (e.g., camping fees). Recreation also places a high value on undeveloped ecosystems because natural areas seem to best provide unique experiences and benefits (Wohlwill 1983).

Nature-based recreation is an interdisciplinary and evolving area of study. It involves the complexities of understanding human behavior in different and often fairly remote outdoor settings. For a variety of reasons, recreation in the wildland–urban interface can be even more complex. In particular, the interface is an easy-to-access natural area for many urban dwellers. That means it can and does provide recreational opportunities for many different types of people. Specifically, lower income and culturally diverse people use interface recreation areas to a greater extent than they use more remote nature-based recreation areas (Dwyer and Gobster 1992). Since the majority of nature-based recreation research has been conducted in wildland recreation settings, the needs and attitudes of these unique groups are, therefore, relatively unknown (Dwyer and Gobster 1992). Although recreation areas in the wildland–urban interface can provide unique recreational experiences to a diversity of people, this high recreational use also has its drawbacks. Environmental impacts and social conflicts are likely to be much more intense and unique in the interface due to the higher use levels and different activities available to a more diverse group of recreationists.

This chapter discusses the benefits of nature-based recreation and how researchers and managers are working to plan and manage for these recreation outputs. It puts nature-based recreation into context by describing some of the recreational activities that people demand from the wildland–urban interface and which organizations are attempting to supply this demand. Then current recreation management approaches for natural areas are described, focusing on zoning, monitoring, and management strategies and tactics. Finally, specific planning issues like financing recreational opportunities and collaborating with diverse constituents are addressed in terms of their relevance to the wildland–urban interface.

10.2 OUTPUTS OF NATURE-BASED RECREATION

"If you build it, they will come" is not always true. In the late 1990s, the Apalachicola National Forest began closing campgrounds and limiting staff at several recreation areas just outside Tallahassee, FL, which has a population of over 150,000 people. One facility was specifically designed in the 1970s for people with disabilities and had a playground, fishing opportunities, a swimming pool, and other recreational facilities. Nearby, a campground was closed, even though it was adjacent to a lake with excellent beach facilities. Surprising to local managers, few people used these sites and the cost of maintaining the facilities simply overwhelmed the benefits of keeping them open (Andrew Colanino, personal communication, May 2001). Since both recreation areas were within 5 mile of Tallahassee, it seemed logical to provide campgrounds and other recreational opportunities. So, why didn't the public, especially local residents, take part in the opportunities provided? The Forest Service is still searching for answers, and it probably will not find those answers until it has a better understanding of its adjacent urban residents, who are most likely to use those facilities.

Developing recreational opportunities for an unknown and mysterious set of visitors is not unique to Florida. Natural resource planners and managers across the nation need to take a closer look at what they are really providing to the public. Recreation planners do not simply provide campgrounds, trails, picnic tables, and other recreational facilities. They provide opportunities for people to benefit in many different ways; therefore, they need to find out what all of their potential visitors (both traditional and nontraditional groups) want from recreation areas in order to help them attain these benefits. Managing for visitors' needs is particularly important in the wildland–urban interface because of the large and diverse populations who use these areas and have many different reasons to take part in recreation.

Recreation is defined as an activity conducted on people's free time to achieve a desired experience that is intrinsically beneficial (Driver and Tocher 1970). This intrinsically beneficial experience, which is considered to be the output of a recreational engagement, is often hard to define and measure — making recreation's outputs much different from other uses of natural resources. For example, when managing for timber production, managers base their success on board-feet. For grazing, productivity is measured in animal unit months. What is the measure of success for recreation? Historically, and in many current cases, recreation managers used numbers of visitors or miles of trail to base their success, but these statistics ignored the reason why people come to natural areas.

According to Driver and Tocher's definition, the output of recreation is a desired experience. The recreational activity is only a means to an end, and the end is something intrinsically beneficial to the person who takes part in the recreational activity (Brown 1984). This places the recreation manager in a unique role. He or she does not directly provide the final output of recreation management (the beneficial experience). The recreation manager can only provide the opportunity for people to attain their own desired experiences by providing them with the opportunity to participate in certain activities within specific settings (Driver and Brown 1983). For example,

FIGURE 10.1 Visitors to Mill Creek Canyon in the Wasatch-Cache National Forest just outside of Salt Lake City, UT, attain diverse benefits from recreation, ranging from physical fitness to enjoying nature. (Photo by Reece Stein.)

someone who goes to a forest to run is really there to experience nature and get exercise. The manager can provide a place for the visitor to run in the forest, but the visitor has the running experience. Even though the initial experience and eventual benefit (physical fitness and experiencing nature) are produced and realized by the individual recreationist, the manager still has an important role in helping to provide that benefit. The manager creates the opportunity for the person to attain physical fitness by designing running trails throughout the forest with diverse topography and providing information about the trails (e.g., location and length), so that the recreationist can plan to work a variety of muscles and move through diverse ecosystems (Figure 10.1).

Although it is difficult to fully understand all the benefits accrued to on-site recreationists, natural resource managers are also being asked to look beyond the sites they manage and identify how management affects and benefits surrounding ecosystems and communities (Grumbine 1994; Driver 1996). Since the early 1990s, researchers and managers in the U.S., Canada, and New Zealand have been taking a benefits-based approach to nature-based recreation planning and management. Benefits-based management (BBM) is a new nature-based recreation management framework that identifies and highlights the multitude of benefits generated from nature-based tourism and outdoor recreation (Anderson et al. 2000). Researchers are using this approach to identify all the benefit opportunities a recreation area might provide, which includes benefits to surrounding communities (e.g., community pride) and ecosystems (e.g., increased biodiversity), as well as to people visiting and

TABLE 10.1
Examples of Nature-Based Recreation and Tourism Benefits

Personal Benefits	
Enjoy nature	Achievement/stimulation
Learn new things	Physical rest
Family relations	Teach/lead others
Reduce tension	Risk taking
Escape physical stress	Risk reduction
Share similar values	Meet new people
Independence	Creativity
Introspection	Nostalgia
Be with considerate people	Agreeable temperatures
Community Benefits	
Attracting tourism dollars to the community	
Preserving/conserving various natural and unique ecosystems	
Experiencing unique outdoor recreation opportunities	
Protecting a natural setting in which your community takes great pride	
Understanding your natural environment	
Feeling that your community is a special place to live	
Feeling a sense of security that the natural environment will not be lost	
Giving a chance for local people to maintain an outdoor-oriented lifestyle	
Environmental Benefits	
Life support	Endangered species/ecosystem benefits
Aesthetics	Religious/philosophical
Scientific	Sources and resources
Historical	

Sources: Driver et al. (1991); Stein et al. (1999); Rolston (1991).

using recreation areas (Table 10.1). Areas that have used a benefits-based approach to provide recreation in the wildland–urban interface include Chandler Park in Detroit, MI, the Bureau of Land Management lands outside of Grand Junction, CO, and the city parks in Portland, OR (Bruns et al.1994).

10.3 WILDLAND RECREATIONAL USE

10.3.1 Recreation Demand

The USDA Forest Service conducts periodic assessments of supply and demand trends for outdoor recreation in the U.S., and the most recent was reported in 1999 (Cordell 1999). Although a recreation assessment specifically for the wildland–urban interface has not been conducted, the USDA results provide a good starting point for understanding the status of recreation in interface areas.

The recreation assessment shows that people take part in various recreational activities throughout the U.S., many of which are likely to occur in wildland–urban interface recreation areas. Four of the most popular recreational activities of people in 1994–1995 were walking (66.7 percent of the U.S. population participating), viewing a beach or waterside (62.1 percent), family gatherings outdoors (61.8 percent), and sightseeing (56.6 percent). Using current recreation participation data, sociodemographic characteristics that influence recreation participation, and logistic regression modeling, the national assessment projected the 10 fastest-growing outdoor recreational activities between 2000 and 2050 (Table 10.2). All 10 activities are likely to have a dramatic effect on the use of the wildland–urban interface.

Recreation managers in the interface have the unique opportunity to accommodate and serve people who traditionally do not participate in nature-based recreational activities in more remote or rural settings. This includes cultural and ethnic minority groups, such as African Americans, Asian Americans, Latinos, and Native Americans. It also includes people classified as the inner-city poor (people who live in cities with at least one million people, have family incomes less than $10,000, received a high school education or less, and are at least 25 years old) and people with disabilities. The wildland–urban interface has great potential to provide urban residents, including these unique groups, with nature-based recreational opportunities (Figure 10.2).

TABLE 10.2
Ten Fastest-Growing Recreation Activities in the United States, 1995–2050

Recreation Activity	Percentage Growth in Number of Users Participating
1. Cross-country skiing	95
2. Downhill skiing	93
3. Visiting historic places	76
4. Sightseeing	71
5. Biking	70
6. Horseback riding	66
7. Nonconsumptive wildlife activities	61
8. Nonpool swimming	58
9. Family gathering	57
10. Hiking	57

Source: Bowker et al. (1999).

The majority of research on visitors to interface recreation areas has focused on the issue of ethnic group participation in recreation. Among other research, studies conducted in California (Carr and Williams 1993; Chavez 1995), the southwestern U.S. (Irwin et al.1990; Floyd et al.1993), Chicago (Floyd and Shinew 1999), and the southeastern U.S. (Johnson et al. 1998; Johnson and Bowker 1999) have begun to provide a theoretically based understanding of minority participation in recreation. Also, Ewert et al. (1993) discussed the overall issue of ethnic participation in recreation in their book *Culture, Conflict, and Communication in the Wildland– Urban Interface*. It is beyond the scope and goals of this chapter to summarize this research, but the important issue of ethnic group visitation to wildland–urban interface recreation areas should be well understood by planners and managers. In this chapter, I examine the issue within a practical context, through the use of existing planning and management frameworks.

As opposed to focusing on a specific region or ethnic group in the U.S., Cordell et al. (1999) examined the factors that prohibited people's participation in recreation. They then analyzed these data according to the ethnic groups' perceived barriers. In general, many people believe they simply do not have enough time or money for recreation. However, when examined at a more specific level, different groups have different reasons for not participating. For example, African Americans and Latinos have a higher tendency than other groups of people to say that inadequate information is a barrier to their participation in outdoor recreation. Hispanic Americans were more likely than the general population and other ethnic groups to list inadequate transportation and crowded areas as barriers. More than a third of the African Americans surveyed listed outdoor pests as a reason for not participating in outdoor recreation. Potential solutions to many of these groups' perceived barriers are addressed at the conclusion of the chapter.

10.3.2 Recreation Supply

Most nature-based recreation sites in the U.S. are managed by government agencies. Federal land management agencies manage much of the land outside western cities,

FIGURE 10.2 Picnickers in Florida's Apalachicola National Forest represent the diverse types of visitors who visit wildland–urban interface recreation areas. (Photo by Shanon Harvey.)

but state, county, and city governments provide the majority of nature-based recreational opportunities in the Midwest and the East. Private landowners rarely open their land for recreation for a variety of reasons; therefore, the discussion focuses on recreation on public lands.

City and county recreation planning has traditionally emphasized family and sports-related recreational activities (Betz et al.1999). However, city, county, and special recreation and park districts do own and manage natural areas in the wildland–urban interface for nature-based recreation, especially when local governments purchase land for conservation. Pinellas County, FL, is a good example of a local government taking on the enormous task of providing recreational opportunities in the wildland–urban interface for both its residents and tourists. In the early 1970s, Pinellas County residents voted to use money generated from a new sales tax in the acquisition of natural areas for conservation, passive recreation (low-impact recreational activities), and environmental education. Due also to skilled leadership and a defined vision, the county now operates one of the largest nature-based recreational programs in Florida.

State governments manage many types of natural areas. The vast majority of areas managed for recreation by state agencies include state parks, but most states also have trail programs, forest systems, fish and wildlife programs, river and water agencies, and wildernesslike systems. These state agencies provide diverse opportunities for residents and tourists that range from recreation-and leisure-related opportunities such as golf courses to nature-based recreation such as nature trails and campgrounds.

Federal land management agencies manage recreation areas in the wildland–urban interface throughout the U.S. In total, the U.S. government manages more than half a billion acres of land (652 million acres) (Table 10.3). The Bureau of Land Management manages the most land (267.6 million acres), but ranks fourth in number of visits (58.9 million visits in 1996). The USDA Forest Service lands have the highest visitation rates, with almost 860 million visits in 1996. One reason these lands receive such heavy visitation is that numerous national forests border some of the nation's largest cities. These Urban National Forests are described more fully in Chapter 18. Examples include the Angeles National Forest, which lies just north of Los Angeles; Roosevelt, Arapaho, and Pike National Forests encroached by Denver's western suburbs; and the Chattahoochee National Forest, which lies less than an hour away from most Atlanta and Chattanooga residents. The USDA Forest Service and Bureau of Land Management also actively manage for a variety of recreational uses (e.g., hunting,

TABLE 10.3
Federal Agency Acreage (Land and Water) and Visitation

Agency	Acres (millions)	Visits in 1996 (millions)
Bureau of Land Management	267.6	58.9
Forest Service	191.6	859.2
Fish and Wildlife Service	90.5	29.5
National Park Service	83.2	265.8
Corps of Engineers	11.6	375.7
Bureau of Reclamation	6.5	38.3
Tennessee Valley Authority	1.0	0.6
Total	652.0	1628.0

Source: Stenger (1999).

backpacking, off-road vehicle driving, and cross-country skiing) and provide easy access to some of the most desired nature-based recreational attractions in the U.S.

Public land management agencies are not providing for outdoor recreational opportunities alone, however. A recent trend in the provision of wildland–urban interface recreational opportunities is the development of partnerships between government agencies, nonprofit organizations, and private landowners to form greenways, trails, and other linear recreational opportunities. The National Park Service (NPS) manages many greenways throughout the U.S., but local and state governments are also active in managing greenways, since these areas are dependent upon strong local participation in their initial creation and ongoing management. For example, residents of Cleveland, OH, began working with state and local governments in the 1960s to protect the Cuyahoga River Valley in northern Ohio from spreading development (Steven R. Davis, personal communication, July 2001). They joined forces with the NPS, which was attempting to establish urban recreation areas throughout the country, and in 1974, the U.S. Congress created the Cuyahoga Valley National Recreation Area. Now known as Cuyahoga Valley National Park, the NPS continues to work with private and local governments to protect 33,000 acres along the northern 22 mile of the Cuyahoga River (Steven R. Davis, personal communication, July 2001).

Although undeveloped natural areas are prized locations for the initiation of greenways and trail programs, urban areas are often looked at as places to restore natural areas to serve as recreation areas in the wildland–urban interface. One of the most important programs helping to develop recreational opportunities both within the interface and throughout rural areas is the Rails-to-Trails program. In brief, the program acquires abandoned railroad corridors and converts those linear areas into trails. Local trail enthusiasts, local and state government agencies, and business leaders usually spearhead the program. In 1999, there were more than 900 rail-trails covering over 9300 miles in the U.S. (Morris 1999).

In the case of greenways and trails, it is often complex and difficult to identify, acquire, and manage natural areas adjacent to development. These efforts require a variety of groups working together to create greenways and trails that are successful both recreationally and ecologically. In northeastern Massachusetts, a regional land trust (Trustees of Reservations) led the effort in protecting land along the Charles River, which empties into Boston Harbor (Flink and Searns 1993). The land trust, the Audubon Society, municipal governments, and the U.S. Army Corps of Engineers protect thousands of acres of wetlands and floodplains bordering the river to create a recreation and conservation corridor extending about 60 mile (Flink and Searns 1993) (See Box 10.1).

Box 10.1

Local and Federal Governments Coming Together to Enhance Recreation in Mill Creek Canyon, UT

Although most public outdoor recreation areas are managed by a single government agency, occasionally different organizations team up to enhance the recreational opportunities offered. This happened in 1994 in Salt Lake City at one of the most popular outdoor recreation areas in the state. Mill Creek Canyon is one of several canyons feeding into the Salt Lake Valley and is used for a variety of purposes. The USDA Forest Service, which manages 80 percent of land in the canyon, keeps recreational use by local residents a top priority (USDA Forest Service 2000). As federal budgets were being slashed, the Salt Lake Ranger District of the Wasatch-Cache National Forest was looking for ways to improve the dilapidated recreation facilities in Mill Creek Canyon (Wondolleck and Yaffee 2000). Trails along the creek were eroding into the water, vegetation in picnic areas was trampled and not growing back, and old bathrooms and picnic shelters were falling into disrepair. Due to the immense demand that Salt Lake County residents placed on the canyon, the Salt Lake District teamed up with the Salt Lake County Commission to devise a new way to generate money to improve the recreational opportunities (Kathy Jo Pollock, personal communication, June 2000).

Although the most direct way of raising revenue was to develop an entrance fee to the canyon, any money the USDA Forest Service, collected would have reverted to the U.S. Treasury, and would not have been used for needed canyon restoration (Kathy Jo Pollock, personal communication, June 2000). Therefore, Salt Lake County took the lead to develop a new entrance fee station to the canyon as a way to raise money for restoration and recreation improvements (Figure 10.3). However, creating a new charge for something that was previously free is never an easy task, and after heavy debate, which lasted more than a year, a memorandum of understanding (MOU) was developed between the Salt Lake County Commission and USDA Forest Service, (Kathy Jo Pollock, personal communication, June 2000).

The MOU called for the county to establish and hire personnel to occupy the fee station. The money collected from fees would then go to the Salt Lake District for improving the recreational opportunities in Mill Creek as well as conserving and protecting Mill Creek's ecological resources (USDA Forest Service 2000). Collected funds were used for recreation, trail, and interpretation facility development; watershed protection; and fish and wildlife habitat protection. Also,

—continued

Box 10.1 — (continued)

this new revenue would be used to build and operate the entrance station, which included redesigning the road to accommodate the station (USDA Forest Service 2000).

In its first year of implementation, the county charged $2 a car or $20 for a season pass and raised $308,000 (Wondolleck and Yaffee 2000). Through the use of a Challenge Cost-Share agreement, Salt Lake County kept $100,000 for maintenance costs and donated the remainder to the USDA Forest Service. Additionally, the Salt Lake District received $308,000 from the federal government as a match to the donated funds, resulting in more than a half a million dollars in revenue that did not exist before the unique partnership was developed (Wondolleck and Yaffee 2000).

Not only has the USDA built new recreation facilities in Mill Creek (Figure 10.4), but the agency has also restored stream banks that were badly damaged due to heavy recreational use. The USDA Forest Service also pays for an interpretive ranger in the canyon who gives guided talks to visitors. Beyond the additional expenditures, the use of an entrance station has improved depreciative behavior in Mill Creek. For example, vandalism and other forms of disorderly conduct dropped 42 percent after the station was installed (Wondolleck and Yaffee 2000).

10.4 RECREATION MANAGEMENT FRAMEWORKS

As stated earlier, it is important to identify the essential outputs (experiences and benefits) that recreation managers believe they can provide to people and to natural ecosystems when planning for recreation in the wildland–urban interface. Evaluating recreation success solely on the development of facilities and high visitation numbers ignores the very reasons why people visit an area and participate in a particular activity. The focus on visitor numbers also ignores the community and environmental benefits potentially derived from recreation areas. Since nature-based recreation is intrinsically an interdisciplinary field, planning and managing for recreation requires thorough integration of the natural and social sciences. Federal public land management agencies played major roles in developing the first recreation planning framework by incorporating research that identified how settings and activities result in desired recreational opportunities. Several of these management frameworks are discussed below in terms of zoning, identifying acceptable levels of change, and management strategies and tactics.

10.4.1 Zoning for Recreation Benefit Opportunities

Representing information spatially is an important technique to manage and plan for human use in any natural area. New geographical information systems technologies have improved the ability of natural resource managers to examine a multitude of variables over large areas (Lang 1998). This spatial representation of data can also be used to improve mapping and planning for recreational use in natural areas.

As researchers better characterized the essential outputs of a recreational engagement, elements of the landscape were identified to help afford those outputs. The Recreation Opportunity Spectrum (ROS) is one system that was developed to help us understand the relationships between landscapes and recreation outputs. It is based on the concept that people choose a specific setting to

FIGURE 10.3 Visitors pay their user fee at the pay station to Mill Creek Canyon in the Wasatch-Cache National Forest jointly managed by Salt Lake County and the USDA Forest Service. (Photo by Reece Stein.)

FIGURE 10.4 New recreational facilities, funded through a recreational user fee program, provide for diverse recreational opportunities in Wasatch-Cache National Forest's Mill Creek Canyon. (Photo by Reece Stein.)

participate in recreational activities in order to realize a desired set of experiences (Driver et al. 1987). ROS offers a framework for understanding the relationships between settings, activities, and people's desired experiences (USDA Forest Service 1982; Manfredo et al.1983; Virden and Knopf 1989; Stein and Lee 1995). It categorizes six setting characteristics, which managers can identify, measure, and manipulate depending on the type of recreational experiences they want to provide in a certain area. These characteristics are access, remoteness, naturalness, social encounters, facilities and site management, and visitor management.

ROS is the first example of a land management framework that truly integrates the complex social needs of recreationists into the physical management of settings. Using the six setting characteristics, the framework describes six different types of recreation zones, ranging from primitive to urban, that could potentially provide different recreational experiences for people (Figure 10.5). Some managers might want to use ROS as a cookbook approach to recreation management since ROS provides detailed setting characteristics for each zone. This would be a mistake. Natural areas are unique unto themselves, and when humans are thrown into the mix, no single set of strictly applied characteristics can provide for optimal solutions. Instead of using a firm set of detailed characteristics, land managers can use the approach as a general guideline in managing a setting to provide the opportunities and benefits they believe are most desired by their customers and constituents.

The Angeles National Forest in southern California is a good example of a wildland–urban interface area that requires flexible recreational opportunity zoning because its adjacency to many urban areas leads to high use. Although many of the Angeles' visitors may be searching for nature appreciation and solitude, establishing pure primitive environments, which ROS identifies as settings that best help people achieve these two experiences, may not be possible in most of this highly used forest. This does not mean people are not attaining experiences of nature appreciation and solitude in the Angeles National Forest. People can attain these primitive-type experiences in areas that have paved roads, high social contacts, and heavy facility development because these types of environments are much less urban than the areas where they work and live. Therefore, it is up to the wildland–urban interface planners to develop their own type of recreation zoning description based on the ROS guidelines, the desires of their visitors, and the setting in which the visitors work and live.

In the early 1990s, the NPS developed the Visitor Experience and Resource Protection (VERP) framework, which gave recreation planners the flexibility to design their own zones (USDI National Park Service 1997). Although VERP is based on ROS and other recreation management strategies, VERP allows recreation managers and planners to be more flexible in their designation of zones. This proves to be particularly helpful in the wildland–urban interface, where managers recognize they have diverse areas that provide unique recreational opportunities. In general, VERP requires managers to determine the resource and social conditions, kinds and levels of visitor use, levels of park development, and types of management activity throughout a recreation area. Armed with this knowledge, managers can then create management zones based on the range of potential visitor experiences and resource conditions that can be accommodated in that recreation area (USDI National Park Service 1997).

ROS and VERP also focus on two other aspects of recreation management particularly important in the wildland–urban interface: maintaining levels of acceptable

FIGURE 10.5 Examples of the six zones described within the Recreation Opportunity Spectrum framework. Top, left to right: (A) primitive (photo by Larry Korhnak); (B) semiprimitive nonmotorized (photo by Larry Korhnak); and (C) semiprimitive motorized (photo by Larry Korhnak). Bottom, left to right: (D) roaded natural (photo by Taylor Stein); (E) rural (photo by Taylor Stein); and (F) urban (photo by Ron Nickerson.)

change in recreation areas and identifying management strategies to maintain the quality of areas within designated standards.

10.4.1.1 Maintaining Levels of Acceptable Change

Often proponents of nature-based recreation and tourism argue that managing the area only for recreational purposes will ensure the area's protection. However, all human use of the environment results in some sort of change — even recreation. Therefore, managers must identify and manage for acceptable levels of change to the environmental and social setting when providing recreational opportunities. For example, the Boundary Waters Canoe Area (BWCA) in northern Minnesota received wilderness protection in 1964, but that did not mean the area was protected from human impact. In the summer of 1969, recreational users left behind an estimated 360,000 pounds of bottles, cans, and other nonbiodegradable refuse (King and Mace 1974). Since 1969, BWCA managers have prohibited bottles and cans in backcountry lakes (Hammitt and Cole 1998).

With an understanding that change will occur, the Limits of Acceptable Change (LAC) framework was developed to help planners define achievable and acceptable desired future conditions for the ecosystem. In other words, LAC allows planners to define how much change is acceptable before the ecosystem is unalterably changed. LAC then helps provide a management strategy to prevent unacceptable conditions (Stankey et al.1985). VERP also uses this concept of managing recreation areas within acceptable change limits. Although LAC was initially developed to manage recreation in wilderness areas, natural resource managers have increasingly used the process to manage more developed recreation areas (Brunson 1997). Both LAC and VERP require managers to develop a baseline assessment of natural resource conditions and visitor experiences, establish desired conditions for a range of management zones, monitor environmental and social indicators to compare affected areas with control areas, and develop a management strategy to address indicators that are below defined standards.

Indicators are the "variables for which objectives are written" (Hammitt and Cole 1998, p. 215). However, even with clear objectives it is sometimes difficult to identify measurable indicators of impact. Examples of environmental indicators might include percentage of vegetation loss, number of pieces of litter, and nitrogen levels in adjacent water bodies. Social indicators should also be developed for recreation areas; however, these are often more difficult to attain and might require managers to survey visitors. They could include the number of people visiting an area, complaints about a specific issue, and conflicts among visitors. If managers are truly attempting to provide visitors with opportunities to attain their desired experiences and benefits, visitors' self-reported attainment of experiences could be used as indicators. For example, the percentage of people who say they learned more about nature would be a useful indicator for an area managed for educational benefits.

Standards are considered to be subjective evaluations of the most appropriate compromise between use and resource protection (Hammitt and Cole 1998). Since LAC was created for federally designated wilderness areas, managers had a clear mandate in which to define standards because federal law required wilderness areas to be managed in their most pristine condition (Brunson 1997). For example, wilderness managers might have defined standards as one piece of litter per primitive campsite or 90 percent of their visitors experienced solitude. In most cases, wildland–urban interface managers are not confined by the strict preservation goals of wilderness areas. This does not make it easier for them. In the interface, managers are faced with the responsibility of deciding how much is too much when it comes to impacts caused by large numbers of visitors with a variety of recreational demands in areas often smaller than designated wildernesses. Identifying both indicators and standards of impact is dependent on the goals and objectives of both management and stakeholders. When zoning an area for different benefit opportunities (whether they are for recreation, conservation, or economic benefits), each zone should have its own indicators and standards. Therefore, managers and planners must have a clear idea of the desired future conditions within the area they are working.

Wildland–urban interface planners and managers should not, however, define desired future conditions by themselves, but they should take the lead in the process. Although they should rely upon their own scientific knowledge of natural systems to develop these standards, interface planners and managers must also collaborate with area stakeholders and constituents to make these difficult decisions. Decision making should incorporate people who have a stake in the future condition of that area and who can potentially help move that area toward and maintain a desirable condition. With a clear understanding of the desired condition, managers can then identify measurable indicators of quality and standards to judge when those indicators show that an area has reached unacceptable conditions. They can also design proactive management to ensure the area does not fall into unacceptable conditions.

Drawing this *line in the sand* might be financially and politically risky. Usually, it is easier to lower standards than actually manage to maintain areas with high standards. Through successful monitoring of the environmental and social indicators, managers do not have to wait for indicators to lapse into unacceptable standards. Instead, they can use a variety of visitor and setting management strategies to maintain quality recreational opportunities, which result in sustainable benefits. If the indicators and standards for a desired condition are initially developed collaboratively with stakeholders, the stakeholders are more likely to understand and support management actions to stay within the standards.

10.4.1.2 Management Strategies and Tactics

As discussed above, recreation managers must understand that recreational use will result in some change to the resource. It is the wildland–urban interface planner and manager's job to manage the visitor and setting to ensure that changes do not exceed acceptable standards. Anderson et al. (1998) developed a handbook to help NPS managers maintain the quality of recreation area resources and visitor experiences. The handbook provides a decision making process based on the LAC and VERP frameworks to help a manager identify appropriate management tactics for specific social or resource impacts. These tactics were categorized into the following management strategies:

1. Modify character of use by controlling where use occurs, when use occurs, what type of use occurs, and how visitors behave.
2. Modify the resource base by increasing resource durability and maintaining/rehabilitating the resource.
3. Increase the supply of recreational opportunities.
4. Reduce use in the entire area, or in problem areas only.
5. Modify visitor attitudes and expectations.

Based on these management strategies, recreation managers should formulate appropriate management tactics to minimize impacts of prime concern. For example, concentrating users into specific areas and educating users as to appropriate behavior are common management tactics used in interface recreation areas. Visitors naturally concentrate around the main attraction of a recreation area (beach or trailhead), an easily accessible location such as the first picnic area off the main road, and travel routes such as trails (Hammitt and Cole 1998). These concentrated areas will generally experience higher levels of impact to the resource and to the visitors' experiences. However, this concentration of impacts does not have to be a bad thing. In fact, many recreation areas are deliberately managed to concentrate specific uses in certain areas. Concentration means impacts are focused in a single area, minimizing impacts in other parts of the recreation area. Managing the site (strategy 2) through the development of facilities and services, strengthening or hardening the site, and removing litter and other problems can minimize impacts while allowing for high recreational use. Educating users (strategy 5) about the recreational opportunities available at a concentrated zone also helps to direct visitors to the designated area and minimize their impact in more sensitive areas or areas designated as low impact.

Paynes Prairie Preserve State Park is a 20,000-acre park just south of Gainesville, FL, and is a good example of a recreation area in the wildland–urban interface that concentrates recreational development in manageable areas and communicates to visitors when and where they can use these opportunities. Visitors are restricted to three short trails into the prairie. The most popular, La Chua Trail, is open from sunrise to sunset and begins only a short distance from a Gainesville neighborhood. The well-maintained trail leads visitors to one of the most popular wildlife viewing areas in the park (Figure 10.6) and to an

FIGURE 10.6 Visitors to Paynes Prairie Preserve State Park watch wildlife at a highly visited wetland just outside of Gainesville, FL. (Photo by Taylor Stein.)

observation tower, which provides a 360° view of the prairie. However, visitors are not allowed to venture beyond that trail without being accompanied by a park ranger on an organized prairie walk. Therefore, recreation-related impacts are restricted to the trail, thus protecting most of the sensitive prairie habitat while allowing the visitor to walk into and view the prairie.

Education is also a fairly common technique used to influence visitor behavior and contribute to positive visitor experiences (Anderson et al. 1998). Much research has been conducted to understand how education can most effectively influence visitor behavior. Swearingen and Johnson (1994) found that visitors were more likely to change their behavior in accordance with information signs in Mount Rainier National Park, south of Seattle, WA, if the sign advocated behavior that visitors found desirable. Also, visitors will more readily accept information that is portrayed as helpful to the visitor experience, as opposed to confining or restraining them (Hultsman et al. 1998). For example, if a trail is closed for ecological restoration, a sign that provides alternative trails and directions to find those trails is more likely to dissuade visitors from using the closed trail than a sign that simply says "Closed Trail. Keep Out!" For more information on effective interpretation strategies, refer to Sam Ham's (1992) book, *Environmental Interpretation: A Practical Guide for People with Big Ideas and Small Budgets*. Also, in this text, Chapter 10 specifically addresses environmental education in the wildland–urban interface.

10.5 RECREATION PLANNING ISSUES IN THE WILDLAND–URBAN INTERFACE

As stated at the beginning of this chapter, managing recreation in the wildland–urban interface is often more complicated than managing rural nature-based recreation due to the large numbers of users, diverse populations, and various environmental impacts and user conflicts often occurring in these areas. Using these characteristics to guide the context of discussion, this section will discuss potential solutions to many of the recreation management problems that interface managers face. Among these are the need for financing quality recreation management in high demand areas and collaborating with diverse groups of people.

10.5.1 Financing Recreation in the Wildland–Urban Interface

Nature-based recreation in the U.S. is a highly subsidized activity. As discussed earlier, most outdoor recreational opportunities are owned and operated by local, state, or federal agencies that either charge small amounts or do not charge the public anything for use of recreational facilities and services on public lands. For a variety of reasons, both the American public and government agencies have assumed that nature-based recreation should be free or provided at a low cost. However, in periods of shrinking government budgets, decision makers must identify methods to sustain recreational opportunities.

Private nature-based recreation operators are much more directly tied to identifying and managing for sustainable economic benefits. The existence of their businesses depends on ensuring that the economic benefits outweigh the economic costs. These private businesses have taken a variety of approaches to sustain their revenues, which include moving beyond the nature-based opportunities to create more grandiose and unique opportunities. For example, Silver Springs, owned by Alta Smart Parks, Inc., is located just outside Ocala, FL, and claims to be one of the state's first ecotourism destinations. The park is built around one of Florida's largest natural springs, which is still a major park attraction. Silver Springs achieved international fame in the 19th century by providing glass-bottom boat tours of the spring and the adjacent Silver River. However, throughout the 20th century, Silver Springs diversified the recreational and tourism opportunities to compete with Florida's more famous theme parks. Along with glass-bottom boat tours, the park now offers animal exhibits, a jeep safari, a jungle cruise, trained animal shows, and a petting zoo. Silver Springs has also ventured into the musical entertainment realm by featuring a variety of performers each year. Due to the park's heavy reliance on developed animal attractions and musical entertainment, it arguably has shifted from a nature park to more of an entertainment and zoological attraction. Silver Springs represents the need of many private nature-based recreation operators to explore diverse and unique opportunities in order to maintain economic viability. Generally, such businesses survive by attracting large numbers of people who are willing to pay high prices to take part in unique recreational opportunities.

In contrast to Silver Springs, public agencies may not have the financial ability to develop a diversity of opportunities to attract high-paying customers. Also, public agencies have management goals and objectives (serving all constituents — not just constituents who can afford high recreation fees, conservation of ecosystems, and managing the resource for timber and other natural resource products), which take precedence over the development of expensive tourism opportunities. However, these agencies still must work within tight fiscal constraints. Three funding options are generally available to natural resource managers working in the wildland–urban interface: developing and increasing user fees, forming partnerships with volunteer groups and businesses, and expanding political and constituency support. John Crompton (1999), in his book, *Financing and Acquiring Park and Recreation*

Resources, and John Loomis and Richard Walsh (1997), in their book, *Recreation Economic Decisions,* examine several strategies for funding public recreation that have varying degrees of applicability to recreation areas in the wildland–urban interface.

10.5.1.1 User Fees

Although recreational user fees can help fund the management of nature-based recreation areas, it is often difficult for any provider to charge a fee for something that was traditionally free. The fees charged in nature-based recreation areas range from small fees per carload of visitors to more expensive fees charged per person entering the area. Generally, user fees should be based on the cost of producing recreational opportunities (Loomis and Walsh 1997). This makes sense for organizations simply trying to cover their costs. However, public agencies must take into account a number of variables that private organizations do not need to consider. As with other public services such as highway maintenance and education, many people expect the government to provide free or inexpensive recreational opportunities on public lands. Also, many groups consider user fees to be a type of *double taxation* since many recreation areas are supported from general tax revenue (Reiling and Kotchen 1996). Finally, since these lands are managed for all citizens, not just citizens who can afford to pay to visit recreation areas, government agencies generally do not want to exclude lower income populations from participating in nature-based recreation on public lands (Manning 1999). User fees inherently force this exclusion.

There are several things to consider when planning to provide new user fees or to increase fees for recreation areas in the wildland–urban interface. Of particular importance is how visitors or potential visitors might react to the fee. Manning (1999) surveyed the literature and found that the following factors most influence recreational use:

1. *Uniqueness of the area.* New or increased user fees are not likely to affect visitation rates for unique areas such as the only freshwater spring within 100 mile of a city; however, if the area with the new fee is similar to other neighboring areas, visitation will likely drop.
2. *Percentage of total cost represented by the fee.* A user fee will reduce visitation rates to an area if that fee represents a high proportion of the person's overall trip cost.
3. *Type of fee instituted.* Daily fees tend to limit visitation rates more than longer-term user fees like seasonal or annual fees.

Finally, when considering implementation of a user fee, recreation planners and managers should understand how to make a user fee more acceptable to their constituency. Research has shown recreational visitors more likely to accept fees if (1) the visitors believe the money charged is retained by the collecting agency and reinvested into the recreation area, (2) the fee is simply an increase rather than a new fee, (3) the visitors are new to the recreation area and did not use the area over a long period before the fee was imposed, and (4) managers provide information that explains the relatively higher costs of visiting alternative recreational facilities such as movies (Manning 1999). Also, as discussed later in this chapter, important decisions, such as the implementation of user fees, should be done in collaboration with the people who use and benefit from the park (Wondolleck and Yaffee 2000) (See Box 10.2).

Box 10.2

Recreation Fee Demonstration Program

In 1996, the U.S. Congress implemented the Recreation Fee Demonstration Program to generate needed revenue for lands managed by the NPS, USDA Forest Service, USDI Fish and Wildlife Service (FWS), and Bureau of Land Management (BLM). The Fee Demonstration Program allowed these agencies to implement and test new fees throughout the U. S. It also helped agencies better understand visitors' reactions to new and higher recreational user fees on public lands.

Many USDA Forest Service, FWS, and BLM recreation areas imposed fees where fees had not existed before and the NPS raised fees in many areas. All of the agencies informed visitors that the additional revenues collected as part of the User Fee Program were mandated to improve and enhance the recreational opportunities in the area where fees were collected (U.S. Department of Interior and U.S. Department of Agriculture 2002).

Results were fairly positive for the program, although challenges have been identified. For instance, the cost of collecting additional fees was often high. Spending money to make money is often necessary, but these costs can be prohibitive. Understanding this challenge early in the process allowed all four agencies to identify ways of reducing collection costs, and they have worked to keep these costs at approximately 20 percent of the fee demonstration revenues (U.S. Department of Interior and U.S. Department of Agriculture 2002).

Overall, the four participating agencies collected $172.8 million of fee demonstration revenue in fiscal year 2001 (U.S. Department of Interior and U.S. Department of Agriculture 2002). This money was designated for a variety of projects, which included the

—continued

Box 10.2 — (continued)

improvement of visitor services, enhancement of wildlife habitat, maintenance and repairs of infrastructure, and development of interpretive materials and signage. The agencies are continuously looking for ways to improve the collection of fees and visitor services. Specifically, the agencies have worked for better coordination among themselves, such as agreements to simplify payments of fees at contiguous sites, and with local organizations, such as joint arrangements with state and local entities to collect fees (U.S. Department of Interior and U.S. Department of Agriculture 2002).

10.5.1.2 Partnerships

Traditionally, the agency managing a recreation area has done it all: staffed the fee stations, enforced rules, picked up trash, sold souvenirs, cleaned restrooms, built trails, and provided educational programs. Cash-strapped agencies can no longer afford to take the lead in every task, especially when there are organizations and private businesses ready to partner with them. Partnering with external organizations is becoming a much more popular activity in providing recreation on public lands (LaPage 1994). For example, the USDA Forest Service contracts with private businesses to maintain many of its recreation sites throughout the U.S. and the NPS has long worked with private concessionaires to manage gift shops, lodges, and restaurants. For example, the Fred Harvey Company began providing visitor services at Grand Canyon National Park in the early 1900s and continues to be the primary concessionaire on the park's south rim.

Partnering with other organizations is not a solution without costs and challenges. In many cases, agency personnel believe they will lose power and control over a particular area when they rely on the help of a partner (Vaske et al. 1995). Public land management agencies must take on social and leadership roles when working with partners to achieve planning objectives for joint management of the land and visitors. This cooperative approach to management is dramatically different from the more technical and scientific jobs traditionally associated with agencies, but can be rewarding when the agencies hire people with the motivation, expertise, and energy to organize and work with partners (See Box 10.3).

Box 10.3

Five Way Partnerships Can Work in the Wildland-Urban Interface

Local, state, and federal land management agencies have partnered with a variety of organizations in the management of recreation areas in the wildland-urban interface. LaPage (1994) described the following types of partnerships that can be and have been used for recreation management:

1. Private Partners and Benefactors: Individuals, families, and even large businesses donate time, money, and labor to recreation and parks management. Often these benefactors have strong emotional ties to the recreation areas, and they will even donate land to a government agency for conservation and recreation purposes (LaPage 1994). In 1999, Evelyn Alexander donated 87 acres of open areas and forestland to Pulaski County, Virginia, just outside Dublin to be used as a county park (Pulaski County n.d.). Randolph Park, named in honor of Ms. Alexander's family, now provides a variety of developed and nature-based recreational activities for local residents on land that otherwise would have been developed for private use (Pulaski County n.d.).
2. Professional Partnerships: Often wildland-urban interface recreation areas are adjacent to or sometimes managed by several agencies. Through interorganizational partnerships, agencies can work together to provide diverse and unique recreational opportunities (LaPage 1994). For example, five different state agencies, two federal agencies, and more than 40 community agencies manage the 230-mil Heritage Trail, which runs the length of New Hampshire (LaPage 1994).
3. Corporate Partners: Corporations and large businesses have ample money to spend on worthwhile projects such as outdoor recreation areas. Donations help companies put something back into communities while providing long-term, positive public relations and needed tax breaks (LaPage 1994). The New Jersey Division of Parks and Forestry is one of the most active public land management agencies in the country working to develop corporate partnerships (New Jersey Division of Parks and Forestry 2000). As of 2000, the agency has partnered with over 18 different corporations to help provide diverse recreational opportunities for over 13 million visitors (New Jersey Division of Parks and Forestry 2000).

—continued

Box 10.3 — (continued)

4. Organizational Partnerships: Many nonprofit groups are active with land management agencies. Both the managers of interface recreation areas and local environmental groups share concern for adequate management and conservation of natural resources in the wildland-urban interface. Interest groups often volunteer time and effort that would otherwise be beyond the financial resources of the public land management agencies (LaPage 1994). For example, the San Jose Conservation Corps works with the Americorps program to involve at-risk youth in programs to improve publicly managed natural areas throughout California (San Jose Conservation Corps 2000). Such programs not only provide benefits to the areas where they operate, but also provide youth with a new appreciation for nature.
5. Friends Groups: Since the early 1900s, citizens have come together to form nonprofit groups to support the mission of specific parks and recreation areas. Not only do these groups actively volunteer their time, labor, and money to help in the beautification, development, and maintenance of natural areas (litter pick-up, facility maintenance, etc.), they also organize funding drives to provide parks with additional revenue for the development and maintenance of new projects (LaPage 1994). In fact, Friends of Rookery Bay in Naples, Florida, not only assists in the development and organization of fishing tournaments and other festivals (Ginger Hinchcliff, personal communication, February 1998) but has also provided money to conduct visitor surveys to improve the overall recreation management of the heavily used estuary (Figure 10.7).

10.5.1.3 Expanding Constituency

According to Crompton (1999, p. 112), "The single most important task in securing additional resources for park and recreation services is to develop and nurture a broader constituency." For recreation managers in the wildland–urban interface, the opportunities to attract new and diverse groups of people to serve are limitless. Also, natural areas managed for recreation can potentially benefit more than just the people who directly participate in recreational endeavors (Driver 1996). A benefits-driven approach to planning will aid in expanding a constituency for a public agency. When local residents, agency stakeholders, and politicians become more aware of these benefits, the political and financial support for the management of a recreation area is likely to increase.

Youth groups often participate in recreational activities in the wildland–urban interface, which aid in child development. For example, the Austin Parks and Recreation Department initiated the Totally Cool, Totally Art program at nine parks (see Crompton (1999) for a detailed description of the program). Specific objectives for the program were developed, and staff highlighted how the program would benefit community youth. Overall, the program helped to provide a safe place for teens to participate in constructive activities and develop a sense of belonging. Specific benefits included increasing teens' knowledge, skills, and interest in art. More long-term benefits were also realized throughout the community. By giving teens something new and constructive to do, the Austin Parks and Recreation Department took some pressure off working parents and, perhaps, city police, who often have to spend time and energy working with teens in confrontational situations. Although the program was situated mainly indoors, it is an example of how recreation professionals used a benefits-based approach to recreation in order to reach out to nontraditional populations (in this case teenagers) and benefit not only the participants but also the community. This approach helped the Austin City Council evaluate the program's success and decide whether to continue to fund it based on its benefits to county citizens (Crompton 1999).

10.5.2 Collaborating with Diverse Groups

Throughout the U.S., public land management agencies are attempting to involve their constituency in decision making. As Tipple and Wellman (1989, p. 24) point out, "The increased public involvement in forest resource management is not a temporary aberration. Instead, it is part of a societal trend toward more direct citizen participation in administrative decision processes." For our purposes here, collaboration will be considered an activity between two or more stakeholders pooling resources to solve a problem or achieve a goal that a single party could not accomplish (Wondolleck and Yaffee 2000). Wildland–urban interface recreation areas inherently have a multitude of existing and potential stakeholders. Hence, collaboration with these stakeholders is key in tackling the numerous issues and problems associated with these areas.

The diversity of people who might use the recreation areas is perhaps one of the most important characteristics of the wildland–urban interface (Dwyer 1995). Often researchers and managers group the many cultures that use recreation areas into broad ethnic categories; however, within any ethnic group there are many subcultures.

FIGURE 10.7 Members of Friends of Rookery Bay (FORB), a local volunteer group, work with coastal managers to replant native trees on a tiny island between the growing cities of Marco and Naples, FL. FORB has developed a strong partnership with estuary managers to ensure that quality environmental and recreational resources are sustained. (Photo by Randy Penn.)

Although an interface recreation area might play host to a large number of Latinos, that ethnic group could be further categorized into Puerto Rican, Mexican, Caribbean, or many other subgroups — each with its own specific beliefs, customs, and attitudes. Therefore, interface managers must focus on collaborative tools and techniques to help them comprehend the diversity of attitudes, beliefs, values, and expectations their particular constituencies possess (Dwyer 1995).

Collaboration can help managers avoid conflicts with people who hold differing, strong opinions about nature and/or recreation (USDA Forest Service 1993). True collaboration is often difficult to achieve because agency organizational structures are designed to protect individual turf and maintain control of decision making (Wondolleck and Yaffee 2000). Collaboration must be considered two-way communication, for which different groups of people require different methods. For example, an afternoon public hearing might be convenient for the people leading the meeting, but it excludes many constituents who cannot afford to take several hours off from work to attend. Also, many people are not used to the structured and formal public participation methods traditionally used by public land management agencies. Weekend open houses at wildland–urban interface recreation areas or evening workshops in residents' neighborhoods attract people who can benefit from these areas and are traditionally not involved in their planning.

Often more subtle changes in methods might be necessary to communicate with and educate diverse groups. As Wondolleck and Yaffee discussed (2000, p. 172): "Particular attention must be given to culture when different languages, practices, or beliefs are at play." They reported on the cleanup of the Russian River in northern California that was led by Rebecca Kress, a resident of Hopland, CA. Ms. Kress took a variety of approaches to attract a diversity of residents to help in the cleanup. She developed a video, recorded a song about the river, and made cleanup T-shirts in both Spanish and English to include the large Latino population that lived in the area (Wondolleck and Yaffee 2000).

10.6 CONCLUSION

Throughout the country, nature-based recreation in the wildland–urban interface is recognized as a valuable and important activity. In many cases, the organizations that manage nature-based recreation sites in the interface have done an excellent job providing for desired recreational opportunities in a natural setting. However, too much of the management of these resources has focused on reacting to demand and unacceptable impacts, rather than proactive planning to provide benefits to visitors, the environment, and surrounding communities.

This chapter showed that recreation can be planned and managed in the wildland–urban interface. Recreation is not something to be allowed simply because there is a demand for specific recreational activities and managers have not made other plans for a particular area. Natural resource planners should work with their local stakeholders to identify a desired future condition for an area, and they should manage that area to achieve that condition. Unplanned and unmanaged recreation can drastically transform an area and lead interface planners and managers down a chaotic, unknown road where they struggle with unacceptable impacts and unhappy, conflicting constituencies.

As with all natural resource products, planning and management frameworks exist to help planners and managers develop quality outputs of recreation in the wildland–urban interface. However, even with these frameworks, there is no way to escape the multitude of social and ecological interconnecting variables that are inherent in nature-based recreation management. When humans enter a natural ecosystem, unexpected changes are likely to occur to ecological processes. Also, people's attitudes and emotions about that area will change. For example, research shows that people tend to have strong attachments to nearby natural areas (Williams and Stewart 1998). When people use wildland–urban interface areas for recreation, this sense of attachment becomes even stronger, and they want to have more influence in the management and planning of those sites (Eisenhauer et al. 2000). Although this might not sound like a benefit to some managers, the fact that people actually care about the site and value it also adds value and importance to the manager's job. If the site is planned carefully and managed with people and the ecosystem in mind, the benefits are endless.

ACKNOWLEDGMENTS

The author would like to thank all those who reviewed the chapter: John Baust of the Florida Division of Recreation and Parks, Ron Nickerson of the University of Minnesota at Mankato, Julie Pennington and Kristina Stephan of the University of Florida, and the book's editors. All reviewers provided helpful comments and made significant improvements to the chapter. Lisa Pennisi of the University of Florida deserves special recognition for her assistance in the initial development of the manuscript and for her valuable recommendations, which dramatically improved the chapter. Finally, the Florida Agricultural Experiment Station provided additional support in this manuscript's development.

REFERENCES

Anderson, D.H., D.W. Lime, and T.L. Wang, 1998. *Maintaining the Quality of Park Resources and Visitor Experiences: A Handbook for Managers,* University of Minnesota Extension Service, St. Paul.

Anderson, D.H., R.G. Nickerson, T.V. Stein, and M.E. Lee, 2000. Planning to provide community and visitor benefits, in *Trends in Outdoor Recreation, Leisure, and Tourism,* W.C. Gartner and D.W. Lime, Eds., CABI Publishing, Wallingford, U.K., pp. 197–211.

Betz, C.J., D.B.K. English, and H.K. Cordell, 1999. Outdoor recreation resources, in *Outdoor Recreation in American Life,* H.K. Cordell (prin. inv.), Sagamore Publishing, Champaign, IL, pp. 39–182.

Bowker, J.M., D.B.K. English, and H.K. Cordell, 1999. Projections of outdoor recreation participation to 2050, in *Outdoor Recreation in American Life,* H.K. Cordell (prin. inv.), Sagamore Publishing, Champaign, IL, pp. 323–350.

Brown, P.J., 1984. Benefits of wildland recreation and some ideas for valuing recreation opportunities, in *Valuation of Wildland Resource Benefits,* G.L. Peterson and A. Randall, Eds., Westview Press, Boulder, CO, pp. 209–220.

Bruns, D.,B.L. Driver, M.E. Lee, D.H. Anderson, and P.J. Brown, 1994. Pilot Tests for Implementing Benefits-Based Management, paper presented at the Fifth International Symposium on Society and Resource Management, Ft. Collins, CO.

Brunson, M.W., 1997. Beyond Wilderness: Broadening the Applicability of Limits of Acceptable Change, in Proceedings — Limits of Acceptable Change and Related Planning Processes: Progress and Future Directions, May 20–22, Missoula, Mont., S.F. McCool and D.N. Cole (comps.), General Technical Report INT-GTR-371, U.S. Department of Agriculture, Forest Service, Rocky Mountain Research Station, Ogden, UT, pp. 44–48.

Carr, D.S. and D.R. Williams, 1993. Understanding the role of ethnicity in outdoor recreation experiences, *Journal of Leisure Research* 251: 22–38.

Chavez, D.J., 1995. Demographic Shifts: Potential Impacts for Outdoor Recreation Management, in Proceedings of the Fourth International Outdoor Recreation and Tourism Trends Symposium and the 1995 National Recreation Resource Planning Conference, May 14–17, 1995, J.L. Thompson, D.W. Lime, B. Gartner, and W.M. Sames (comps.), University of Minnesota, College of Natural Resources and Minnesota Extension Service, St. Paul, pp. 252–255.

Cordell, H.K. (prin. inv.), 1999. *Outdoor Recreation in American Life,* Sagamore Publishing, Champaign, IL.

Cordell, H.K., B.L. McDonald, R.J. Teasley, J.C. Bergstrom, J. Martin, J. Bason, and V.R. Leeworthy, 1999. Outdoor recreation participation trends, in *Outdoor Recreation in American Life,* H.K. Cordell (prin. inv.), Sagamore Publishing, Champaign, IL, pp. 219–321.

Crompton, J.L., 1999. *Financing and Acquiring Park and Recreation Resources,* Human Kinetics, Champaign, IL.

Driver, B.L., 1996. Benefits-driven management of natural areas, *Natural Areas Journal* 16: 94–99.

Driver, B.L. and P.J. Brown, 1983. Contributions of behavioral scientists to recreation resource management, in *Behavior and the Natural Environment — Human Behavior and the Environment: Advances in Theory and Research,* Vol. 6, I. Altman and J.F. Wohlwill, Eds., Plenum Press, New York, pp. 307–339.

Driver, B.L, and S.R. Tocher, 1970. Toward a behavioral interpretation of recreational engagements with implications for planning, in *Elements of Outdoor Recreation Planning,* B.L. Driver, Ed., University Microfilms Michigan, Ann Arbor, pp. 9–29.

Driver, B.L., P.J. Brown, G.H. Stankey, and T.G. Gregoire, 1987. The ROS planning system: evolution, basic concepts, and research needs, *Leisure Sciences* 9: 201–212.

Driver, B.L., H.E.A. Tinsley, and M.J. Manfredo, 1991. The paragraphs about leisure and recreation experience preference scales: results from two inventories designed to assess the breadth of the perceived psychological benefits of leisure, in *Benefits of Leisure,* B.L. Driver, P.J. Brown, and G.L. Peterson, Eds., Venture Publishing, State College, PA, pp. 263–286.

Dwyer, J.F., 1995. Multicultural Values: Responding to Cultural Diversity, in Proceedings of the Fourth International Outdoor Recreation and Tourism Trends Symposium and the 1995 National Recreation Resource Planning Conference, May 14–17, 1995, J.L. Thompson, D.W. Lime, B. Gartner, and W.M. Sames (comps.), University of Minnesota, College of Natural Resources and Minnesota Extension Service, St. Paul, pp. 227–230.

Dwyer, J.F. and P.H. Gobster, 1992. Recreation opportunity and cultural diversity, *Parks and Recreation* 27: 22–33, 128.

Eisenhauer, B.W., R.S. Krannich, and D.J. Blahna, 2000. Attachments to special places on public lands: an analysis of activities, reason for attachments, and community connections, *Society and Natural Resources* 13: 421–441.

Ewert, A.W., D.J. Chavez, and A.W. Magill, Eds., 1993. *Culture, Conflict, and Communication in the Wildland–Urban Interface,* Westview Press, Boulder, CO.

Flink, C.A. and R.M. Searns, 1993. *Greenways: A Guide to Planning, Design, and Development,* L.L. Schwarz, Ed., Island Press, Washington, DC.

Floyd, M.F. and K.J. Shinew, 1999. Convergence and divergence in leisure style among whites and African Americans: toward an interracial contact hypothesis, *Journal of Leisure Research* 31: 359–384.

Floyd, M.F., J.H. Gramman, and R. Saenz, 1993. Ethnic factors and the use of public outdoor recreation areas: the case of Mexican Americans, *Leisure Sciences* 15: 38–98.

Grumbine, R.E., 1994. What is ecosystem management? *Conservation Biology* 8: 27–38.

Ham, S.H., 1992. *Environmental Interpretation: A Practical Guide for People with Big Ideas and Small Budgets,* North American Press, Golden, CO.

Hammitt, W.E. and D.N. Cole, 1998. *Wildland Recreation: Ecology and Management,* John Wiley and Sons, New York.

Hultsman, J.,R.L. Cottrell, and W.Z. Hultsman, 1998. *Planning Parks for People,* Venture Publishing, State College, PA.

Irwin, P.N.,W.C. Gartner, and C.C. Phelps, 1990. Mexican-American/Anglo cultural differences as recreation style determinants, *Leisure Sciences* 12: 335–348.

Johnson, C.Y. and J.M. Bowker, 1999. On-site wildland activity choices among African Americans and White Americans in the rural South: implications for management, *Journal of Park and Recreation Administration* 17: 21–39.

Johnson, C.Y., J.M. Bowker, D.B.K. English, and D.Worthen, 1998. Wildland recreation in the rural South: an examination of marginality and ethnicity theory, *Journal of Leisure Research* 30: 101–120.

King, J.C. and A.C. Mace Jr., 1974. Effects of recreation on water quality, *Journal of Water and Pollution Control Federation* 46: 2453–2459.

Lang, L. 1998. *Managing Natural Resources with GIS,* Environmental Systems Research Institute, Redlands, CA.

LaPage, W.F., 1994. *Partnering for Parks: To Form a More Perfect Union,* National Association of State Park Directors, Tallahassee, FL.

Loomis, J.B. and R.G. Walsh, 1997. *Recreation Economic Decisions: Comparing Benefits and Costs,* Venture Publishing, State College, PA.

Manfredo, M.J., B.L. Driver, and P.J. Brown, 1983. A test of concepts inherent in experienced based setting management for outdoor recreation areas, *Journal of Leisure Research* 15: 263–283.

Manning, R.E., 1999. *Studies in Outdoor Recreation: Search and Research for Satisfaction*, Oregon State University Press, Corvallis.

Morris, H., 1999. Rails-to-trails, in *Outdoor Recreation in American Life: A National Assessment of Demand and Supply Trends,* H.K. Cordell (prin. inv.), Sagamore Publishing, Champaign, IL, pp. 140–143.

New Jersey Division of Parks and Forestry, 2000. NJ Corporate Partners, New Jersey Department of Environmental Protection, Division of Parks and Forestry, Trenton, http://www.state.nj.us/dep/forestry/njcorp.htm. [Accessed: July 15, 2000.]

Pulaski County, n.d. Randolph Park, Pulaski County, Dublin, Virginia, http://www.rootsweb.com/~vapulask/randolph/indexr.html. [Accessed: July 15, 2000.]

Reiling, S.D. and M.J. Kotchen, 1996. Lessons Learned from Past Research on Recreation Fees, in Recreation Fees in the National Park Service: Issues, Policies, and Guidelines for Future Action, A.L. Lundgren, Ed., Minnesota Extension Service Pub. No. BU-6767, Cooperative Park Studies Unit, Department of Forest Resources, University of Minnesota, St. Paul, pp. 49–69.

Rolston, Holmes, III, 1991. Creation and recreation: environmental benefits and human leisure, in *Benefits of Leisure,* B.L. Driver, P.J. Brown, and G.L. Peterson, Eds., Venture Publishing, State College, PA, pp. 393–403.

San Jose Conservation Corps, 2000. SJCC & EP. Americorps, San Jose Conservation Corps, San Jose, CA, http://www. sjcc-ep.org/acorps.htm. [Accessed: July 15, 2000.]

Stankey, G.H., D.N. Cole, R.C. Lucas, M.E. Petersen, and S.S. Frissell, 1985. *The Limits of Acceptable Change (LAC) System for Wilderness Planning*, General Technical Report INT–176, U.S. Department of Agriculture, Forest Service, Intermountain Forest and Range Experiment Station, Ogden, UT.

Stein, T.V., D.H. Anderson, and D.Thompson, 1999. Identifying and managing for community benefits in Minnesota state parks, *Journal of Park and Recreation Administration* 17: 1–19.

Stein, T.V. and M.E. Lee, 1995. Managing recreation resources for positive outcomes: an application of benefits-based management, *Journal of Park and Recreation Administration* 13: 52–70.

Stenger, R., 1999. Trends in visits to federal areas, in *Outdoor Recreation in American Life,* H.K. Cordell (prin. inv.), Sagamore Publishing, Champaign, IL, pp. 282–284.

Swearingen, T.C. and D.R. Johnson, 1994. Keeping visitors on the right track: sign and barrier research at Mount Rainier, *Park Science* 14: 17–19.

Tipple, T.J. and J.D. Wellman, 1989. Life in the fishbowl: public participation rewrites public foresters' job descriptions, *Journal of Forestry* 87: 24–27, 30.

U.S. Department of Agriculture (USDA), Forest Service, 1982. *ROS Users Guide*, Department of Agriculture, Forest Service, Washington, DC.

U.S. Department of Agriculture (USDA), Forest Service, 1993. The Power of Collaborative Planning: Report of the National Workshop, FS-553, Department of Agriculture, Forest Service, Washington, DC.

U.S. Department of Agriculture (USDA), Forest Service, 2000. Salt Lake County — Wasatch-Cache National Forest Partnership: Mill Creek Canyon Protection and Management Operations and Maintenance Plan, Department of Agriculture, Forest Service, Wasatch-Cache National Forest, Salt Lake City, UT.

U.S. Department of Interior (USDI), National Park Service, 1997. *VERP: The Visitor Experience and Resource Protection (VERP) Framework: A Handbook for Planners and Managers*, Department of the Interior, National Park Service, Denver Service Center, Denver, CO, http://www.nps.gov/planning/verp/handbook.pdf. [Accessed: August 1, 2000.]

U.S. Department of Interior and U.S. Department of Agriculture, 2002. Recreation Fee Demonstration Program: Progress Report to Congress, fiscal year 2001, Department of Interior and Department of Agriculture, Washington, DC.

Vaske, J.J., M.P. Donnelly, and W.F. LaPage, 1995. Partnerships for the 21st century: a return to democracy, *Journal of Recreation and Park Administration* 13: i–ii.

Virden, R.J. and R.C. Knopf, 1989. Activities, experiences, and environmental settings: a case study of recreation opportunity spectrum relationships, *Leisure Sciences* 11: 159–176.

Williams, D.R. and S.I. Stewart, 1998. Sense of place: an elusive concept that is finding a home in ecosystem management, *Journal of Forestry* 98: 18–23.

Wohlwill, J.F., 1983. The concept of nature: a psychologists' view, in *Behavior and the Natural Environment — Human Behavior and the Environment: Advances in Theory and Research,* vol. 6, I. Altman and J. F. Wohlwill, Eds., Plenum Press, New York, pp. 5–37.

Wondolleck, J.M. and S.L. Yaffee, 2000. *Making Collaboration Work: Lessons from Innovation in Natural Resource Management*, Island Press, Washington, DC.

Part IV

Conserving and Managing Forests for Ecological Services and Benefits

11 Ecological Assessment and Planning in the Wildland–Urban Interface: A Landscape Perspective

Wayne C. Zipperer
Southern Center for Wildland–Urban Interface Research & Information, USDA Forest Service Southern Research Station, Gainesville, Florida

CONTENTS

A thing is right when it tends to preserve the integrity, stability, and beauty of the biotic community. It is wrong when it tends otherwise.

—Aldo Leopold

11.1 INTRODUCTION

The day starts like any other with one exception, a request to evaluate the effects of a proposed residential development in your management district. Development has occurred in adjacent districts, but not in yours. You realize that the proposal represents more than just one action, it represents the first of a series of actions that can alter the ecological integrity, the management of natural resources, and the aesthetics of the landscape. The simple action of evaluating a development plan confronts you with three questions: (1) How does the current proposal affect the structure and function of the site and adjacent areas? (2) What areas need to be conserved or protected to minimize environmental effects from future development? (3) How will these areas be protected (e.g., legally) from future development?

Of course, the land is privately owned, and a private landowner has the right to develop his or her lands in compliance with federal, state, and local environmental laws and regulations. Does this mean that managers need to resign to the fact that development will occur? Quite the contrary, as land managers, we can provide critical information and insights into the development process by identifying important sites — ecological, physical, and cultural — within the landscape and by providing guidelines to landowners to minimize environmental and cultural degradation of those sites. In addition, we can provide guidelines to local decision makers who develop policy for land-use decisions. Without this input, development will continue to degrade the environment, alter social structure, and change the aesthetic beauty of the landscape.

My intention is not to provide an exhaustive review of the extensive literature on methods to protect the environment from development (e.g., Duerksen et al. 1997; Foresman et al. 1997; Dale et al. 2000) but rather to use the literature to provide natural resource managers and land-use planners with some basic guidelines to begin to evaluate the effect of development on natural systems. This chapter is also not a cookbook with recipes to achieve specific

1-56670-602-5/05/$0.00+$1.50
© 2005 by CRC Press LLC

outcomes, but rather it emphasizes concepts for developing specific recommendations that can be tailored for individual conditions. The chapter has three sections — "Why Ecosystems?," "Why Landscapes?," and "Tomorrow's Landscapes Today" — to address the first two questions facing a land manager. To address question three, the manager or planner can propose to (1) purchase the property, (2) purchase development rights, (3) propose tax incentives, and (4) regulate land use. These options are discussed thoroughly in Chapters 4, 5, and 6 of this book and in the Southern Wildland–Urban Interface Assessment (Macie and Hermansen 2002), and will not be discussed here. Sections "Why Ecosystems?" and "Why Landscapes?" introduce ecosystem management as a holistic approach to land-use management decisions and to evaluate the effect of development on natural systems at a landscape scale. The third section, "Tomorrow's Landscapes Today," applies ecosystem management to identify key physical, ecological, and cultural sites in the landscape, to evaluate proposed development, and to minimize negative effects from future development.

11.2 WHY ECOSYSTEMS?

In 1992, the USDA Forest Service adopted an ecosystem approach to multiple-use management (Overbay 1992). The approach was proposed and accepted by the agency as a means to shift focus from sustaining production of particular goods to sustaining the viability of physical, ecological, and social systems (Kaufmann et al. 1994). Since this policy shift, ecosystem management has been the guiding principle for management decisions in the Forest Service as well as other federal, state, and local natural resource management agencies.

Why ecosystems? An ecosystem refers to a spatially and temporally explicit place that includes all the organisms, the abiotic environment, and their interactions (Likens 1992). Unlike population and community approaches to management, which focus on the interactions of individuals and species, an ecosystem approach focuses on the interactions — flows and processes — among physical, ecological, and social components. Hence, the ecosystem is a functional unit where physical, ecological, and social components interact (Farina 2000), and an ecosystem approach to management accounts for these interactions and flows, and structure that influences them.

Ecosystems are open systems. Energy (e.g., photosynthesis, herbivory, and predation), organisms (e.g., migration, foraging and breeding, and diurnal and seasonal movements), and matter (e.g., nutrients, water, sediments, and heavy metals) flow into, within, and out of the system. Therefore, an ecosystem influences and is influenced by neighboring ecosystems (Likens 1992). For example, in the past, land managers considered individual management units as being ecologically independent of each other rather than as integrated parts of a larger ecological system. By altering a management unit, the manager not only affects the flows and processes occurring within the unit but also the flows into and out of the unit and adjacent units. Furthermore, when we consider each unit independently, we cannot assess the cumulative effects of management actions on individual units at a scale of the larger system. An ecosystem approach takes into account the effect of management activities on a site and on adjacent sites.

Ecosystems also are dynamic; that is, they change over time. These changes alter physical structure or composition and the flow of energy, organisms, and matter. An ecosystem approach acknowledges that change is a characteristic of the system and that there is not a "balance of nature" (Botkin 1990).

Initially, ecologists excluded humans from natural systems (Pickett et al.1997). Today, however, ecologists recognize that humans and their socioeconomic systems are a significant component of all ecosystems. Because physical, ecological, and social components are interdependent, a holistic or ecosystem approach to land-use decisions enables the equitable evaluation of components and their interactions (McCormick 1998). (See Christensen et al.[1996] and McCormick [1998] for excellent overviews of the components and principles of ecosystem management.) Traditionally, land-use decisions focused principally on economic factors at the expense of biophysical and other social and cultural elements. An ecosystem approach to land-use decisions acknowledges biophysical and social complexities of ecosystems and the importance of maintaining those complexities to meet the needs for goods and services used by humans for the current and future generations (Christensen et al.1996; McCormick 1998).

An ecosystem, however, is far too complex for humans to manage as a unit. So, why use an ecosystem approach to decision making? In practice, ecosystem management is more a way of thinking to acknowledge and account for the species diversity as well as physical, ecological, and social patterns and processes (Yaffee et al.1996). Grumbine (1994) offers five management goals to sustain ecological integrity under ecosystem management (Table 11.1). These goals recognize the importance of maintaining biodiversity and ecological processes and incorporating humans and their activities into the decision making process. When a land-use decision is contemplated, an ecosystem approach enables us to assess the effect of development not only on populations and biotic communities but also on biophysical and social components and on the flow of energy, species, and matter in the system. Further, an ecosystem approach enables us to evaluate the effects across ownership and management boundaries; thus, we are able to inventory and evaluate cumulative effects on the landscape. For example, watershed protection is an ecosystem approach

TABLE 11.1
Ecosystem Management Goals to Sustain Ecological Integrity

- Maintain and protect habitat for viable populations of all native species.
- Represent, within protected areas, all native ecosystem types across their natural range of variability.
- Maintain evolutionary and ecological processes (i.e., disturbance regimes, hydrological processes, nutrient cycles, species migrations).
- Manage over periods of time sufficient to maintain the evolutionary potential of species and ecosystems.
- Allow for human use and occupancy at levels that do not result in ecological degradation.

Source: Grumbine (1994).

to planning. By working within the boundaries of a watershed, which often encompass many political and managerial jurisdictions, watershed managers measure hydrologic inputs and outputs and assess, individually and collectively, how existing and proposed land uses affect water quality and quantity.

11.3 WHY LANDSCAPES?

To evaluate the effect of urbanization on physical, ecological, and social patterns and processes, a perspective that is greater than the ecosystem and encompasses the spatial interactions among ecosystems is needed (Turner et al. 2001). A landscape scale provides the opportunity to view the spatial connectedness of ecosystems and assess the cumulative effects of land-use decisions on physical, ecological, and social components (Dale et al. 2000). A landscape, however, connotes different meanings for different people. To some, a landscape may represent a pastoral scene or a planted garden. Ecologically, a landscape is a heterogeneous area composed of a cluster of interacting ecosystems that are repeated in similar form throughout (Forman and Godron 1986). For example, an agricultural landscape is composed of agricultural fields and buildings, hedgerows, and woodlots (Figure 11.1). Similarly, urban landscapes are composed of streets, buildings, and managed greenspaces. Regardless of how a landscape is defined, every landscape has three components: structure, function, and change (Forman and Godron 1986). Structure refers to the types of structural elements that you see on the landscape and their spatial arrangement. Function refers to the flow of energy, materials, and organisms within and through ecosystems. Change refers to modification of structural and flow attributes over time. Development causes change, and a landscape perspective enables managers and planners to ascertain not only the potential effects of urbanization on an ecosystem but also the effects on adjacent ecosystems (Turner 1989). A landscape perspective also accounts for the collective incremental changes by humans and provides the ability to assess their cumulative effects on the ecosystems comprising that landscape (Farina 2000).

FIGURE 11.1 Aerial photograph of an agricultural landscape depicting different patch types — remnant forest, hedgerow, field, building, and transportation.

So, how do we link ecosystem management and a landscape perspective with the issue of changing land use? Looking at an aerial photograph of an agricultural landscape, for example, we can identify different structural elements based on their morphology: agricultural fields and buildings, forests, and hedgerows (Figure 11.1). These homogeneous areas represent structural units called patches, and collectively these patches form the landscape mosaic (Forman and Godron 1986). A patch can also be defined by its functional attributes such as how it is used by a species or its role in an ecosystem process. For example, a riparian patch is characterized structurally as vegetation along streams and functionally as a zone for removing nitrogen. By viewing a landscape as a mosaic of structural and functional patches, we can define how energy, species, and materials are distributed and flow across a landscape. In addition, characterizing the landscape by structural units enables us to assess how the landscape changes spatially and temporally. Subsequently, we can ask how the proposed development plan affects spatial distribution and flow within and among patches. Further, by conducting "what if" scenarios, we can assess future losses of patches to development and the effect on the ecological integrity of the landscape (Forman and Collinge 1996; White et al. 1997). An example would be a new road: a transportation patch. A road fragments a habitat, which creates new edges and disrupts migration patterns; increases storm runoff, which alters stream biota and stability; and serves as a conduit for invasive species, which alter habitat structure. Without a landscape perspective, these cumulative effects could not be assessed.

The concept of defining a landscape by homogeneous patches is not foreign to land managers and planners. Natural resource managers have used terms such as community and forest types to describe a forest landscape. Similarly, planners use land use to designate areas with similar structural and functional attributes. Regardless of the classification system, each unit is based on structural or functional attributes that distinguish it from adjacent units. So, why use the word "patch" rather than some other common terms? First, the term "patch" simplifies terminology across different disciplines; second, the ecological concept of patch dynamics allows one to move from one spatial or temporal scale to another; and third, it is applicable to physical, ecological, and social components of the ecosystem (Farina 2000; Pickett et al. 2001).

11.3.1 Scale

Like forest-type delineation, patch delineation is scale dependent. Scale refers to the spatial and temporal dimensions of an object or process being studied or managed (Forman 1995). Scale contains two components: grain, the finest resolution of the data being collected or mapped; and extent, the areal size of the management site or the duration of the proposed action (Turner et al. 2001). An example of grain is from land cover. An area may be defined rather coarsely as forest or more finely as evergreen or deciduous or even more finely as a sugar maple forest type. Grain resolution (patch delineation) depends on the proposed objectives. For example, patch delineation of a bear habitat would be different from delineation of a butterfly habitat. Examples of extent are the forest being managed and the watershed where a proposed development might occur. Unfortunately, an array of scales is needed to define the complexities of ecosystem processes (Turner et al. 2001), and the manager/planner must pick the scale that best meets his or her objectives. To assess the effects of a proposed development plan, scale needs to have a resolution sufficient to capture population and community characteristics of the area and ecosystem processes such as species migrations, water flow, and disturbance patterns. A scale commonly used by county planners is land use/cover (grain) within a watershed (extent).

11.3.2 Patch Configuration

Patch size, shape, isolation, orientation, and spatial arrangement have significant influence on the distribution and flow of energy, species, and materials in a landscape. See Forman (1995) for an in-depth discussion of each attribute. For example, larger patches may have greater spatial heterogeneity (e.g., structural and environmental conditions) and support larger populations of species for longer periods of time than smaller patches (Arnold 1995). Similarly, smaller patches have greater edge to area ratios and subsequently greater edge effect than larger patches. Edge effect, which can be detrimental to interior dependent species, is the biophysical environment created at the interface between two patches. This edge effect creates edge habitat. In a forest patch, edge habitat is drier because of increased solar radiation and wind, has higher predation and parasitism rates, and may have greater biodiversity than interior habitat (Saunders et al. 1991). The width of edge habitat is species dependent. For example, for forest trees, edge habitat is about 30 m (Levenson 1981), but for some birds it may exceed 500 m (Wilcove et al. 1986), although 100 m seems to be an appropriate width (Temple 1986). Edge effect may be tempered by including a buffer between the core habitat needing protection and the actual edge. One concept used to minimize edge effect is a multiple-use module (MUM) (Harris 1984; Noss and Harris 1986) (Figure 11.2). The MUM contains a core of protected area, a zone of low utilization (e.g., recreation), and a zone of intense utilization (e.g., agriculture and development). These zones can be established through existing zoning regulations and ordinances at the town, county, and state level. In fact, zoning regulations and ordinances can be developed to minimize fragmentation and biodiversity degradation.

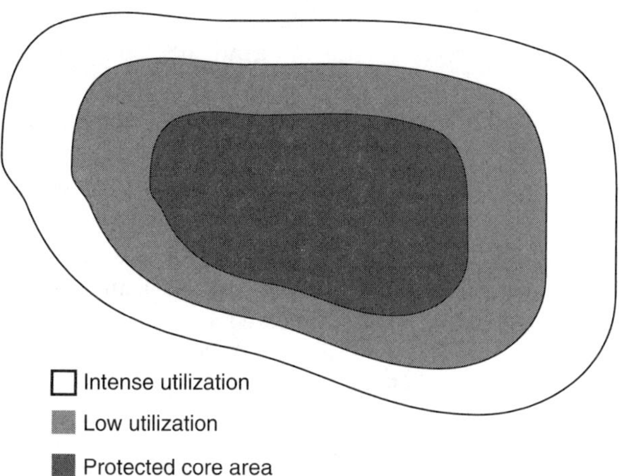

FIGURE 11.2 Graphic illustration of the core conservation area being protected by zones of low and intense utilization. (Adapted from Noss and Harris 1986.)

Forest interior habitat tends to be shadier, cooler, and moister, and possesses a greater density of mesic plant species than an edge habitat (Ranney et al. 1981). The amount of interior habitat depends on patch shape and size. A long, elongated patch may have no interior habitat or an insufficient amount of interior habitat to support interior species. By comparison, a patch of similar size but having a regular or circular shape may have an interior habitat if it is larger than 5 ha (Figure 11.3) (Levenson 1981). This does not mean that all protected patches need to be circles or squares. Elongated patches can connect patches aiding in the dispersal of species, and lobes and extensions from patches add to shape complexity and may influence the movement of organisms (Forman 1995).

Patch isolation significantly affects the movement and dispersal of organisms. Considerable discussion has focused on the functionality of corridors (e.g., Simberloff and Cox 1987; Beier and Noss 1998). Corridors need to be thought of as an element of connectivity rather than just linear habitats and designed to meet the needs of species being managed (Farina 2000). In general, habitat patches that are closer, linked, or occur in a hospitable landscape matrix allow species to disperse more freely among patches and support species for longer periods of time than patches that are distant from one another or occur in inhospitable landscapes (Arnold 1995). Likewise, large patches may be preferred habitats to conserve and protect, but smaller patches, distributed across the landscape, may serve as stepping-stones across a hostile environment and improve connectivity among patches (Forman and Collinge 1996). For example, in urban landscapes, green spaces and belts often link patches of natural habitat. Forman and Collinge (1996) call these smaller patches outliers and support their uses in conservation plans (Figure 11.4).

11.3.3 Disturbance

Geomorphology and other abiotic conditions (e.g., climate, topography, soils, moisture availability), biotic interactions (e.g., competition, herbivory, predator–prey, exotic species), and natural and human disturbances create patches and alter their spatial arrangement on a landscape (Farina 2000; Turner et al. 2001). This section focuses only on attributes of natural and human disturbances. Natural disturbances include windstorms (e.g., hurricanes, tornadoes, thunderstorms), fire, floods, and insect and pathogen outbreaks. Examples of disturbance attributes include severity (intensity), magnitude (spatial — size and shape), frequency (number of events), and return interval (temporal) (Pickett and White 1985; Turner et al. 2001). Collectively, all disturbance types and their attributes describe the disturbance regime for a landscape (Pickett 1998). It is important to note that it is unknown at what time a particular spot will undergo a natural disturbance; however, what is known is that such events will occur at some point in time (Bormann and Likens 1979; Denslow

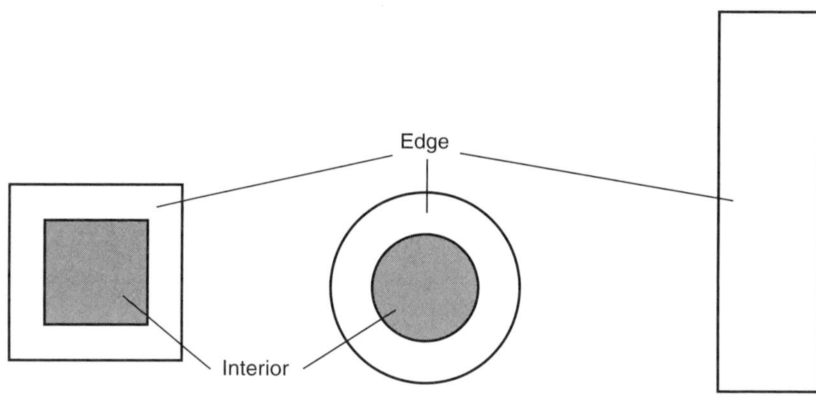

FIGURE 11.3 Illustration of the effect of different patch shape on interior habitat.

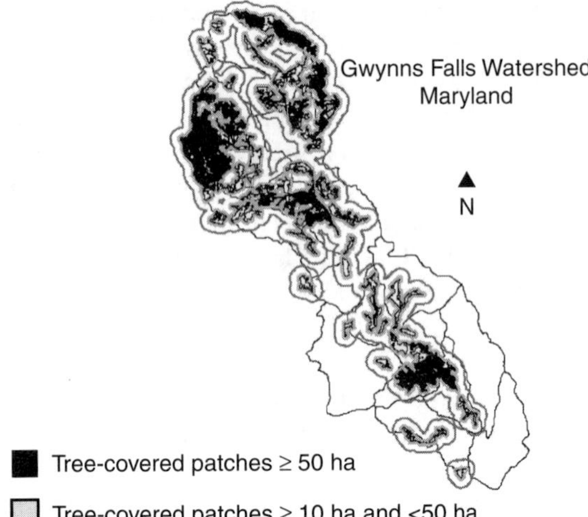

FIGURE 11.4 Distribution of tree-covered patches in an urbanizing watershed, illustrating the importance of including stepping-stones in landscape planning.

1980). Because disturbances will occur, managers and planners need to consider how development will affect the disturbance regime and how disturbances affect the development. For example, the coastal plain landscape contains fire-dependent ecosystems and is also a zone of rapid development. With development, fire suppression occurs to protect properties, but fuels still accumulate. Consequently, when a fire does occur, it is often a conflagration rather than a low-intensity surface fire characteristic of those ecosystems (see Chapter 13). Managers and planners need to account for fire by establishing prescribed burning regimes and proposing firewise landscaping and construction (Monroe 2002).

Natural disturbances create spatial heterogeneity, the landscape mosaic of patches, and changing the disturbance regime will alter this mosaic (Clark 1986; Pickett 1998). Humans alter disturbance regimes through their activities. Fires are suppressed, rivers are dammed, and streams are channelized. These actions directly and indirectly cause shifts in species composition of communities and alter the movement of energy, species, and matter through the system. With fire suppression, short-lived fire-dependent species are being replaced by long-lived mesic species. In addition, nutrient cycling is drastically altered (Stuart 1998). No longer is there a flush of nutrients after a fire. With fire suppression, nutrients reside in live vegetation and dead biomass over a longer period of time. When a fire does occur, its intensity may be so great that textural and chemical composition of soils can be altered. Organic matter burns to a greater depth, reducing the nutrient holding capacity of the soil; soils become hydrophobic (unable to absorb water); and nutrients are volatilized (Stuart 1998).

Because of their effect on ecosystems and humans and their property, large, infrequent disturbances are of particular concern to natural resource managers and land planners (Dale et al. 1998). Dale et al. (1998) identify three management options to plan for this type of disturbance: (1) manage the system prior to the disturbance; (2) manage the disturbance; and (3) manage the system after the disturbance. By managing the system prior to a disturbance, managers can minimize the effects of the disturbance on management goals. For example, reducing fuel load rather than suppressing fires diminishes the severity of the fire when it does occur. Managing a disturbance is often motivated by human desire to lessen effects on life and property (Dale et al. 1998). Such actions are often costly and may result in greater damage than if no management took place. Again, fire provides an excellent example. The disturbance is controlled with suppression, but control is only temporary. A conflagration can still occur, destroying personal property and altering ecosystem structure and function. Efforts to manage a site after a large, infrequent disturbance can also be problematic by creating undesirable plant communities, altering community development, and introducing nonnative species (Dale et al. 1998). An ecosystem and landscape approach to land-use planning enables managers and planners to identify, protect, and maintain viable populations of native species and native ecosystem types and their processes across their natural range of variability. So, when a disturbance occurs in a region, natural populations and processes are represented, thus available to begin a recovery cycle.

Urbanization is a disturbance. Urbanization, however, is different from natural disturbances. With urbanization, land features (e.g., streams and forests) become linear because of roads, ownership, and management practices. Urbanization creates patches that have more regular shapes, smaller sizes, and more diverse types. In addition, landscape changes are more permanent and natural processes (succession and nutrient cycling) are suppressed or altered. These differences alter landscape structure and function and subsequently change the distribution and flow of energy, species, and materials across a landscape.

11.4 TOMORROW'S LANDSCAPES TODAY

The wildland–urban interface is a zone of rapid transformation of natural habitat to urban land use. Urbanization directly and indirectly affects natural ecosystems (McDonnell et al. 1997). The most obvious direct effects are deforestation and fragmentation. Deforestation creates new forest edge, simplifies edges, decreases forest interior habitat, and increases patch isolation (Saunders et al. 1991; Zipperer 1993). Examples of indirect effects include urban heat island effect, soil hydrophobicity (White and McDonnell 1988), introductions of exotic species (Reichard and White 2001), air pollution (Lovett et al. 2000), and altered disturbance regime (Pickett 1998).

As managers, we must try to minimize urban effects on the natural ecosystems to sustain the goods and services provided by them (Christensen et al. 1996). In his book, *The Seven Habits of Highly Effective People*, Stephen Covey (1989) suggests that we "begin with the end in mind." As managers, we need not only think about how a landscape will be structurally and functionally changed after an area has been developed but also how to plan for future events. What should the future landscape look like with continual development? What features are important? What features can be lost? Although final decisions about the future landscape depend on land-use regulations and the goals and objectives of landowners, land managers can provide critical information to decision makers on how the landscape functions. By identifying critical elements of the landscape that contain significant structural and functional attributes before development occurs, the elements might be protected and environmental degradation minimized (Forman and Collinge 1997).

Some managers may state: "just tell me what to save." Unfortunately, there are no pat answers or solutions to conserving critical landscape elements. Each situation is unique. A number of concepts can be applied to each development scenario, but the final decision needs be made within the context of the landscape being developed and planning objectives. Harris (1984) proposes four critical landscape questions of patch importance that we can use to define tomorrow's landscapes:

1. What patches are strategically located with respect to the function and integrity of the overall landscape?
2. What patches make a specific contribution to biodiversity in terms of genetics, endemic species, greater species richness, or ecotypes?
3. What patches are more susceptible to development?
4. Does a patch and its linkages fit into the landscape pattern and processes?

An ecosystem approach to decision making enables a manager or a planner to answer these questions.

Hunter (1990) proposes a two-filter approach — a macro- or coarse-scale filter and a micro- or fine-scale filter — to answer the landscape questions and to begin defining tomorrow's landscape. At the broad-scale filter, the land manager assesses the patch configuration and ecological processes, and the context in which they occur. Fine-scale filters identify site differences from a physical, ecological, and cultural perspective (LaGro 2001).

11.4.1 Landscape Filters

At the coarse-scale level, we map out landscape structure and function. Current geographical information software (GIS) and other specialty software such as FRAGSTATS (McGarigal and Marks 1995) can aid in quantifying patches by their size, edge-to-area ratio, shape, interior habitat, and nearest neighbor of similar size or habitat. The coarse-scale filter also needs to include a temporal component to account for the seral stages of ecosystems and the effects of disturbances. Ephemeral patches need to be identified and mapped because they may provide critical habitat for some species (Smallidge and Leopold 1997). Although the application of a GIS would aid in the analyses, its use is not a prerequisite for the assessment. Assessments can be conducted, for example, on 7.5-min U.S. Geological Survey topographic maps or aerial photographs. What is important is to map patches composing the landscape mosaic (see Diaz and Apostol 1992) and the ecological processes — the movement of energy (e.g., food webs, water flowing downhill), organisms (e.g., seasonal and breeding migrations), and materials (e.g., hydrology, nutrient cycling, sediments). By doing so, we can begin to assess how development may remove significant patches and disrupt ecosystem processes. For example, within a watershed, important hydrologic sources (e.g., headwaters, seeps, springs, streams, aquifers), riparian habitats, and wetlands can be identified and mapped to evaluate how urbanization may alter the flow of water across and within the watershed.

Although patch importance is determined by management or planning objectives, importance links landscape structure, function, and change to achieve the goals of ecosystem management (Table 11.1). For example, variable source areas in water movement include significant locations such as riparian areas, headwaters, and seeps. Similarly, migration corridors for mammals, reptiles, and amphibians reflect landscape connectivity. The intersection of contrasting habitats indicates a unique habitat feature used by a variety of species. Each of these structural and functional attributes represents a set of structural and functional elements that need to be identified before landscape alterations occur.

Duerksen et al. (1997) identifies strategies for patch selection (Table 11.2). These strategies generally follow concepts for refuge design (Figure 11.5) (Simberloff 1997):

- Large patches will hold more species than a smaller patch.
- Assuming the patches are of the same habitat type, a large patch is preferable to several small patches.
- If only small patches are available, they should be clustered and preferably linked rather than linear or disconnected.
- Reduce the edge to interior ratio to minimize edge effect.

TABLE 11.2
Criteria and Principles to Select Significant Patches within a Landscape

1. Select and maintain large, intact patches of native vegetation, prevent fragmentation by development, and establish MUM to negate development along edges.
2. Establish priorities for protecting biodiversity and maintaining ecological processes in protected areas.
3. Protect not only threatened and endangered species but also rare landscape elements. Divert development toward "common" patches and landscape attributes.
4. Reduce patch isolation by maintaining connectivity through creating a hospitable landscape matrix, stepping-stones of habitat, and corridors.
5. Select patches to create riparian buffers around headwater streams, streams, and variable source areas.
6. Establish patch redundancy to protect from disturbances.
7. Maintain or reestablish disturbance regime.

Source: Adapted from Duerksen et al. (1997).

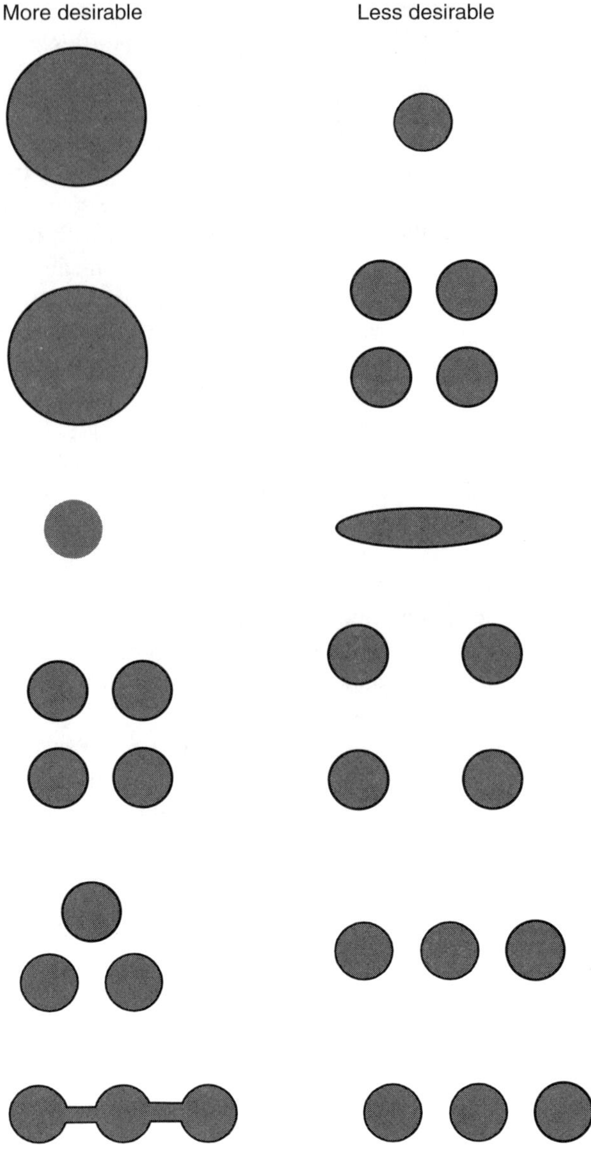

FIGURE 11.5 Desirable and less desirable patch configuration for refuge design. (From Simberloff (1997). With permission.)

These strategies have been successfully applied to land-use decisions in the Colorado Front Range to protect large, unfragmented patches of natural habitat, maintain native biodiversity and ecosystem structure and functions, and maintain connectivity (Duerksen et al. 1997).

How large is large when selecting patches? It depends on the objectives of the selection process, disturbance regime, and landscape context. For example, if an objective is to maintain or restore a viable population of a large predator, then significantly larger patches are needed than if the objective is to maintain forested habitat for carrion beetles. In general, larger animals need larger spaces to breed and survive than smaller animals (Holling 1992). Nevertheless, even if large mammals are absent from the regional fauna, the largest possible patches should be selected for conservation to minimize the loss of a species or community type to large, infrequent disturbances (Pickett and Thompson 1978). Selection should include not only the desirable patch type but also its seral stages (Harris 1984).

Context plays an important role in evaluating which patch to keep and which to develop. Context refers to where the patch occurs within the landscape and what surrounds the patch. In general, land managers will encounter some variation of three context types: forest, agricultural, and urban (Zipperer et al. 1990). Each context type differs in opportunities for conservation and protection of biodiversity and ecosystem processes. Forest context offers the greatest flexibility to identify significant patches. By comparison, options in agricultural and urban contexts may be limited and depend on the extent of previous patterns of deforestation and fragmentation. For example, large patches necessary to maintain large mammal species may not be available in urban and agricultural contexts. Context also influences patch importance. What may have been an unimportant patch in a forest context could be a significant patch in an urban or agricultural context because of the absence of other patch types. Within these deforested landscapes, patch occurrence can significantly affect its importance with respect to species presence and dispersal across a landscape (Andrén 1994).

Although larger patches are often favored over smaller patches with similar habitat value, both Forman and Collinge (1996) and Hunter (1990) argue for including smaller patches of natural habitats in the landscape design. In agricultural and urban contexts especially, smaller patches provide ecological benefits by protecting rare habitats and species outside the large patches; enhancing connectivity between large patches by providing "stepping-stones" for species movement; and enhancing heterogeneous conditions throughout the landscape (Forman and Collinge 1996).

Once the coarse-level assessment has been completed, patches can be prioritized by their attributes. For example, in a forest context, a score of 1 may be given to patches <50 ha, whereas a score of 5 may be given to patches >1000 ha. By comparison, a score of 1 may be given to patches <1 ha, and a score of 5 for patches >100 ha in an urban landscape. A similar scoring range can be developed for each of the other measured attributes, such as riparian habitat, headwater area, corridors, and unique spatial arrangements. Using a spreadsheet, we can sum, average, weight, or use the maximum value to identify and rank patches based on their structural and functional significance (see Duever and Noss 1990; White et al. 1997).

Once scoring has been completed for attributes deemed important at the coarse-scale level, a fine-scale-level assessment needs to be conducted to identify intrinsic differences among patches. LaGro (2001) identified important physical, ecological, and cultural attributes of site content (Table 11.3). Each of these categories can be expanded to meet objective needs. Duever and Noss (1990) provide an expanded list of ecological elements that can be used to answer the following questions: what patches make a specific contribution to biodiversity in terms of genetic, endemic species, greater species richness, or ecotypes, and what patches are more susceptible to development (Table 11.4)? From their list, Duever and Noss (1990) developed a scoring protocol to rank patches by their ecological importance. For example, for vulnerability to future development, they scored a patch as 1 if protection was guaranteed by deed restriction, easement, or established regulatory authority and as 5 if the patch was slated for development or had no significant regulatory protection. Using the scoring approach, Duever and Noss (1990) conducted "what if" scenarios to determine whether rankings would change under different land management decisions. The final resolution of the assessment depends on objectives and data availability.

11.4.2 Examples

The conservation of significant habitats is not new, but linking conservation strategies with land development decisions has only recently been acknowledged as an important step toward creating sustainable landscapes (Cohan and Lerner 2003). To illustrate the evaluation process, I will use two terrestrial examples — the Highlands Region of New York and New Jersey and Alachua County, FL.

TABLE 11.3
Examples of Physical, Ecological, and Cultural Attributes that May Be Included when Inventorying Site Content

Physical
Topography
Elevation
Slope
Aspect
Geology
Serpentine
Caves, ledges, escarpments
Hydrology
Surface water
Ground water
Aquifer recharge
Thermal springs
Wetlands
Hazards
Earthquakes
Volcanos
Landslides
Soils
Permeability
Erosion potential
Textural/chemical composition
Depth to water table
Depth to bedrock
Ecological
Threatened and endangered species
Federal and state listings
Unique community types
Significant wildlife habitat
Breeding/nesting, foraging
Cultural
Historic
Buildings, meeting locations, burial grounds, gardens
Circulation/use and transportation
Roads, trails, paths
Perceptual amenities
Viewsheds
Human populations
Native
Ethnic

Source: LaGro (2001).

The Highlands of New York–New Jersey (1.5 million acres) is part of a geomorphic province called the Reading Pong that stretches from northwest Connecticut to east-central Pennsylvania (Figure 11.6) (van Diver 1992). The Highlands, although only an hour from Manhattan, N Y, is renowned for its biological diversity, unique ecological communities, and significant cultural sites. In addition, over 11 million people use the water resources of the Highlands and more than 14 million individuals visit the region annually.

Human population growth threatens this region. Just in the past decade, human population levels increased by 11.5

TABLE 11.4
Possible Criteria Used to Rank the Importance of Patches within a Landscape

Vulnerability: How vulnerable is the patch to being developed? Is the patch protected through deeds and conservation easements; is it owned by indi viduals willing to develop; or does it occur on a good developable site? Is the patch vulnerable to the initiation of disturbances?

Rarity: Does the patch contain rare plants and animals? Is the patch a rare community type? Is the patch community listed by the state's Heritage Program?

Connectedness: Is the patch connected to other elements of the landscape? Is it isolated from large parcels of land; is it part of a natural corridor; or does it serve as a stepping-stone between two significant habitats?

Completeness: Does the patch represent ecological communities with a full complement of species? If species are missing, can neighboring sites be used as a source for colonizing individuals? How disturbed or degraded is the site? Is the patch large enough to contain different seral stages and representations from different types of disturbances?

Manageability: Manageability can be viewed from two perspectives: management for products and management to maintain the ecological integrity of the site. If the patch is degraded, can it be restored? Are sites too small to restore a complement of species and natural processes?

Nature-oriented human use potential: Is the patch suitable for passive recreation? Is it accessible for recreational use, or is it aesthetically pleasing?

Source: Duever and Noss (1990).

percent to 1.4 million individuals. In October 2000, Congress authorized a study of urbanization effects on the region. One of four goals of the regional study was to identify significant areas to conserve and protect (Phelps and Hoppe 2002). To accomplish this objective, criteria were selected and importance weighted for each of the five critical resources (Table 11.5). One criterion for water resources was the presence of an aquifer (coarse filter) and weights were given based on the type of aquifer (fine filter) (Hatfield et al. 2003). Data were mapped to a 30 m grid and each grid cell was assigned a value from 1 to 5 for each criterion of a resource. To create the final resource map that depicted critical areas, the authors assigned a cell's value based on the maximum value of a criterion used to evaluate that resource. For example, to evaluate biological conservation, individual criteria could be scored as 2 for critical animal habitat, 3 for critical plant habitat, and 2 for significant vegetation community. The cell's final value for biological

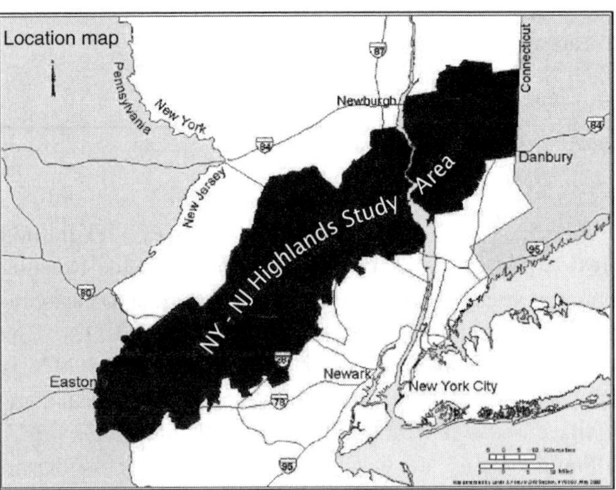

FIGURE 11.6 Location of the Highlands Regional Study in New York and New Jersey.

conservation would be 3. This approach enabled the authors to evaluate resources individually and collectively as well as regional and local patterns.

Like the Highlands Region, Alachua County, FL, has rapidly increased in population. During the past decade, Alachua County's human population grew by 20 percent to more than 218,000 individuals. Also, like the Highlands Region, Alachua County has a number of unique geological and ecological sites. Recognizing the ecological and social importance of conserving or protecting these sites, county planners, commissioners, and environmentalists created a program in 2000 called Alachua County Forever. With voter-approved funds, the program purchases unique properties or their development rights. Each recommended parcel of land is evaluated, scored, and prioritized based on six categories (coarse filters) and 26 criteria (fine filters) (Table 11.6). Each criterion is scored from 1 to 5, with 1 being the least beneficial and 5 being the most beneficial. For the environmental and human categories, scores are averaged and multiplied by a weight of 1.3333. The acquisition and management categories are also averaged and weighted by a factor of 0.6667. By June 2003, 158,669 acres had been identified as significant and over 65,000 acres of land had been purchased.

Landscape assessments often focus on ecological and physical components of an ecosystem to maintain ecological integrity. In working landscapes, social and cultural components also play an important role in defining integrity (LaGro 2001). In both examples, social and cultural attributes were assessed. In the Highlands study, the assessment evaluated two cultural resources: farmland and recreation. The recreation critical element included viewsheds and cultural and historical sites. In the Alachua County Forever assessment, economic and management factors were evaluated.

Obviously, a significant amount of information is needed to assess current and future development propos-

TABLE 11.5
Resources, Criteria, and Rationale Used to Identify Areas for Conservation and Protection in the New York–New Jersey Highlands Region

Resource/Criteria	Rationale
Water	
Aquifer	Provides groundwater for drinking water supply wells
Wellhead protection zone	Immediate source of groundwater for public water supply
Riparian zones including streams with 150-ft buffer	Buffers surface water systems from nonpoint pollution, overland runoff, and soil erosion
Headwater streams	Sources for surface waters, sensitive to pollution
Steep slopes: > 15%	Soil erosion source
Wetlands	Surface waters important to flood and pollution control
Biological Conservation	
Critical animal habitat	Habitat important for wildlife populations, including threatened and endangered species
Critical plant habitat	Habitat important for plant populations, including threatened and endangered species
Significant natural vegetation communities	Intact and rare communities of native vegetation
Recreation and Open Space	
Trails with buffers	Access for humans to experience nature
Scenic viewsheds	Accessible viewpoints to enjoy scenic beauty
Visible ridgetops	Accessible viewpoints to enjoy scenic beauty from valley roadways
Existing parks	Public investment
Historical, cultural, and recreational resource areas with 150-ft buffer	Significant historical or cultural resource
Recreational waters and shoreline	Major recreational areas
Farmland	
Cultivated lands	Active agriculture
Preserved farmlands	Public investment
Forest Resources	
Forest stewardship lands	Active forest management
Contiguous forest tracts	Forest management efficiency

Source: Hatfield et al. (2003).

als within a landscape. Information technology provides access to a variety of databases containing information on flora and fauna distribution, movement of species, and disturbance regimes in a region (see Cooperrider et al. 1999 for data sources). Further information can be gathered from discussions with local residents and other land managers. The assessment cannot be done overnight. It requires time to conduct appropriate assessment, interpret the information, and build political support for the evaluation. However, once the time has been invested, maps can be periodically updated to reflect current landscape structure, evaluate any proposed human activities on the landscape, and reassess patch importance. Without the assessment, evaluations are only guesses with anecdotal information.

11.5 CONCLUSION

Ecological assessment provides the manager with information on characteristics that are needed to maintain the physical, ecological, and social processes required for healthy ecosystems and for delivering ecosystem goods and services (Kaufmann et al. 1994). Landscapes are composed of a mosaic of patch types and ownership types. Land-use decisions are based on ownership. Ecological decisions are based on patch types and the movement of energy, organisms, and materials in the landscapes. Ecosystem management provides the avenue to link ecological and land-use decisions and assesses how development will alter the landscape. Returning to Harris's four questions, the proposed assessment provides a means to identify strategically important patches with respect to the landscape function, patches significantly contributing to biodiversity, patches susceptible to development, and a patch's importance to the overall landscape. So, when a request for site development needs to be evaluated, the manager can provide scientific-based information on potential benefits and costs of the proposed action, and use the information to propose alternatives.

TABLE 11.6
Categories and Criteria Used by the Alachua County (FL) Forever Program to Evaluate Unique Ecological and Geological Sites

Category	Criterion
Protection of water resource	Whether the property has geologic/hydrologic conditions that would easily enable contamination of vulnerable aquifers that have value as drinking water sources
	Whether the property serves an important groundwater recharge function
	Whether the property contains or has direct connections to lakes, creeks, rivers, springs, sinkholes, or wetlands for which conservation of the property will protect or improve surface water quality
	Whether the property serves an important flood management function
Protection of natural communities and landscapes	Whether the property contains a diversity of natural communities
	Whether the natural communities present on the property are rare
	Whether there is ecological quality in the communities present on the property
	Whether the property is functionally connected to other natural communities
	Whether the property is adjacent to properties that are in public ownership or have other environmental protections such as conservation easements
	Whether the property is large enough to contribute substantially to conservation efforts
	Whether the property contains important, Florida-specific geologic features such as caves or springs
	Whether the property is relatively free from internal fragmentation from roads, power lines, and other features that create barriers and edge effects
Protection of plant and animal species	Whether the property serves as documented or potential habitat for rare, threatened, or endangered species or species of special concern
	Whether the property serves as documented or potential habitat for species with a large home range
	Whether the property contains plants or animals that are endemic or near-endemic to Florida or Alachua County
	Whether the property serves as a special wildlife migration or aggregation site for activities such as breeding, roosting, colonial nesting, or over-wintering
	Whether the property offers high vegetation quality and species diversity
	Whether the property has a low incidence of nonnative invasive species
Social and human values	Whether the property offers opportunities for compatible resource-based recreation, if appropriate
	Whether the property contributes to urban green space, provides a municipal defining greenbelt, provides scenic vistas, or has other value from an urban and regional planning perspective
	Average for environmental and human values
Management issues	Whether it will be practical to manage the property to protect its environmental, social, and other values (examples include controlled burning, exotic removal, maintaining hydroperiod, etc.)
	Whether this management can be completed in a cost-effective manner
Economic and acquisition issues	Whether there is potential for purchasing the property with matching funds from municipal, state, federal, or private contributions
	Whether the overall resource value justifies the potential cost of acquisition
	Whether there is imminent threat of losing the environmental, social, or other values of the property through development and/or lack of sufficient legislative protections (this requires analysis of current land use, zoning, owner intent, location)
	Whether there is an opportunity to protect the environmental, social, or other values of the property through an economically attractive less-than-fee mechanism, such as a conservation easement
	Average for acquisition and management values

Source: Alachua County Environmental Protection Department (2003).

REFERENCES

Alachua County Environmental Protection Department, 2003. Land conservation, http://environment.Alachua-county.org/Land_Conservation/. [Date accessed: December 2003.]

Andrén, H., 1994. Effects of habitat fragmentation on birds and mammals in landscapes with different proportions of suitable habitat: a review, *Oikos* 71: 355–366.

Arnold, G.W., 1995. Incorporating landscape pattern into conservation programs, in *Mosaic Landscapes and Ecological Processes,* L. Hansson, L. Fahrig, and G. Merriam, Eds., Chapman & Hall, London, pp. 309–337.

Beier, P. and R.F. Noss, 1998. Do habitat corridors provide connectivity? *Conservation Biology* 12: 1242–1252.

Bormann, F.H. and G.E. Likens, 1979. Catastrophic disturbance and the steady-state in northern hardwood forests, *American Scientist* 67: 660–669.

Botkin, D.B., 1990. *Discordant Harmonies: A New Ecology for the Twenty-First Century,* Oxford University Press, New York.

Christensen, N.L., A.M. Bartuska, J.H. Brown, S. Carpenter, C.D'Antonio, R. Francis, J.F. Franklin, J.A. MacMahon, R.F. Noss, D.J. Parson, C.H. Peterson, M.G. Turner, and R.G. Woodmansee, 1996. The report of the Ecological Society of America Committee on the Scientific Basis for Ecosystem Management, *Ecological Applications* 6: 665–691.

Clark, J.S., 1986. Coastal forest tree populations in a changing environment, southeastern Long Island, New York, *Ecological Monographs* 56: 259–277.

Cohan, J.P. and J.A. Lerner, 2003. *Integrating Land Use Planning and Biodiversity,* Defenders of Wildlife, Washington, DC.

Cooperrider, A.Y., L. Fox, III, L.R. Garrett, and N.T. Hobbs, 1999. Data collection, management, and inventory, in *Ecological Stewardship: A Common Reference for Ecosystem Management,* Vol. III, W.T. Sexton, A.J. Malk, R.C. Szaro, and N. C. Johnson, Eds., Elsevier Science, Oxford, U.K., pp. 603–627.

Covey, S.R., 1989. *The 7 Habits of Highly Effective People: Restoring the Character Ethic,* Simon and Schuster, New York.

Dale, V.H., S. Brown, A. Haeuber, N.T. Hobbs, N. Huntly, R.J. Naiman, W.E. Riebsame, M.G. Turner, and T.J. Valone, 2000. Ecological principles and guidelines for managing the use of land, *Ecological Applications* 10: 639–670.

Dale, V.H., A.E. Lugo, J.A. MacMahon, and S.T.A. Pickett, 1998. Ecosystem management in context of large, infrequent disturbances, *Ecosystems* 1: 546–557.

Denslow, J.S., 1980. Patterns of plant species diversity during succession under different disturbance regimes, *Oecologia* 46: 18–21.

Diaz, N. and D. Apostol, 1992. Forest Landscape Analysis and Design, R6-ECO-TP-043-92, U.S. Department of Agriculture, Forest Service, Seattle.

Duerksen, C.J., D.L. Elliott, N.T. Hobbs, E. Johnson, and J.R. Miller, 1997. Habitat Planning: Where the Wild Things Are, PAS 470/471, American Planning Association, Chicago.

Duever, L.C. and R.F. Noss, 1990. A computerized method of priority ranking for natural areas, in *Ecosystem Management: Rare Species and Significant Habitats,* R. S. Mitchell, C. J. Sheviak, and D. J. Leopold, Eds., New York State Museum, Albany, pp. 22–33.

Farina, A., 2000. *Landscape Ecology in Action,* Kluwer Academic Publishers, Dordrecht, Netherlands.

Foresman, T.W., S.T.A. Pickett, and W.C. Zipperer, 1997. Methods for spatial and temporal land use and land cover assessment for urban ecosystems in the greater Baltimore-Chesapeake Region, *Urban Ecosystems* 1: 201–216.

Forman, R.T.T., 1995. *Land Mosaics,* Cambridge University Press, New York.

Forman, R.T.T. and S.K. Collinge, 1996. The 'spatial solution' to conserving biodiversity in landscapes and regions, in *Conservation of Faunal Diversity in Forested Landscapes,* R.M. DeGraaf and R.I. Miller, Eds., Chapman & Hall, New York, pp. 537–568.

Forman, R.T.T. and S.K. Collinge, 1997. Nature conserved in changing landscapes with and without spatial planning, *Landscape and Urban Planning* 37: 129–135.

Forman, R.T.T. and M. Godron, 1986. *Landscape Ecology,* John Wiley, New York.

Grumbine, R.E., 1994. What is ecosystem management, *Conservation Biology* 8: 27–38.

Harris, L.D., 1984. *The Fragmented Forest,* University of Chicago Press, Chicago.

Hatfield, C.,R. Lathrop, and D. Tulloch, 2003. Land Resources, in New York–New Jersey Highlands Regional Technical Report, M.C. Hoppe, Ed., U.S. Department of Agriculture, Forest Service, Northeastern Area State and Private Forestry, Newtown Square, PA, pp. 45–219.

Holling, C.S., 1992. Cross-scale morphology, geometry and dynamics of ecosystems, *Ecological Monographs* 62: 447–502.

Hunter, M.L., Jr., 1990. *Wildlife, Forests, and Forestry,* Prentice-Hall, Englewood Cliffs, NJ.

Kaufmann, M.R., R.T. Graham, D.A. Boyce, Jr., W.H. Moir, L. Perry, R.T. Reynolds, R.L. Bassett, P. Mehlhop, C. B. Edminster, W.M. Block, and P.S. Corn, 1994. An Ecological Basis for Ecosystem Management, General Technical Report RM-246, U.S. Department of Agriculture, Forest Service, Rocky Mountain Forest and Range Experiment Station, Fort Collins, CO.

LaGro, J.A., Jr., 2001. *Site Analyses: Linking Program and Concept in Land Planning and Design,* John Wiley and Sons, New York.

Levenson, J.B., 1981. Woodlots as biogeographic islands in Southeastern Wisconsin, in *Forest Island Dynamics in Man-Dominated Landscapes,* R.L. Burgess and D.M. Sharpe, Eds., Springer-Verlag, New York, pp. 1340.

Likens, G.E., 1992. *The Ecosystem Approach: Its Use and Abuse,* Ecology Institute, Olendor/Luhe, Germany.

Lovett, G.M., M.M. Traynor, R.V. Pouyat, W. Zhu, and J.W. Baxter, 2000. Atmospheric deposition to oak forests along an urban-rural gradient, *Environmental Science and Technology* 34: 4294–4300.

Macie, E.A. and L.A. Hermansen, Eds., 2002. Human Influences on Forest Ecosystems: The Southern Wildland–Urban Interface Assessment, General Technical Report SRS-55, U.S. Department of Agriculture, Forest Service, Southern Research Station, Asheville, NC.

McCormick, F.J., 1998. Principles of ecosystem management and sustainable development, in *Ecosystem Management for Sustainability,* J. D. Peine, Ed., Lewis Publishers, New York, pp. 3–22.

McDonnell, M.J., S.T.A. Pickett, P. Groffman, P. Bohlen, R.V. Pouyat, W.C. Zipperer, R.W. Parmelee, and K. Medley, 1997. Ecosystem processes along an urban-to-rural gradient, *Urban Ecosystems* 1: 21–36.

McGarigal, K. and B.J. Marks, 1995. FRAGSTATS. Spatial Analysis Program for Quantifying Landscape Structure, General Technical Report PNW-GTR-351, U.S. Department of Agriculture, Forest Service, Portland, OR.

Monroe, M.C., 2002. Fire, in Human Influences on Forest Ecosystems: The Southern Wildland–Urban Interface Assessment, E. A. Macie and L.A. Hermansen, Eds., General Technical Report SRS-55, U.S. Department of Agriculture, Forest Service, Southern Research Station, Asheville, NC, pp. 133–150.

Noss, R.F. and L.D. Harris, 1986. Nodes, networks, and MUM's: preserving diversity at all scales, *Environmental Management* 10: 299–309.

Overbay, J.C., Ed., 1992. Ecosystem management, Speech delivered at the National Workshop on Taking an Ecological Approach to Management, Salt Lake City, UT, U.S. Department of Agriculture, Forest Service, Washington, DC.

Phelps, M.G. and M.C. Hoppe, 2002. New York–New Jersey Highlands Regional Study, NA-TP-02-03, U.S. Department of Agriculture, Forest Service, Northeastern Area State and Private Forestry, Newtown Square, PA.

Pickett, S.T.A., 1998. Natural processes, in *Status and Trends of the Nation's Biological Resources,* Vol. 1, M.J. Mac, P.A. Opler, C.E.P. Haecker, and P.D. Doran, Eds., U.S. Geological Survey, Washington, DC., pp. 11–35.

Pickett, S.T.A., M.L. Cadenasso, J.M. Grove, C.H. Nilon, P.V. Pouyat, W.C. Zipperer, and R. Costanza, 2001. Urban ecological systems: linking terrestrial ecological, physical, and socioeconomic components of metropolitan areas, *Annual Review of Ecology and Systematics* 32: 127–157.

Pickett, S.T.A., W.R. Burch, Jr., S.E. Dalton, T.W. Foresman, J.M. Grove, and R.A. Rowntree, 1997. A conceptual framework for the study of human ecosystems in urban areas, *Urban Ecosystems* 1: 185–201.

Pickett, S.T.A. and J.N. Thompson, 1978. Patch dynamics and the design of nature reserves, *Biological Conservation* 13: 27–37.

Pickett, S.T.A. and P.S. White, Eds., 1985. *The Ecology of Natural Disturbance and Patch Dynamics,* Academic Press, New York.

Ranney, J.W., M.C. Bruner, and J.B. Levenson, 1981. The importance of edge in the structure and dynamics of forest islands, in *Forest Island Dynamics in Man-dominated Landscapes,* R.L. Burgess and D.M. Sharpe, Eds., Springer-Verlag, New York, pp. 67–96.

Reichard, S.H. and P.S. White, 2001. Horticulture as a pathway of invasive plant introductions in the U.S., *Bioscience* 51: 103–113.

Saunders, D.A., R.J. Hobbs, and C.R. Margules, 1991. Biological consequences of ecosystem fragmentation: a review, *Conservation Biology* 5: 18–32.

Simberloff, D., 1997. Biogeographic approaches and the new conservation biology, in *The Ecological Basis of Conservation,* S.T. A. Pickett, R.S. Ostfeld, M. Shachak, and G.E. Likens, Eds., Chapman & Hall, New York, pp. 274–284.

Simberloff, D. and J. Cox, 1987. Consequences and costs of conservation corridors, *Conservation Biology* 1: 63–71.

Smallidge, P. J. and D.J. Leopold, 1997. Vegetation management for the maintenance and conservation of butterfly habitats in temperate human-dominated landscapes, *Landscape and Urban Planning* 38: 259–280.

Stuart, J.D., 1998. Effect of fire suppression on ecosystems and diversity, in *Status and Trends of the Nation's Biological Resources,* Vol. 1, J.M. Mac, P.A. Opler, C.E.P. Haecker, and P.D. Doran, Eds., U.S. Geological Survey, Washington, pp. 45–47.

Temple, S.A., 1986. Predicting impacts of habitat fragmentation on forest birds: a comparison of two models, in *Wildlife 2000: Modeling Habitat Relationships of Terrestrial Vertebrates,* J. Verner, M.L. Morrison, and C.J. Ralph, Eds., University of Wisconsin Press, Madison, pp. 301–304.

Turner, M.G., 1989. Landscape ecology: the effect of pattern on process, *Annual Review of Ecology and Systematics* 20: 171–197.

Turner, M.G., R.H. Gardner, and R.V. O'Neill, 2001. *Landscape Ecology in Theory and Practice,* Springer-Verlag, New York.

van Diver, B.B., 1992. *Roadside Geology of New York,* Mountain Press Publishing Company, Missoula, MT.

White, C.S. and M.J. McDonnell, 1988. Nitrogen cycling processes and soil characteristics in an urban versus rural forest, *Biogeochemistry* 5: 243–262.

White, D., P.G. Minotti, M.J. Barczak, J.C. Sifneos, K.E. Freemark, C.F. Satelmann, A.R. Kiester, and E.M. Preston, 1997. Assessing the risks to biodiversity from future landscape change, *Conservation Biology* 11: 349–360.

Wilcove, D.S., C.H. McLellan, and A.P. Dobson, 1986. Habitat fragmentation in the temperate zone, in *Conservation Biology: The Science of Scarcity and Diversity,* M.E. Soulé, Ed., Sinauer Association, Sunderland, MA., pp. 237–256.

Yaffee, S.L., A.F. Phillips, I.C. Frentz, P.W. Hardy, S.M. Maleki, and B.E. Thorpe, 1996. *Ecosystem Management in the U.S.: Assessment of Current Experience*, Island Press, Washington, DC.

Zipperer, W.C., 1993. Deforestation patterns and their effects on forest patches, *Landscape Ecology* 8: 177–184.

Zipperer, W.C., R.L. Burgess, and R.D. Nyland, 1990. Patterns of deforestation and reforestation in different landscape types in central New York, *Forest Ecology and Management* 36: 103–117.

12 Managing Hydrological Impacts of Urbanization

Larry V. Korhnak and Susan W. Vince
School of Forest Resources and Conservation, University of Florida

CONTENTS

What we do on the land is mirrored in the water.

—Anonymous

12.1 INTRODUCTION

As the earth's water cycles through the land, it supports, defines, and integrates the landscape's physical and biological parameters. In turn, the land flavors the water's chemistry and determines the paths and timing of the water's return to the atmosphere. Globally, one half of the precipitation falling on land is cycled through plants (Jackson et al. 2000).

Forested ecosystems serve as a critical linkage in the water cycle, but they have been and continue to be intensely altered by humans. Since the arrival of European settlers in the 1600s, forestland cover in the northeastern U.S. has been reduced by 70 percent (Joyce et al. 2001). After the 1920s, forest cover increased slowly with the abandonment of cropland and re-growth of cutover areas. However, forest cover is now on the

1-56670-602-5/05/$0.00+$1.50
© 2005 by CRC Press LLC

decline in many regions, largely due to sprawling urban growth. Conversion of forestland to urban uses in the U.S. was estimated to be 10.3 million acres from 1982 to 1997 (USDA Natural Resources Conservation Service 2000).

Managing water in the wildland–urban interface requires an understanding of the forest water cycle and an awareness of how human activities change this cycle. Water is an essential community resource that must be managed as carefully as shelter, transportation, energy, and food. However, the nature of water makes communal management especially challenging. The location, timing, and amount of water falling from the sky cannot be controlled. Water falls on and moves across all manner of political, economic, and ethnic boundaries, making centralized management difficult. Often the most challenging step in resource management is finding common ground, but what more common ground can we have than knowing that water makes life possible?

In this chapter, we describe the forest water cycle and outline the impacts of urbanization on hydrology, water quality, and aquatic and wetland ecosystems. We then offer some examples of promising and innovative approaches to prevent and mitigate these changes. Restoration of hydrological functions in already highly urbanized areas is difficult and often economically unfeasible, but opportunities exist in the wildland–urban interface to more wisely manage our water resources.

12.2 FOREST WATER CYCLE

Forested watersheds supply about two thirds of the U.S. runoff and the National Forests alone provide drinking water for more than 3400 communities in 33 states (Ryan and Glasser 2000). Forests are increasingly valued and protected for their clean water, flood protection, and other ecological functions. Many of these values are coupled to the manner in which forests cycle water.

12.2.1 Evapotranspiration

On average, two thirds of the precipitation entering U.S. forests is returned to the atmosphere through the processes of interception, transpiration, and evaporation from the soil and surface waters. Collectively, these processes are termed evapotranspiration (ET).

The landscape is bombarded with the energy of quadrillions of raindrops, but the protective cover of forests intercepts both the rain and its kinetic energy. Intercepted rainfall is held by leaf surface tension and a portion is evaporated back to the atmosphere. How much of the water is returned depends on climatic factors and also the type of vegetation; conifers have interception losses ranging from about 20 to 40 percent of total precipitation and hardwoods from about 10 to 20 percent (Zinke 1967). When leaf storage capacity is filled, drops grow until their weight exceeds surface tension and they fall to the next lower leaf or the ground.

Evaporation from saturated surface soils, like water evaporation from rivers, lakes, and ponds, is limited mainly by the supply of energy and the availability of drier air. However, as the soil dries, movement of water vapor up through the soil is impeded and evaporation is significantly reduced. Tree roots can bridge the dry soil barrier and reach deeper sources of water than other types of vegetation. Soil water is absorbed through the roots and transmitted through the xylem to replace water vaporized in the leaves. This physiological process, called transpiration, is responsible for most of the water evaporated from the forested landscape. For example, Hewlett (1982) reported that transpiration accounted for 78 and 63 percent of the total ET for mature hardwood and white pine forests, respectively. Forests have greater transpiration rates than other vegetation types because their roots are deeper, their leaves are positioned higher and so allow for greater water vapor loss, and leaf surface areas may be up to 10 times the corresponding ground area.

12.2.2 Infiltration and Storage

Forest soils are protected by a layer of live and dead vegetation that inhibits erosion, preserves the porous structure of the soil, and detains surface water, thereby increasing infiltration of water into the soil. Additionally, forest soils teem with roots, nematodes, earthworms, insects, and moles, whose activities promote a porous matrix. Water storage by vegetation and litter can be important for small, frequent storms but is relatively minor for large precipitation events. In contrast, soil storage capacity can be very high. For example, a 1-m thickness of typical Florida flatwoods soil can hold about 60 cm of water or about 45 percent of average annual precipitation. Between storms, storage capacity is renewed by ET.

12.2.3 Storm Flow

Forest soils generally have infiltration and surface storage capacities that exceed the intensity of most rainfall events, and therefore overland storm flow is infrequent. When it occurs, most surface runoff is generated where rain falls on saturated soils or impervious surfaces such as rock outcroppings. Saturated soils in and near stream channels and wetlands produce the first storm flow. As the storm continues, rain infiltrates and percolates to fill soil storage, and the saturated area (termed the variable source area) expands up the stream banks and into intermittent and ephemeral channels.

Most of the storm flow from forests travels at least part of its journey by the slower subsurface route. Storage and the tortuous pathways of water through the leaves, litter, and soil delay precipitation from reaching the stream.

Therefore, stormwater inputs will be spread over time and stream discharge rates will increase gradually, resulting in a lower peak discharge and less chance of flooding than in nonforested watersheds (Figure 12.1).

After a storm ceases, water slowly continues to make its way downslope, thus sustaining groundwater inputs. In forested watersheds of the eastern U.S., about 70 percent of total stream flow occurs between storms as a slow release of groundwater (base flow) (Hewlett 1982). Not only does groundwater keep streams from drying up, it is the source of 40 percent of the U.S. public supply (Council on Environmental Quality 1996).

12.3 EFFECTS OF URBANIZATION ON THE FOREST WATER CYCLE

12.3.1 HYDROLOGY

The leaves, roots, and soil of forests form an enormous net that captures precipitation and releases most of it back to the atmosphere. Removal of forest vegetation, reductions in soil infiltration, and removal of watershed storage significantly alter the paths and speed of the water cycle.

When forests are converted to other types of vegetation, increased runoff volume is consistently observed (Sahin and Hall 1996). Grasses and other vegetative covers frequently have lower ET rates and so more of the precipitation is available to produce runoff. Deforestation contributes to the hydrological impacts of development in the wildland–urban interface. For example, in rural King County, WA, forest clearing by small landowners for lawns and pasture was observed to have significant effects on watershed hydrology (Booth et al. 2002).

The hydrological effects of deforestation are particularly severe and difficult to reverse when coupled with loss of water infiltration into the ground. Impervious surfaces are extraordinarily problematic: as their coverage increases, stream quality declines. In many cases, physical and biological degradation of streams has been observed where impervious surfaces cover 10 percent or less of the watershed (see the review by Center for Watershed Protection 2003). Until the 1980s, the objective of stormwater managers was to collect and transport runoff away from developed areas as quickly as possible. Impervious areas, such as roofs, gutters, curbs, and storm sewers, were connected to speed the exit of runoff and bypass opportunities for infiltration or detention.

Additional decreases in infiltration result from soil compaction, which accompanies urban activities (reviewed by Schueler 2000a). Pit et al. (1999) observed that infiltration rates of urban soils were reduced between 5 and 10 times. Less infiltration results in more runoff, and Wignosta et al. (1994) found that compacted soils produced from 40 to 60 percent of the runoff from a developed catchment. Other urban conditions that reduce the penetration of water include loss of the soil's protective canopy, removal of leaf litter, reduction in soil organic matter and fauna, and mechanical disturbances that alter soil structure.

Blocked and impeded infiltration prevents precipitation from being stored in the soil and then slowly released. Water running over smooth impervious surfaces can reach a stream in minutes as opposed to the hours, days, and even months that infiltrated rain may take. Therefore, as a watershed becomes increasingly urbanized, the shape and magnitude of the hydrograph (a graph of discharge rate over time) change (Figure 12.2). Peak discharge rate is amplified and it occurs more quickly after the precipitation event. The total volume of runoff increases, in some cases by up to 16 times that from natural areas (Schueler 1995). The transformation of the forest water cycle is

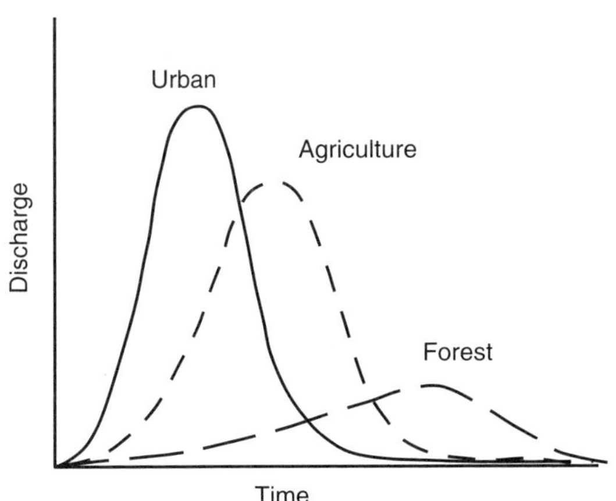

FIGURE 12.1 Generalized hydrographs of different land uses. Landscapes with reduced infiltration and storage have more runoff with higher peak flows occurring more quickly after a precipitation event. (Adapted from Beaulac and Reckhow 1982.)

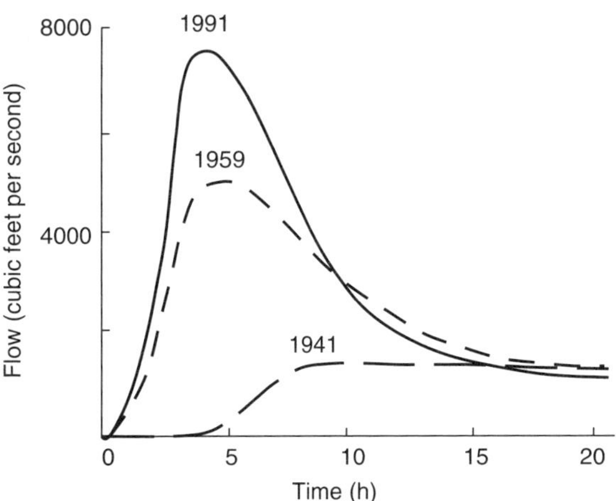

FIGURE 12.2 Hydrographs of Brays Bayou (Houston, TX, metropolitan area) as land use changed. Note how the total flow and peak flow increased as the watershed was developed. (Adapted from Bedient and Huber 1992.)

profound: as impervious cover increases, surface water runoff greatly increases, and groundwater recharge and ET decrease (Figure 12.3).

Increased runoff from urbanized landscapes contributes to flooding losses in the U.S., estimated to be $5 billion annually in the 1990s (Pielke and Downton 2000). Urbanization causes flooding to become more frequent, especially for small- and medium-sized precipitation events. For example, Hollis (1975) found that flooding that would normally happen once a year would occur 10 times a year after urbanization, and flooding events that happened once every 10 years would occur on average every 5 years.

The reduction in infiltration can also reduce the groundwater available for dry weather flow (base flow), as noted in many urbanizing watersheds, including Atlanta (Ferguson and Suckling 1990) and the Maryland Piedmont (Klein 1979). Urbanization of Long Island, N Y, from 1940 to1970 reduced groundwater contribution to a stream from 86 to 22 percent (Simmons and Reynolds 1982). Hunt and Steuer (2001), using the computer model MODFLOW, predicted that future development in an urbanizing Wisconsin watershed would cause a 57 percent decline in groundwater recharge that would dry up the stream between storms. The absence of water can be even more disastrous than too much water, especially for aquatic life.

12.3.2 Water Quality

Forest ecosystems require minimal inputs of chemicals, even when forest products are a goal. On the other hand, to sustain agricultural and urban systems, nutrients, pesticides, herbicides, metals, and hydrocarbon fuels are concentrated on the landscape. Irrigation, rainfall, and snowmelt, aided by urban drainage systems, wash these pollutants into lakes, rivers, coastal waters, and groundwater. This diffusely distributed pollution is called nonpoint source, and it is now the largest water quality problem in the U.S. (U.S. EPA 2000a). Of the total pollution load to our nation's waters, nonpoint sources contribute 90 percent of nitrogen, 90 percent of the fecal coliform bacteria, 70 percent of the biochemical oxygen demand, 70 percent of the oil, 70 percent of the zinc, 66 percent of the phosphorus, 57 percent of the lead, and 50 percent of the chromium (Thompson et al. 1989).

According to the 1998 U.S. National Water Quality Inventory, urban nonpoint sources are the leading cause of water quality impairment for coastal waters and major contributors to impairment of lakes and rivers (U.S. EPA 2000b). Over 1.2 billion pounds of pesticides are applied in the U.S. each year (U.S. EPA 2002), and home and garden sales account for about 20 percent of total U.S.

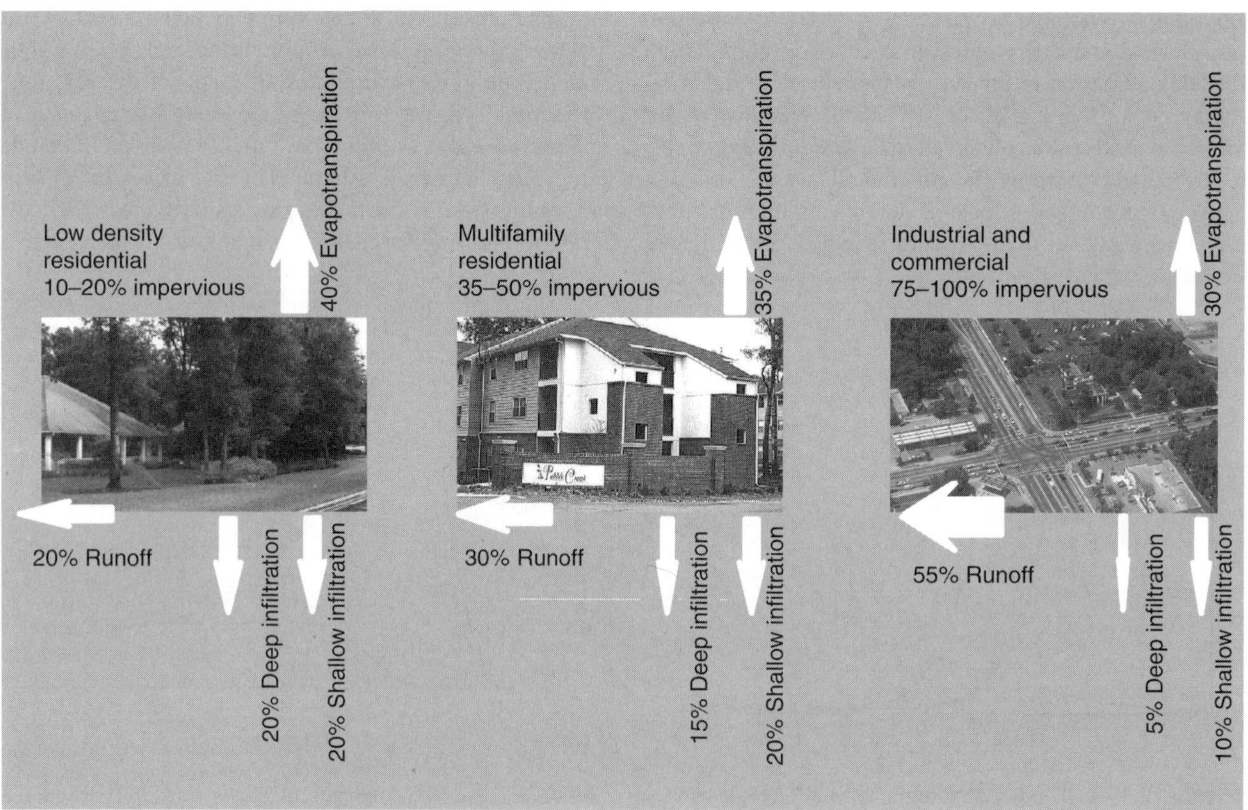

FIGURE 12.3 As impervious area increases, ET and groundwater recharge decrease and surface runoff increases. (Adapted from U.S. EPA 1993.)

sales. So it is not surprising that the U.S. Geological Survey's National Water Quality Assessment (NAWQA) program found at least one pesticide in most shallow wells and in almost every water and fish sample collected from streams in urban and agricultural watersheds (USGS 1999). All urban stream samples had concentrations of insecticides that exceeded at least one guideline established to protect aquatic life. Sediment from urban streams was frequently polluted with persistent organic toxins and heavy metals. These compounds are concentrated in fish tissue, and one or more organochlorine compounds were detected in 97 percent of fish samples collected at urban sites (USGS 2001). In contrast, pesticides are seldom found in stream water, sediments, or fish tissue in forested watersheds (Anderson et al. 1996; Munn and Gruber 1997).

Exports of the nutrients nitrogen and phosphorus and their concentrations in runoff are also related to land use (Table 12.1). The U.S. is one of the world's leading consumers of fertilizers. Nitrogen and phosphorus concentrations in agricultural and urban surface waters frequently exceed levels that cause excessive algal growth, reduce recreational value, and decrease biodiversity. Our love of the green lawn is particularly problematic. Lawns can comprise the largest surface area in urban watersheds and can be the largest sources of phosphorus in runoff (Waschbusch et al. 1999).

Although many pollutants are directly applied to urbanized landscapes, a significant portion is delivered in rainfall. The Clean Air Acts of 1970 and 1990 have reduced sulfur emissions by 38 percent, but nitrogen emissions continue to rise (Driscoll et al. 2001). Twenty to thirty percent of the nitrogen entering the Chesapeake Bay is derived from atmospheric transport from an air shed that expands far beyond its watershed boundaries (Linker et al. 1996).

Point sources of pollution, such as wastewater treatment plants, also significantly degrade water quality in urbanized watersheds. Municipal point sources are the leading cause of impairment of U.S. estuaries (U.S. EPA 2000b). Each person who inhabits the forested landscape increases the demand for water resources and increases the pollutant load. Public water use in the U.S. is about 700 l per day per person (Solley et al. 1998), but only about a liter is actually consumed as drinking water. The majority is used for waste flushing and washing.

TABLE 12.1
Approximate Export of Phosphorus and Nitrogen from Lands with Different Uses

Land Use	Phosphorus Export (kg/ha/year)	Nitrogen Export (kg/ha/year)
Forest	0.2	2.5
Agriculture (row crops)	2.8	9.4
Urban	1.2	5.6

Source: Beaulac and Reckhow (1982).

Fewer than 20 percent of U.S. major cities had any wastewater treatment in 1925 (Linsley and Franzini 1972). However, the passage of the Clean Water Act in 1972 and more than $61 billion from the Municipal Construction Grant Program (Freeman 1990) spurred the construction of approximately 17,000 wastewater treatment plants. Removal efficiencies vary among systems, but even advanced treatment plants discharge an effluent with nutrient levels that can disrupt ecological relationships in receiving waters. Furthermore, as the population grows, improvements in treatment efficiency will be offset by the increasing amount of wastewater that needs to be treated. For example, loading of biochemical oxygen demand is projected to exceed the historical maximum, experienced in 1968, by the year 2025 (U.S. EPA 2000c).

Sparse populations and lower land values often make the wildland–urban interface a target for the disposal of biosolids (the solid materials separated from wastewater) and municipal solid waste. Over 5.3 million metric tons per year of dry weight biosolids are generated annually in the U.S. (National Academy of Sciences 1996). Because biosolids contain beneficial nutrients and organic matter, and other disposal options are limited, they are frequently applied to rural farms and forests as soil amendments. However, biosolids also contain pathogens, high levels of metals, and other pollutants. Urban garbage is another pollutant that is frequently dumped in rural areas. The U.S. has an estimated 55,000 municipal solid waste landfills, of which approximately 75 percent are polluting groundwater (Jones-Lee and Lee 1993). Liners and leachate recovery systems improve the safety of new landfills, but leaks are possible and eventually liners will fail (U.S. EPA 1988).

Many rural communities lack wastewater treatment plants and rely on decentralized systems (usually septic tanks) for the treatment of human wastes. Approximately 30 percent of U.S. households use septic systems, producing an estimated 5.5 billion gallons of effluent annually (U.S. EPA 1999). High water tables, thin soils, unsuitable geology, poor maintenance, and excessive density of septic tanks can result in septic effluent harming both humans and the natural environment. Septic tank effluent has concentrations of total nitrogen ranging from 25 to 60 mg/l and total phosphorus concentrations ranging from 10 to 30 mg/l (Tchobanoglous and Burton 1991). Much of the effluent nitrogen is converted to highly mobile nitrate in the septic drain field. Primary production of coastal waters is often nitrogen limited, and septic systems have been identified as the largest contributor of nitrogen loading to some estuarine waters (e.g., Valiela et al. 1997). Septic tank effluent also contains pathogenic bacteria and viruses, which can contaminate rural water supplies (USGS 1992). On rural home sites, the well and septic drain field may be less than 100 ft apart, and frequently one aquifer serves the dual role of providing drinking water and receiving wastewater.

Wastewater disposal in the wildland–urban interface is not only directly responsible for environmental degradation, but in many cases it is also the *de facto* controller of land use. That is, because building permits in the interface often depend on septic tank permits. When Wisconsin's legislators revised their private on-site wastewater treatment systems rule (COMM 83), allowing the use of new types of on-site systems in shallower soils, they increased the amount of land that could be developed from 15.5 to 22.4 million acres (Jaskula and Hohn 2002). Although the rule was not intended as land-use regulation, it had that consequence.

12.3.3 Aquatic and Wetland Ecosystems

12.3.3.1 Physical Alterations of Stream Habitat

Urbanization of forested watersheds results in the physical and biological degradation of receiving water bodies, especially streams. Although some physical alterations of streams are directly caused, many are indirectly due to hydrological changes in the watershed. Reductions in infiltration and storage far from the stream can dramatically energize the forces that shape stream channels. Urbanization increases the frequency of flood events that are primarily responsible for determining channel geomorphology (Dunne and Leopold 1978).

A stream is in dynamic equilibrium when it can transport sediment delivered by the watershed and, at the same time, maintain its physical characteristics. Urbanization disturbs stream equilibrium by increasing sediment inputs, increasing flood frequency and flow volumes, and changing stream channel gradients. In the early stages of watershed urbanization, construction and the removal of stream bank vegetation deliver sediment to a stream channel faster than the stream can transport it. The channel fills (aggradation), burying important bottom habitat. The filled channel can no longer carry as much water and may overtop the banks more often. As more impervious area is added to the watershed, flow volumes increase and stream sediment is washed away (degradation), enlarging and downcutting the channel and often isolating it from the floodplain. In instances of severe incision, unstable slopes result in bank collapse, which further increases sediment transport downstream (Figure 12.4).

Construction, drainage, and transportation activities are frequently responsible for direct stream impacts. Construction removes protective vegetative cover and disturbs soil. In the U.S., about one million construction projects begin each year, disturbing an estimated five million acres of land (U.S. Census Bureau 2001). The resulting erosion can produce 10–100 times more sediment than natural areas (up to 50,000 t/km^2/year) (Novotny and Olem 1994). Sediment smothers stream bottoms, depletes hydrological storage, and is frequently a carrier of other pollutants.

Stream channels are often straightened, widened, deepened, lined with hard materials, cleared of large woody debris, cleared of forest cover, and mined, dammed, and culverted. The magnitude of these changes is enormous. The length of channelized rivers in the U.S. could reach the moon, there are over 75,000 dams, and two thirds of the riparian forest has been converted to nonforest land uses (Council on Environmental Quality 1996). Stream straightening and in-stream structures such as culverts can have a permanent effect on stream slope. For example, undersized and improperly installed culverts commonly slow water and cause aggradation on the upstream side; the increased slope then results in erosion on the downstream side. In severe cases, the destabilizing grade change can migrate considerable distances upstream and downstream (Riley 1998). Large woody debris have

FIGURE 12.4 Comparison of a stream channel in a relatively undisturbed watershed (left) with a channel in a watershed with over 50 percent impervious area (right). Increased runoff increases the energy for sediment transport, eroding the stream channel and disconnecting it from its floodplain. (Photos by Larry Korhnak.)

important biological and hydrological functions in a stream ecosystem, but they are often scarce in heavily urbanized areas. They are washed out of streams by increased flooding, left stranded above the channel banks due to incision, and are not replenished due to forest clearing on the stream banks (Booth 1990; Finkenbine et al. 2000). Removal of the riparian forest canopy can also cause urban streams to be colder in winter and warmer in summer than forested streams (Paul and Meyer 2001).

12.3.3.2 Biological Changes in Streams

Streams in urbanized landscapes are likely to differ considerably from those in forested settings due to altered flow regimes, enlarged channels, greater levels of turbidity, nutrients and toxins, lower levels of oxygen, and fewer sources of organic fuel for aquatic food webs. These modifications eliminate critical habitat for some organisms and generally impair the ability of urban streams to support aquatic life.

Numerous studies have documented the biological degradation of urban streams (see the review by Paul and Meyer 2001). The response of many types of aquatic organisms to urbanization is similar. Species diversity of algae, macrophytes, invertebrates, and fish decreases as watershed development increases. Both the number of species and how evenly the species are represented may decline. Bottom-living (benthic) invertebrates in particular have been well studied. Some, such as mayflies and stoneflies, are relatively intolerant of pollution and other disturbances and so their relative abundance decreases with increasing urbanization. Local extirpation of some types is not uncommon. At the same time, more tolerant invertebrates like midges increase in dominance.

Knowledge gained from these studies and those by Karr (1981) on fish communities has led to the creation of multimetric biologic criteria or indices of biological integrity (IBI) for diagnosing the health of aquatic ecosystems. They provide indications of cumulative impacts of multiple stressors with both short- and long-term effects and are particularly useful for gauging the impacts of watershed development. For example, IBI scores measured at 1000 stream sites across Maryland indicated that one half of the streams were in poor or very poor condition (Boward et al. 1999). Biological integrity decreased with increasing urban land use, and nearly all sites with greater than 30 percent of the catchment in urban land use (approximately 15 percent impervious area) had poor to very poor IBI scores. Numerous other IBI studies have concluded that invertebrate as well as fish diversity progressively decline with increasing impervious area (reviewed by Center for Watershed Protection 2003).

12.3.3.3 Wetland Loss and Degradation

Wetlands are intimately connected with surface and groundwater flow systems in the landscape. Their structure and functions, for example, how they process nutrients, are largely determined by hydrology: how deeply and how frequently they are flooded, the duration of flooding, and the source and rate of water flow. Accordingly, hydrological changes in urbanizing watersheds result in wetland alterations (see the review by Ehrenfeld 2000). Species richness of native plants often decreases while the number of weedy, nonnative plant species increases (e.g., Ehrenfeld and Schneider 1991). Wetland functions may also be disrupted. In the Puget Sound lowlands of Washington, wetlands in urbanized watersheds support fewer species of amphibians than do wetlands in less developed watersheds (Richter and Azous 2001). Many wetlands filter and store nutrients and pollutants, enhancing the quality of water passing through. However, upland development has modified nitrogen cycling processes in some wetlands, such as Atlantic white cedar swamps, potentially causing them to be sources rather than sinks of nitrogen (Zhu and Ehrenfeld 1999).

Urbanization also causes direct changes in wetlands. Despite increased regulation and a federal policy of "no net loss," wetlands continue to disappear and urbanization is often implicated. Between 1982 and 1992, 40 percent of isolated wetlands in Portland, OR, were lost; of these, 46 percent were replaced by developments (Holland et al. 1995). Besides outright destruction, wetlands in urbanizing areas are commonly altered by ditching, diking, or draining.

The landscape consequences of direct and indirect effects of urbanization are a significant reduction in the extent of wetlands, selective loss of some wetland types, and degradation of those remaining (Bedford 1999). In Portland, seasonally flooded wetlands were lost at a higher rate than wetlands with more permanent water (Holland et al. 1995), and the residual wetlands were dominated by introduced plant species, especially aggressive invaders (Magee et al. 1995). The loss in area and variety of wetlands in an urbanized watershed is likely to result in the diminishment of the hydrological and ecological services we expect, such as water quality protection, flood attenuation and storage, and provision of wildlife habitat.

12.3.4 Water Supply

The forest water cycle is also altered by withdrawals of water for human needs, which frequently occur at the expense of ecological needs. Competition between water users and uses is certain to intensify as both the human population and water use per person increase and as the ability of urbanizing watersheds to recharge underlying groundwater decreases. Total water withdrawals in the U.S. are more than 1.5 billion cubic meters per day (Solley et al. 1998).

Almost 40 percent of those withdrawals are for irrigation, and most of that water is lost to evaporation. Farming is an important land use in many interface regions, and irrigation issues can be critical, especially in the western states. For example, irrigation demands on the Klamath River (Oregon–California) were determined to be detrimental to several fish protected by the Endangered Species Act. In 2000, the Bureau of Reclamation was ordered to reduce irrigation flows by 90 percent, leaving 1200 farmers without water (Schoch 2002). Protesting farmers obtained a presidential order to reopen the irrigation gates, which resulted in more than 12,000 salmon deaths.

Groundwater resources in rural areas are particularly important, supplying over 85 percent of local drinking water needs. However, these resources are also coveted by large cities, and both natural and human rural communities can have their life-supporting water sucked out from beneath them. For example, groundwater withdrawals by the city of Green Bay, WI, have lowered regional water tables by as much as 131 m and threaten to dry up wells in rural communities (Axness et al. 2002). Green Bay and its suburbs are working to solve their water crisis but have not sought input from the rural communities that are affected.

Well fields are often located outside cities to prevent contamination. Heavy pumping lowers groundwater levels, and surface water can also be affected depending on the degree of its hydrological connection with underlying aquifers. Springs, rivers, lakes, and wetlands can dry up when poorly confined aquifers are pumped. This is exactly what happened in the Tampa, FL, area, where poorly confined aquifers were under a pumping demand of about 1 million cubic meters per day (Southwest Florida Water Management District 1996). The ecological costs were so severe that Tampa has turned to more expensive desalination to supply its water needs.

12.4 WAYS TO PREVENT AND MITIGATE HYDROLOGICAL IMPACTS OF URBANIZATION

Urbanization of forested landscapes alters the pathways and rates of the water cycle and threatens the integrity of aquatic ecosystems. In densely urbanized areas, these impacts are difficult and very costly to offset, and restoration of predevelopment hydrology may be impossible. More opportunities exist at the wildland–urban interface to avoid and mitigate the hydrological changes commonly associated with development, particularly through proactive planning and design.

Here we will briefly describe the concepts and key elements of watershed planning, an approach essential to preventing and managing the hydrological consequences of land development. Then we will present three general strategies for managing water at the wildland–urban interface: (1) protecting forestlands and wetlands that provide critical hydrological functions; (2) limiting impacts of new development by reducing impervious surfaces and controlling sources of pollutants; and (3) mitigating unavoidable consequences of urbanization. Some examples and applications of each are given to illustrate the variety of new techniques and approaches. However, a feature common to all is the aim to move from "end-of-the-pipe" fixes to solutions that treat the underlying causes of the hydrological changes.

12.4.1 WATERSHED PLANNING AND MANAGEMENT

A watershed is an area of land that drains water, sediment, and dissolved materials to a common water body. Watersheds range in size from a few to several million km^2, from small, local drainages (subwatersheds) to large river basins. The hydrologically defined limits of a watershed rarely correspond with political, jurisdictional, and land ownership boundaries. During the past two decades, planners and resource managers have come to recognize that the watershed is the most appropriate geographic unit on which to focus efforts to protect our water resources.

A watershed-based planning approach to solving water issues at the wildland–urban interface is needed for many reasons. Although water resources in the interface sustain local and distant human communities as well as natural ecosystems, they are often over-appropriated and poorly managed. Realistic assessments are needed of the water resources and their capacity to sustain both off-stream and in-stream uses. Watershed-wide planning and management are required to match the demands of the various users to the available resources. The threats to water supplies and water bodies are often the cumulative impact of countless assaults, some from discrete sources but many from dispersed, nonpoint sources; some threats are short term while others are sustained. A watershed-based approach can provide the necessary framework for assessing, integrating, and managing these complex influences.

The aim of the watershed planning process is to produce a blueprint — a long-term strategy — for protecting water resources and human health. To do this, it must integrate ecological, social, and economic considerations into decision making. Planners and other participants in the process must consider all the direct and indirect factors that affect water quantity and quality, set appropriate management goals, and decide on the suite of actions and instruments, from land-use zoning to wastewater management, to best achieve these goals. Ideally, the watershed approach results in the implementation and continued use of measures that collectively safeguard its aquatic resources.

Since the 1980s, many watershed planning efforts have been initiated throughout the U.S. Some have been successful, others not. Reports by the EPA (e.g., U.S. EPA 2001), scientific committees (National Research Council 1999), and nongovernmental organizations (Schueler and Holland 2000) have highlighted the successes and shortcomings of these attempts and made recommendations for improvement. Key elements of an effective watershed planning and management program were identified:

- *Early involvement of all stakeholders, public and private.* Any agency, organization, or individual that is affected by the decisions made in the watershed planning process is a stakeholder. Involvement of all stakeholders is needed to improve cooperation and coordination across political, institutional, and cultural boundaries and to obtain support for designing and employing appropriate management strategies.
- *Increased public awareness and education.* Residents need to know about their watershed, how their daily activities can adversely affect water resources, and what they can do to help protect aquatic resources. Public support is necessary for developing feasible management goals and for implementing actions.
- *Watershed assessment.* Characterization of the watershed, including physical attributes, present and future land use, and linkages among land and water components, is necessary to identify the causes and sources of water resource problems. The information collected through monitoring and research can also be used to help prioritize management goals, to design management strategies, and to develop performance measures with which to evaluate the effectiveness of implemented actions.
- *Development of a watershed plan.* Broad policy goals and more explicit objectives are articulated and prioritized, and then translated into specific management practices (for a detailed example of the process, see Blau's [2002] description of watershed planning in the San Francisco Bay Area). A watershed management plan documents these steps and guides public and private actions, both of which are needed — at spatial scales from the backyard to the watershed — to address the complex water resource issues.
- *Implementation.* Effective watershed plans do not sit on the shelf; actions identified in the plan must be implemented. This involves obtaining adequate funding and technical assistance, and having an appropriate administrative structure (e.g., a watershed council) and the mandate to ensure that the work is carried out, sustained, and evaluated.
- *Monitoring and evaluation.* Good watershed plans are dynamic. Ongoing monitoring supplies feedback on how aquatic resources are responding to the plan's management practices and provides a basis for making adjustments to the plan.

12.4.2 Preserving the Hydrological Services of Forests and Wetlands

The cornerstone of an effective watershed management plan is protection of the hydrological functions of forestlands and wetlands. These ecosystems cleanse water, ensure a steady supply, and control flooding. Making certain that these natural services are retained is often less costly than replacing them with technological fixes such as water filtration plants. Preservation of forestlands and wetlands yields other ecological and social benefits as well, including provision of wildlife habitat and open space.

12.4.2.1 How Much and Where

Planners and other stakeholders must decide on the appropriate mix and location of land uses within a watershed. A useful approach is to treat forests and wetlands like public utilities and ask what capacity (land area, in this case) is needed to meet the needs and performance requirements of the human population. Booth et al. (2002) used this method to estimate the extent of forest cover that should be retained in rural Washington watersheds. First they set a hydrological performance criterion: they empirically determined that stream channels begin to degrade when peak stormwater flow exceeds a certain rate. Then they used a hydrological model to ascertain that in these rural watersheds, where impervious cover is about 4 percent, forest cover of 65 percent is needed to avoid surpassing this threshold. Mitsch and Gosselink (2000) summarized the results of case studies that estimated the percentage of wetland in landscapes required to accomplish certain hydrological functions, usually water quality improvement through nutrient removal. They concluded that an optimal amount is about 3–7 percent in temperate-zone watersheds. At a more local level, CITYgreen software developed by American Forests (2003) allows users to calculate hydrological (e.g., reduced stormwater flow) and economic benefits of increasing tree coverage in urban areas.

Location, as well as extent, influences the hydrological functions and values of forests and wetlands. For example, because of their landscape position, riparian buffers and wetlands are especially important for protecting the water quality of adjacent streams and lakes. Watershed planners need to identify and rank critical lands for protection. Based on their potential for providing hydrological services, these areas should be given highest priority:

- *Aquifer recharge areas.* These are upland regions where water percolates through the soil and resupplies the aquifer, the source of drinking water for many communities.
- *Riparian buffers.* Described in more detail in a following section, these areas are critical for maintaining overall stream health.

- *Small-sized (headwater) streams and their associated wetlands.* Often these are most vulnerable to urban degradation, but they are particularly influential in regulating the water quality of the entire watershed (National Research Council 2001).

12.4.2.2 How to Preserve

Acquisition generally affords the most protection to critical lands. The recent success of referenda that fund land acquisition demonstrates that the public is willing to pay, especially if it is informed about the ecological and hydrological benefits (The Trust for Public Land 1998). However, the expense makes it unlikely that land acquisition will play more than a small part in protecting the natural hydrological services of forests and wetlands. The American Water Works Association found in 1991 that only 2 percent of watershed lands in the U.S. were owned by water utilities (The Trust for Public Land 1998). Other tools for protecting lands, such as conservation easements, are becoming more important and are discussed in detail in Chapters 5 and 6. In general, these tools allow lands to remain in private ownership and the owners are compensated for keeping the land undeveloped, for example through lowered taxes.

Communities may employ a variety of means to ensure the preservation of sufficient forests and wetlands in their watershed (see Box 12.1). Besides purchasing development rights, downstream beneficiaries may offer other incentives to upstream forest landowners. For example, they may provide technical and financial assistance to make forestry more economically viable. Regulatory actions, including zoning (see Chapter 6), may also play a role, and the regulatory approach has been particularly important in preserving wetlands.

Section 404 of the 1972 Clean Water Act laid the groundwork for protecting wetlands by prohibiting the discharge of materials into them, and it also led to many state and local laws guarding wetlands from development. Increasingly, the focus has evolved from saving wetland areas to saving wetland functions (National Research Council 2001). Wetlands in a sea of urbanization do not function in the same ways or provide the same services as wetlands in undisturbed settings. Preserving the hydrological services of a wetland requires not only that the wetland be protected from alteration but also that an adequate natural buffer around the wetland be guaranteed.

12.4.2.3 A Special Case: Riparian Buffers

Streamside forests are among the most diverse and productive ecological systems on earth (Naiman and Décamps 1997). The riparian zone, which stretches from the low water level through the floodplain and continues

Box 12.1

New York City Water Supply

The eight million residents of New York City (NYC) depend on upstate forested watersheds to provide safe drinking water. However, most of the lands are privately owned and conversion to other land uses has deteriorated water quality. Pollution from development in watersheds east of the Hudson River has contaminated 10 percent of NYC's water supply and the Environmental Protection Agency (EPA) has required the construction of a filtration plant (scheduled for completion in 2007). Development in watersheds west of the Hudson River threatened the rest of the city's water supply, and NYC faced the prospect of spending $8 billion to build an additional filtration system and $300 million a year to operate it. In order for NYC to avoid the expense of another filtration plant, the EPA required that the city acquire environmentally sensitive land in the supply watersheds, adopt strong watershed regulations, and institute and maintain a comprehensive watershed protection program.

In the early 1900s, NYC's Board of Water Supply began its relationship with upstate communities by notifying all landowners that title to their property would be vested in NYC and that they would be given only a 10-d notice to move (Weidner 1974). This legacy of distrust spurred towns to quickly organize a "Coalition of Watershed Towns" to oppose NYC's newly proposed watershed regulations. In particular, landowners perceived the proposed rules regulating agriculture and forestry as forcing them out of business and actually increasing economic pressure to subdivide and develop their lands (Watershed Forest Ad Hoc Task Force 1996). The agricultural and forestry communities formed task forces to explore nonregulatory alternatives to protect water quality. NYC responded by including and funding many of their ideas in the 1997 New York City Watershed Memorandum of Agreement (MOA). Total funding for the MOA was approximately $350 million. The majority of the funds were designated for buying environmentally sensitive lands, with additional funds for programs to encourage sustainable forest management. Thus far, EPA has determined that New York's investment in the ecological services of its water supply watersheds has yielded results that avoid the costs of technological filtration.

Programs in NYC's watershed encourage sustainable forestland use by funding forestry plans and Best Management Practices (BMPs), sponsoring economic development opportunities, and providing a variety of educational programs for landowners and loggers. Landowners willing to commit to a 10-year forest

—continued

Box 12.1 — (continued)

management plan can get up to an 80 percent reduction of their property taxes.

The urban/rural partnership in the NYC watershed is justifiably hailed as a national model of conflict resolution and watershed management. In many instances, NYC's investment in the ecosystem services of upstate watersheds not only benefits city dwellers but also protects the values of rural landowners by preserving open space, wildlife, and small farms. However, some issues need further resolution. For example, local governments that absorb the impacts of forestland tax exemptions feel unfairly burdened. Additionally, the transfer of septic permitting authority from local governments to NYC has caused concern among landowners because they feel that the more restrictive rules have effectively condemned their land.

upslope to merge with the uplands, encompasses a wide range of soil types, soil moisture gradients, disturbance regimes, and topography. Riparian forests are important regulators of the flow of nutrients, water, organic material, and organisms among landscape elements. They fuel streams with dissolved and particulate organic matter and enhance and stabilize stream habitats with their roots and large woody debris.

Riparian forests intercept and remove pollutants and sediment that enter from the uplands, and so they act to buffer streams from anthropogenic watershed changes. Because of their crucial ecological position, conservation and restoration of riparian forests has become an integral component of watershed management. For example, in an effort to enhance water quality in Chesapeake Bay, 2090 km of forested riparian buffer have been restored and the goal is to restore an additional 1140 km by the year 2010 (Chesapeake Bay Program 2002).

The ability of forested riparian buffers to successfully remove nonpoint source pollutants is influenced by a number of factors, including the detention time of water passing through, the rate of infiltration, the flow paths of groundwater, the physicochemical conditions of the soil, and the availability of carbon sources. Different conditions optimize removal of different pollutants. For example, nitrogen is most readily eliminated in the dissolved form (nitrate) under reduced soil conditions, whereas phosphorus is trapped more efficiently in the particulate form under oxidative conditions. Thus, restored riparian buffers are usually structured in zones, with three zones being frequently recommended (Figure 12.5) (e.g., see U.S. EPA 1995).

The most upland riparian zone (Zone 3) is designed to spread out runoff and remove sediment (Figure 12.6). To be effective, buffers must first slow and disperse the movement of stormwater. This promotes sediment settling and prevents channelized flow from rapidly short-circuiting buffer treatment. Grass-vegetated strips on shallow slopes can accomplish these functions, but if the impervious source area is large, engineered detention and spreading structures may also be required. Generally, the recommended width of Zone 3 is 6 to 30 m, but large upslope source areas, steep slopes, and fine sediments may require much wider buffers for effective sediment removal (Klapproth and Johnson 2001).

A vigorously growing managed forest 6 to 150 m wide (Zone 2) is recommended downslope of Zone 3 (Figure 12.5). The required width depends on the pollutant load, slope, soils, vegetation, and levels of disturbance. Sites with slopes greater than 10 percent, poor infiltration, and active recreational uses may be unable to remove large pollutant loads. The major function of Zone 2 is to provide the environment for pollutant removal through sedimentation, filtration, cation exchange, and plant uptake. In forested and agricultural landscapes, selective removal of trees is recommended to maintain the forest in a vigorous growth stage.

Ideally, the zone at the land–water interface (Zone 1) is an undisturbed mature forest at least 10 m wide (Figure 12.5). This forest stabilizes stream banks, cools water temperature, and contributes to stream productivity and habitat. Although nutrient uptake by a mature forest may be balanced by loss to litter, Zone 1 provides suitable conditions for other nutrient removal mechanisms. Often, high water tables and anoxic carbon-rich soils promote denitrification, the process by which microorganisms convert nitrate to nitrogen gas (which escapes to the atmosphere). The degree of removal depends on the extent of interaction between nitrate-enriched groundwater from the uplands and denitrifying riparian zones (Gold et al. 2001). Hydrological modifications to the watershed outside of the riparian area can severely impair a buffer's nitrogen removal efficiency. For example, increases in stream runoff reduce base flow and degrade stream channels. Both factors can lower water tables in the streamside riparian zone, causing groundwater to flow under instead of through the subsurface organic soils required for denitrification.

Although buffers with a fixed design and set zone widths are easier to administer, objectives are best accomplished by flexible design matched to site characteristics, mitigation goals, and cultural sustainability. Unfortunately, buffer design in urban areas is often borrowed from different ecoregions, based on agricultural standards, or set by political expediency, and water quality benefits are unrealistically assumed to occur (Herson-Jones et al. 1995). Furthermore, urban riparian buffers face intense disturbance pressures that may compromise their effectiveness (Figure 12.7). Buffer design must be coupled with provisions for proper establishment and long-term management if significant benefits are to be realized and sustained.

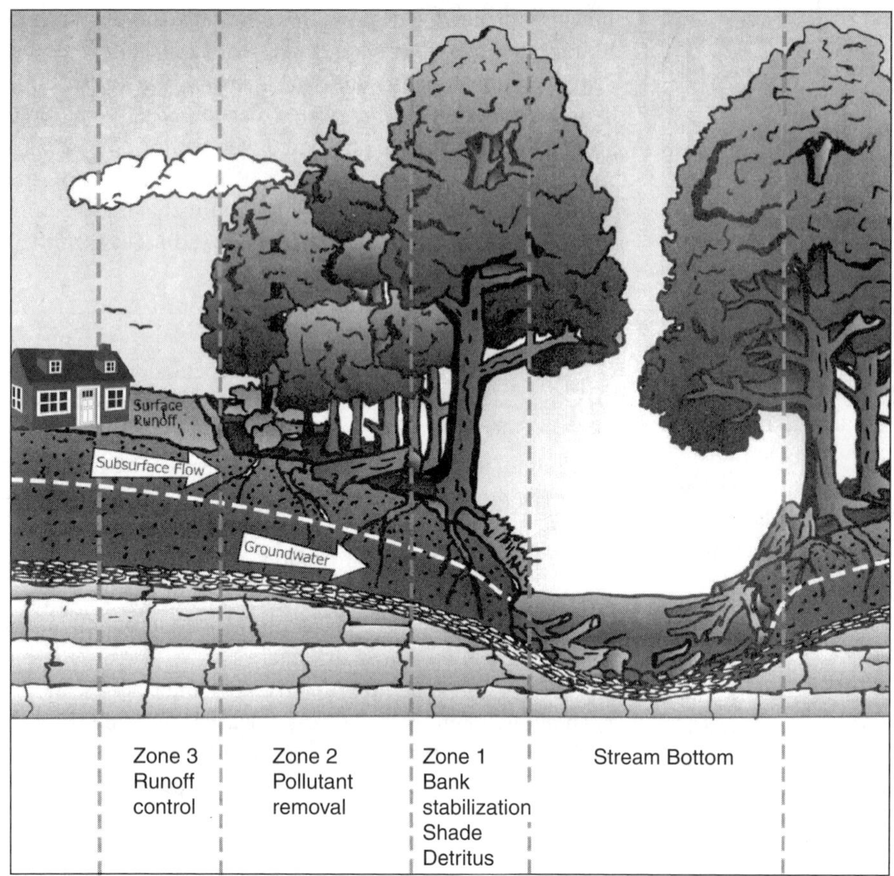

FIGURE 12.5 Riparian buffers can be more effective in enhancing water quality if they are designed with different functional zones. (Adapted from Welch 1996.)

FIGURE 12.6 A grass-vegetated buffer zone separates a residential development from a riparian forest. Its main purpose is to spread out concentrated urban runoff. (Photo by Larry Korhnak.)

Both the riparian area and the entire watershed will shape stream characteristics. For example, good lateral buffers will not be adequate to maintain biological integrity if the upstream riparian corridors are highly disturbed; on the other hand, surprisingly high biological integrity has been observed in some very impervious watersheds where most of the upstream riparian corridor was intact (Horner et al. 1997). However, there are limits to what riparian buffers alone can accomplish. Research in agricultural landscapes has shown that riparian buffers are much more effective in preserving and restoring stream health if they are combined with BMPs throughout the watershed (Wang et al. 2002). The watershed approach is even more critical in urban landscapes where pollutant

FIGURE 12.7 Judy Okay of the Virginia Department of Forestry observes the most frequent type of riparian buffer disturbance — conversion of natural vegetation into lawn by bordering homeowners. Encroachment can be limited by clearly marking the buffer boundaries. (Photo by Larry Korhnak.)

stresses on buffers are higher, but buffer efficiency is often reduced by design limitations and frequent disturbance. Forested riparian areas can be effective in maintaining healthy stream conditions in urban watersheds with low to moderate development, but they are often overwhelmed when impervious cover reaches 25 percent (Meyer and Couch 1999; Morley 2000).

12.4.3 Limiting the Hydrological Impacts of Development

The consistent relationship between impervious cover and altered hydrological conditions, from increased runoff to reduced water quality, signifies the importance of minimizing paved surfaces when planning and constructing developments. However, water quality in urbanized watersheds declines not only because water moves off the land more rapidly but also because more nutrients and pollutants are added. Local actions by individuals and organizations to control sources of pollution are cumulative at the larger watershed level and can help to limit the hydrological impacts of land development.

12.4.3.1 Reducing Impervious Cover

Much of the urban landscape in the U.S. consists of habitat designed for cars. On average, two thirds of the impervious cover in an urbanized watershed is pavement associated with transportation — roads, parking lots, driveways, and sidewalks (Schueler 1995). The key, then, to minimizing paved areas in the wildland–urban interface is to reduce the size of the transportation networks. This can be accomplished at various spatial scales, from the regional level to local parking lots.

At the regional level, concentrating growth within existing urban centers limits sprawl and, therefore, the extension and proliferation of roads. People live closer to their workplaces and shops, and fewer additional roadways are required to meet their transportation needs. Chapters 5–7 discuss some tools for promoting infill development and controlling sprawl.

Within residential subdivisions, a new approach to site design called clustering, open space, or conservation design promotes more compact development (see Chapter 7 for more details and an example). In exchange for retaining undeveloped areas, the developer is allowed to build a greater density of homes elsewhere on the site. Often the purpose is to protect sensitive areas such as wetlands and to provide open and green space for recreation and aesthetics, but an additional benefit is reduced imperviousness of the site. Zielinski (2002) compared conventional subdivisions and what might have been had the developers used open space designs (Table 12.2). Although the overall number of homes is the same, an open space design clusters the homes on smaller lots, requiring fewer and shorter roads than the conventional design. The consequence in many cases is a marked decrease in impervious cover and projected stormwater runoff (Table 12.2).

In any type of residential development, the extent of paved surfaces can be reduced by narrowing the streets, shortening driveways, providing fewer sidewalks (e.g., on only one side of the street), eliminating cul-de-sacs, and substituting permeable pavements or gravel on driveways and other appropriate areas. Several publications give

TABLE 12.2
Estimated Impervious Cover and Stormwater Runoff from Conventional Subdivisions and their Redesigned, Open Space Alternatives

		Impervious Cover in the Subdivision		
Residential Density	Original Lot Size (acres)	Conventional Design (%)	Open Space Design (%)	Reduction in Stormwater Runoff (%)
Low	3–5	5.4–8	3.7–5	20–23
Medium	1/5–1	13–29	7–20	24–66
High	</= 1/8	23–35	20–21	8–31

Source: Adapted from Zielinski (2002).

detailed analyses of the effects of specific changes on impervious cover and how to alter local ordinances to allow these modifications (Schueler 1995; Center for Watershed Protection 1998).

Streets are the primary component of impermeable cover in a residential subdivision, but a significant obstacle to reducing their cover has been local codes that specify road width according to federal highway standards rather than to actual volume of traffic (Schueler 1995). Frequently, streets can be narrowed to accommodate the flow of vehicles and still satisfy public safety concerns. Similarly, local regulations generally call for commercial parking lot size to meet peak demand rather than more characteristic use, and developers often exacerbate the problem by providing even more spaces than required (Schueler 1995). Parking lots are the greatest source of impervious cover in commercial developments — exceeding rooftops — and so downsizing them is critical to limiting hydrological impacts of development. Reducing the number of parking spaces to meet normal parking demand and making various design changes, such as narrowing the driving aisles and providing smaller stalls for compact cars, can substantially shrink parking lots (Zielinski 2000). Overflow lots covered with grass or porous pavement can hold excess vehicles on peak demand days.

12.4.3.2 Controlling Sources of Pollution

People's use of chemicals and water pollution are connected. For example, in the Seattle area, pesticides with the largest sales were detected in all urban and suburban streams that were tested, while no pesticides were detected in the undeveloped reference stream (Voss et al. 1999). Methods of mitigating chemical pollution once it washes off the land are expensive and frequently ineffective. Thus, preventing pollution by reducing the amount of pollutants entering a watershed is a strategy that merits increased attention.

Restricting nitrogen inputs to developing watersheds is particularly important because nitrogen runoff is linked to the degradation of coastal ecosystems, and nitrate, a common form of nitrogen in freshwater, can be harmful to human health. On Cape Cod, MA, nitrogen is delivered to suburban and semirural watersheds mainly by atmospheric deposition, and then by wastewater disposal and fertilizer application (Valiela et al. 1997) (Figure 12.8). In contrast, the nitrogen leaving the watersheds and entering coastal waters derives predominantly from wastewater. Atmospheric nitrogen is efficiently intercepted by natural vegetation (primarily forests), but much of the nitrogen introduced as fertilizer and especially wastewater escapes removal and exits the watersheds via groundwater. These findings point to effective strategies for controlling pollution sources in the wildland–urban interface: retaining large areas of natural vegetation, limiting fertilizer application, and improving the treatment of wastewater.

On Cape Cod, as in many interface landscapes, most fertilizers and pesticides are applied to residential lawns and golf courses, not to agricultural fields. One town, Falmouth, has sent a brochure to every homeowner, explaining the connection between lawn fertilization and water quality impairment and describing better lawn care practices to reduce nitrogen leaching. However, education

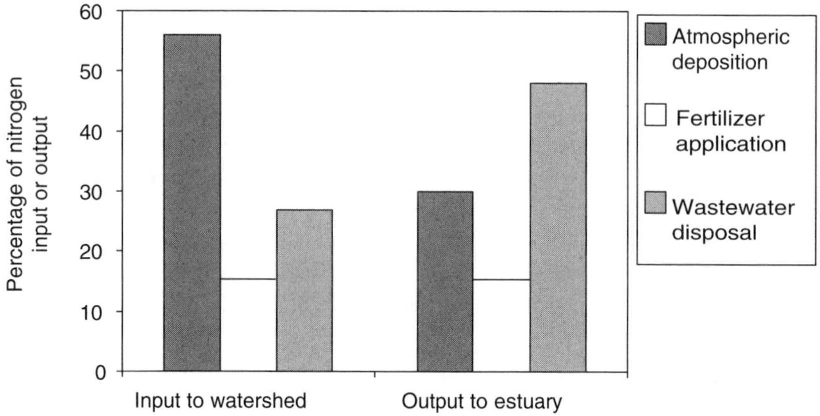

FIGURE 12.8 Primary sources of nitrogen to the watershed of Waquoit Bay, MA, and losses to the receiving estuary. (Adapted from Valiela et al. 1997.)

by itself does not necessarily lead to changed behavior (see Chapter 9). Cooperative extension personnel in Prince William County, VA, found that many residents were already aware and concerned about the link between lawns and water quality, but they had a number of reasons for not adopting practices that would reduce the impacts, including the desire to have a lush green lawn (Schueler 2000b). Therefore, many communities have initiated more active programs to encourage landowners to adopt landscaping that uses fewer pollutants and produces less runoff. Florida Yards and Neighborhoods and the BayScapes program use extensive outreach and demonstration gardens to promote attractive alternatives to lawns (e.g., native shrubs and ground covers) that require fewer inputs of water and chemicals. (Note that in fire-prone regions, attention should also be given to reducing fire risk. See Chapter 13.) Some programs such as one in Prince William County use hands-on training to convince homeowners that good-looking lawns are possible even when using practices that improve water quality (Schueler 2000b). These include testing the soil prior to fertilization and fertilizing only when and in the amount needed, recycling lawn clippings back onto the yard, and practicing integrated pest management.

In extreme cases, restrictive laws can replace voluntary source control. The worsening eutrophication of many of the nation's freshwaters in the 1960s eventually resulted in phosphorus being eliminated from detergents in many states. Initially, the proposed ban stimulated heated scientific, legal, and political debate, but now the link between the eutrophication of freshwaters and phosphorus is so well established that the phosphorus ban is being extended to the fertilization of residential lawns. For example, Minnesota law SF 1555 prohibits phosphate in lawn fertilizer in the Twin Cities metropolitan area and limits its content to 3 percent in other areas of the state.

In rural and semirural areas, such as Cape Cod and much of the wildland–urban interface, most homes are beyond the reach of public sewer lines and instead use on-site systems for the disposal of wastewater. Conventional septic systems consist of a belowground holding tank that receives the raw wastewater and a drainage or leaching field where pollutants in the partially treated wastewater are removed by adsorption and microbial degradation. The effluent then disperses into surrounding soils and travels down gradient, ultimately entering groundwater and receiving water bodies.

The retention efficiency of on-site systems, and therefore the amount of nutrients and pollutants escaping treatment, varies enormously. One survey found that nitrogen retention ranged from 10 to 90 percent and averaged 46 percent (Valiela et al. 1997). Many factors account for this wide variation and the outright failure of some systems (Schueler and Holland 2000). Septic systems perform poorly when too much wastewater is moving through; they must be sized properly to match the input. Performance of on-site systems declines with age, and regular maintenance is needed (e.g., septic tanks should be periodically pumped out). The soil in the drainage field must be unsaturated and of appropriate texture and organic content to facilitate movement of the effluent and removal of pollutants. At some sites, unsuitable soil conditions necessitate alternatives to conventional septic systems, such as mound sand filters.

Rural communities face a challenge in ensuring that the cumulative impact of hundreds or thousands of septic systems does not compromise their water resources. Management tools include setback rules to prevent the installation of septic systems too close to wells and surface waters and zoning to limit the density of on-site systems, perhaps to as few as one per 4 ha (Canter and Knox 1985). Special restrictions can be applied to areas with shallow soils, steep topography, and unprotected aquifers. Communities should inspect new on-site systems for proper installation and functioning, and identify and eliminate older, failing ones. Unfortunately, although many of these responsibilities are mandated to local governments, the financial resources and expertise needed to fulfill these duties are often lacking.

Many rural communities have on-site septic systems that function badly because of age or because they were poorly designed and located at the outset, and new developments with clustered small lots may be unsuitable for individual septic systems. Centralized wastewater treatment plants may not be a solution to these problems because they are too costly to build and operate, and residences are too dispersed. In some cases, multiuser decentralized wastewater treatment technology can improve community wastewater treatment and fill the gap between single household systems and centralized wastewater treatment plants. Decentralized systems use existing septic tanks to pretreat wastewater from several homes or businesses. Then a low-cost collection system transports the wastewater to a more effective treatment unit that can include aerobic digestion and disinfection.

Replacing septic systems with a centralized wastewater plant may be the only option in some situations, but it may have some unintended impacts. For example, wastewater treatment plants concentrate effluent, the disposal of which may overwhelm and degrade a valued surface water resource, and centralized plants frequently result in large transfers of water from one watershed to another. Also, waste disposal is often the physical factor limiting population density, and once more growth becomes possible, it is hard to stop.

12.4.4 Mitigating the Hydrological Consequences of Development

Stormwater management has long been required in urbanizing landscapes because land development inevitably results in increased surface runoff. Traditionally, the aim was to solve on-site drainage problems by moving stormwater off-site as quickly as possible, usually through gutters and culverts to streams. Often this merely transferred the problems downstream, and therefore a variety of BMPs have been designed and implemented to better manage the stormwater generated by developed areas.

The goals and designs of detention ponds (Figure 12.9), the most common type of constructed BMP, have greatly evolved during the past two decades (Wang 2002). Originally, the ponds were intended solely for flood control. They capture, detain, and slowly release stormwater runoff from developed areas, preventing peak discharges from exceeding predevelopment levels. More recently, the goal has extended to water quality improvement, and many jurisdictions require that detention ponds are sized and configured to hold water longer and so enhance pollutant settling.

Although detention ponds can be very effective in controlling floods such as the 10-year storm, they do not adequately mitigate many of the other hydrological changes generated by development (Booth et al. 2002; Ferguson 2002). The ponds dampen the maximum rate of runoff, but they allow significant increases in runoff volume and duration of flow, contributing to stream erosion. Most of the water eventually runs out of the ponds or evaporates, and little infiltrates the soil to recharge groundwater. Moreover, detention ponds that dry out between runoff events are limited in their ability to trap pollutants (Winer 2000). Other types of BMPs, such as retention ponds and infiltration basins, perform better in removing pollutants, but like detention ponds they often are "end-of-pipe" facilities located at the base of drainage areas and have limited capacity to maintain predevelopment hydrological processes. Maxted and Shaver (1999) found that stormwater ponds were ineffective in preventing the physical degradation of streams when watershed impervious cover reached 20 percent, and none of the ponds in their study prevented a shift to pollution-tolerant macroinvertebrates. In fact, the ponds' influences on temperature and organic carbon cycling may have contributed to the biological declines in the streams they were intended to protect.

New approaches for managing stormwater have a broader set of goals, including preventing downstream flooding, protecting stream channels and water quality, and maintaining groundwater recharge and base flow. Constructed wetlands are increasingly used for pollution control as well as moderation of stormwater flow rates. An array of integrated techniques, collectively termed low-impact development practices, promote on-site water treatment and infiltration at the source of surface runoff, thereby resupplying groundwater and precluding water quality and quantity problems downstream.

12.4.4.1 Constructed Wetlands

Wetlands are commonly referred to as nature's kidneys because of their ability to remove contaminants from inflowing water and thereby protect downstream water resources. A number of physical, biological, and chemical processes interact to promote this capacity (DeBusk and DeBusk 2001). Many incoming contaminants are attached to particulate matter, and wetlands provide highly favorable conditions for the settling of these sediments: shallow water, slow water flow, and filtering action by plant stems and leaves. The vegetation ingests some of the dissolved pollutants, but the flooded soils in wetlands are

FIGURE 12.9 This typical detention pond captures flashy stormwater runoff from roadways and parking lots and more slowly releases it through an outflow control structure. (Photo by Larry Korhnak.)

particularly conducive to their removal, either by adsorption onto the soil particles or by uptake and transformation by microbes.

The efficiency with which a wetland removes contaminants depends on many factors, including input concentrations, the depth and flow rate of the water (hydraulic loading), and the type, setting, and design of the wetland. Since the 1970s, water managers have constructed wetlands specifically to treat wastewater effluent and stormwater runoff. Kadlec and Knight (1996) give a detailed discussion of design considerations for these wetlands. In general, a successful design ensures intimate contact between the contaminated water and the wetland soil: the inflowing water must be shallow, it must disperse throughout the wetland, and it must be retained long enough for contaminant removal processes to take place.

Designing constructed wetlands for stormwater treatment is more difficult than for wastewater treatment. Wastewater is delivered in pipes and its volume and rate of inflow can be controlled, but stormwater flow is highly variable and unpredictable. To provide effective treatment, the stormwater wetland must accommodate the maximum discharge as well as function during the dry periods between storm events. Also, stormwater tends to carry much higher sediment loads than wastewater and greater concentrations of heavy metals, hydrocarbons, and other pollutants that wash off parking lots and roads. Nevertheless, stormwater wetlands have been shown to be very effective at reducing the loadings of a wide variety of contaminants (Tilley and Brown 1998).

One way to increase the effectiveness of stormwater wetlands is to incorporate them in a treatment train. In this approach, different water treatment units are assembled in sequence in order to optimize pollutant removal. For example, Lake Greenwood, a stormwater treatment system constructed in Orlando, FL, in 1988, includes a baffle box where coarse sediments and litter are trapped and three aerated detention ponds that permanently hold water (Figure 12.10). About one half the area of the first pond consists of shallow shelves that were planted with wetland vegetation. The baffle box is periodically cleaned out, helping to maintain high removal rates of contaminants such as phosphorus (Carla Palmer, personal communication, September 2003).

Lake Greenwood also illustrates that stormwater systems can be designed and constructed to provide multiple benefits, including the creation of greenspace within urban areas and educational and recreational opportunities (Campbell and Ogden 1999). Wildlife habitat may be another benefit of treatment wetlands, although some caution is necessary. While treatment wetlands are highly effective in trapping many contaminants, some pollutants (e.g., some of the heavy metals) may become accessible (and potentially toxic) to wetland inhabitants. Most states do not permit the use of natural wetlands for wastewater and stormwater treatment because of potential toxicity problems and the likely alterations in wetland structure due to changed hydrology.

Wetlands can be constructed on-site to treat and dampen stormwater outflows from residential and commercial developments and regionally to capture larger flows from urban watersheds. A watershed-level approach is needed to decide on the size, placement, and total area of wetlands necessary to adequately treat runoff and prevent flooding. Tilley and Brown (1998) provided a start in that direction by designing a hierarchical network of stormwater wetlands for urbanized watersheds in south Florida.

FIGURE 12.10 Lake Greenwood, a constructed stormwater management system in Orlando, FL, illustrates an attractive, multifunctional alternative to detention ponds. (Photos by Larry Korhnak.)

12.4.4.2 Low-Impact Development Practices

While many BMPs are effective at cleaning runoff and reducing peak discharge rates, they do not address the fundamental problems of excess surface water generation and insufficient groundwater recharge in urbanized watersheds. A new approach tackles these problems at the source — where water hits paved and other impervious surfaces. The objective is to treat the water where it falls and to direct it back into the ground, restoring the natural hydrological pathways of infiltration and ET. Because this is likely to result in diminished (or even eliminated) surface runoff, downstream impacts of land development may be avoided.

The Department of Environmental Resources in Prince George's County, MD (PGCDER), has pioneered this approach to stormwater management, called low-impact development (LID), and produced a design manual for national use (PGCDER 1999). LID is a site design strategy that aims to mimic or restore the predevelopment hydrological functions of the landscape. A number of tactics are used in combination to achieve this goal, including:

- Conservation of critical natural features and permeable soils
- Reduction and disconnection of impervious surfaces
- Lengthening of water flow paths and slowing of runoff
- Distribution throughout the site of integrated, microscale management practices to reduce and cleanse stormwater.

Many of these measures are not new or unique to LID. What sets LID apart from conventional approaches to stormwater management is that water is not drained off-site and conveyed to streams or centralized facilities, but instead is managed and treated locally — on the building lot — by a variety of dispersed, small-scale techniques.

Some of the practices, such as vegetated rooftops, store and evaporate precipitation where it falls, reducing the volume and rate of stormwater discharge. Rain barrels collect water that washes off rooftops, saving it for reuse on lawns and gardens. The overflow may be filtered and directed to depressions on the lot that are specifically engineered to hold back and treat water. Stormwater from hard roofs, parking lots, and driveways is also routed to the depressed areas, called bioretention cells, and/or to infiltration trenches. LID designs exclude conventional curb and gutter systems, which store and move stormwater underground, and instead convey water through open grass channels and wetland swales. A bioretention area (Figure 12.11) typically consists of grass buffer strips to slow and filter inflowing water, a ponding area to temporarily store the water, a conditioned soil bed covered with mulch to support physical filtering and biological uptake of pollutants, vegetation to take up and transpire water, and an underlying bed of sand to enable infiltration

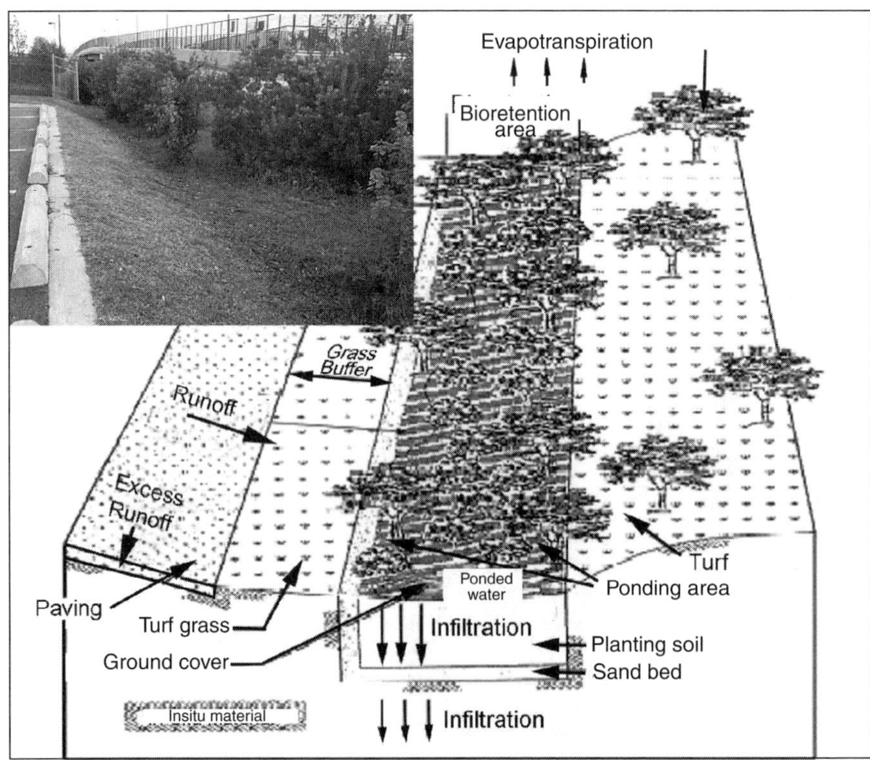

FIGURE 12.11 A bioretention area is a vegetated depression designed and engineered to capture, retain, purify, and infiltrate stormwater runoff. (Adapted from PGCDER 1993).

of treated water. An infiltration trench is another LID design feature that reduces the volume of site runoff by promoting water penetration into the ground.

LID techniques can be applied to residential and commercial lots of various sizes and to parking lots (see Box 12.2). Originally created for new developments, the practices are also being applied to urban retrofits. Clar (2002) documents a variety of LID applications and case studies, and the Low-Impact Development Center website (www.lowimpactdevelopment.org) provides more information. However, LID is not suitable for every site. Steep slopes, impermeable soils, and a shallow water table are conditions that may prevent the use of LID techniques. Water that infiltrates the soil can sometimes drain to the groundwater table and, therefore, practices that promote infiltration may not be appropriate for sites that have chemical storage areas and other activities that can contaminate the land and the groundwater below. In most cases, conventional BMPs (e.g., detention ponds) may be required in addition to LID practices in order to handle large storm events.

The long-term performance of LID practices is unknown. A review of available data on LID projects (U.S. EPA 2000d) found that bioretention areas and grass swales were highly effective in reducing runoff volume and removing pollutants, but noted that data were scarce and the projects were young. The review also observed that continued high performance depends on regular maintenance of these practices. The soil needs to be tested and replaced as necessary, perhaps every 5–10 years. Grass swales need to be mowed periodically, and bioretention areas need regular mulching, pruning of woody vegetation, and replacement of dead or diseased plants.

Maintenance is required for all stormwater management facilities but is viewed as a thornier problem for LID practices because they are installed on private lots and homeowners must shoulder the responsibility (U.S. EPA 2000d). Ensuring that LID practices are attractive and educating homeowners about the benefits of these techniques may help enlist the participation of residents. Also, homeowner associations can be a mechanism for providing long-term maintenance.

Although the use of LID techniques is growing, the vast majority of new developments still depend on conventional stormwater management. There are a number of obstacles to widespread adoption of LID, including homeowner misperceptions (e.g., holding water on-site inevitably causes structural damage), lack of familiarity with LID techniques on the part of developers and engineers, and local development regulations that forbid some LID practices such as open roads without curbs and gutters. Better education, changed ordinances, and incentives (e.g., Maryland's offer of stormwater credits for using LID techniques in stormwater management (Clar 2002)) may help to overcome these obstacles.

Box 12.2

Greening a Parking Lot

The Southwest Florida Water Management District incorporated LID features into the parking lot of the Florida Aquarium in Tampa and studied their effectiveness in retaining and treating runoff (Rushton and Hastings 2001). The overall design was a stormwater treatment train that treated water passing from the parking lot, where microscale management practices were applied, through a forested strand and then to a small detention pond, which discharged into Tampa Bay. The 4.65 ha asphalt and concrete parking lot was modified by installing end-of-island bioretention cells and bioretention swales around the perimeter. Within the lot, eight basins (two of each type) were constructed to examine the effects of paving materials and the presence of bioretention swales between parking rows on runoff and pollutant loadings (Figure 12.12). The four treatment types were asphalt with no swale, asphalt with a swale, concrete with a swale, and permeable pavement with a swale. For 2 years, rainfall and flow from each of these basins were measured following storms, and water quality samples were collected.

The addition of bioretention swales greatly reduced the rate and volume of stormwater generated by the parking lot basins (Figure 12.13). Asphalt and concrete basins with swales discharged about 40 percent less runoff than those without swales, and replacement with permeable pavement decreased runoff even more, by another 40 percent. These treatments also reduced the loads of most pollutants, including nitrate, ammonia, and a variety of heavy metals. The basins with both permeable pavement and swales consistently achieved the greatest reduction in pollutant loads (Figure 12.13). Only phosphorus loading showed an increase, perhaps due to grass clippings from swale maintenance. In any case, little of this nutrient flowed into the parking lot's receiving water, Tampa Bay. Thanks to the stormwater treatment train, only one of 20 monitored storms produced discharge from the detention pond.

12.4.4.3 Stream Restoration

Often one of the most visible effects of urbanization is stream degradation. Collapsing banks, loss of riparian vegetation, flooding that threatens property, and the disappearance of favored fish can mobilize adjacent landowners and community groups and agencies to take actions to restore predevelopment stream conditions.

The first task in many restoration projects is to reestablish the stream's channel stability, which is its ability to carry water and sediments without being physically

FIGURE 12.12 A bioretention swale installed between parking rows in the Florida Aquarium parking lot. (Photo by Betty Rushton.)

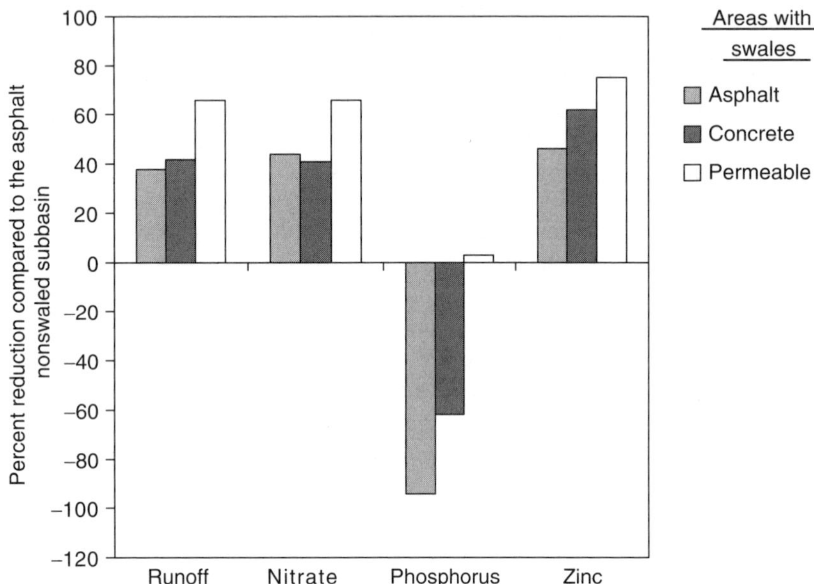

FIGURE 12.13 Runoff and pollutant reduction by three types of pavement with installed bioretention swales in the Florida Aquarium parking lot. (Adapted from Rushton and Hastings 2001.)

degraded. Successful restoration projects identify, mitigate, and/or design for the root causes of the channel instability. However, all too often stream "restoration" consists of treating the local symptoms of disequilibria by channelizing and hard armoring the stream (Figure 12.14). These methods not only destroy ecosystem functions, but often they merely transfer energy downstream where the problem will reoccur.

An important foundation of many stream restoration projects is adjusting the channel dimensions to convey bankfull storms through the channel with enough energy to convey sediment but not enough to erode the stream channel. A channel that is too narrow will erode, and one that is too large will accumulate sediment. Bankfull discharge occurs when water fills the banks and just begins to spread onto the floodplain. These events normally occur every 1 to 2 years, and over time they are the dominant force that shapes channel morphology. Most streams are ungauged, so hydrological models are essential for estimating bankfull flows. Channel dimensions and sinuosity can then be modeled after reference stream sections with the help of regional bankfull curves (e.g., see

FIGURE 12.14 Stabilizing stream channels by hard armoring degrades ecosystem functions and transfers destructive energies downstream. (Photo by Larry Korhnak.)

Rosgen's (1994) stream classification and Doll et al. (2002)). Key elements of many redesigned channels are reconnection of the channel to the floodplain, widening of the floodplain, regrading of stream banks to stable angles, and reduction of hydraulic gradients by reestablishment of meanders to straightened stream sections.

Excessive sediment delivery to the stream from outside sources is most effectively controlled with proper watershed BMPs. Maintaining natural vegetation in the riparian zone is especially critical. Increased stormwater can be mitigated with BMPs such as detention ponds; however, in many instances these will reduce peak flow but not stormwater volumes. The result is longer periods of lower but relatively high-energy erosive stream flows. When watershed-based approaches fail to mitigate the impacts of development, in-stream measures may also be needed.

As the biological values of streams have become better understood and appreciated, in-stream restoration techniques have shifted from "hard" structural approaches to "soft" bioengineering approaches. An important goal of soft restoration strategies is to stabilize the stream channel by revegetating the banks rather than replacing stream bank vegetation with hard materials such as riprap. While the restored vegetation is maturing, degraded stream banks are shielded by energy-absorbent organic materials like brush piles, willow stakes, tree revetments, and fabricated flexible logs made from coconut fibers. In some cases, stream energies need to be deflected from eroding banks by logs or root wads projecting into the stream (Figure 12.15).

Practical constraints limit both the number and scope of stream restoration efforts. Priority should be determined by the probability that the project will restore ecological functions. Brown (2000) found that fewer than 60 percent of urban stream restoration practices fully achieved even limited objectives for habitat enhancement. Isolated restoration projects surrounded by highly disturbed conditions are unlikely to succeed. For example, although large woody debris are known to provide critical in-stream habitat, merely resupplying this component to highly urbanized streams has been found to be ineffectual in reversing biological degradation (Larson et al. 2001).

Restoration efforts in the wildland–urban interface may have a greater chance of success because more of the supporting systems are intact. However, success may be short-lived if future development alters the conditions for which the project was designed. Thus, future plans for watershed development need to be considered when setting restoration priorities and designing stream restoration projects.

Fortunately, land in the wildland–urban interface is often available to accommodate expanded stream zones. However, supportive landowners are just as important. Stream restoration professionals in North Carolina have found that many successful projects begin with a few cooperating landowners with strong interests in improving stream conditions (Jennings et al. 2001). In the South Fork Mitchell River Watershed, restoration efforts began with one cattle rancher's stream. Success with that project catalyzed a community response and 20 other landowners requested assistance. Stream restoration may cost hundreds of dollars a linear foot, so it is important that legal mechanisms, such as conservation easements, are enacted to protect and maintain the investment. Stream restoration design and construction is a multidisciplinary task: teams of engineers, hydrologists, biologists, and extension educators are needed to get the job done. Riley (1998) and the Federal Interagency Stream Restoration Working Group (2001) provide good foundation references for exploring this expansive topic.

FIGURE 12.15 Eroding banks can be protected with soft armoring such as root wads. The streambank in the foreground, protected by root wads, survived high flows caused by Hurricane Isabel (September 2003), while the unprotected bank in the background was severely eroded. (Photo by Larry Korhnak.)

12.5 CONCLUSION

Forests are a crucial link in the cycle of water through the biosphere. Their complex physical structure and biological processes moderate episodic flows from the atmosphere. Forests shield and grip the land, and water in forested watersheds encounters fewer pollutants to leach and transport. For humans, some important benefits of the relationship between water and forests are less flooding and a more sustained flow of clean water.

The urbanization of forested watersheds reverses these benefits: flows and erosive forces are amplified, and pollutants are dispersed across the landscape or directly mixed with water in domestic and industrial uses. Expensive technology is required to protect human health and life-supporting resources when forest ecosystem services are lost. Other species that have less flexible links to forested watersheds are frequently harmed at very low levels of urbanization (less than 10 percent impervious area), and watershed restoration practices may only provide limited protection up to 25 percent impervious area.

In many ways, the wildland–urban interface is the front line in the battle for preserving the sustainability of forested watersheds, the natural ecosystems they support, and the ecological services they provide to human society. Knowledge is accumulating about the hydrological, physical, and chemical dangers resulting from the replacement of forests with impervious surfaces and high-maintenance vegetation. A number of promising strategies for preventing or mitigating these threats are being pursued, including identification and defense of critical landscape areas, limits on the spread of impervious area, restriction of pollutant supplies, and purification and infiltration of surface runoff at its source. Watershed-based planning encompasses all these elements and provides the framework for effective control of the hydrological changes wrought by urbanization. However, a critical key to success is educating the public about the importance of forested watersheds and their vulnerability to careless development and transforming this knowledge into broad-based individual and political action.

REFERENCES

American Forests, 2003. CITYgreen, http://www.americanforests.org/productsandpubs/citygreen/. [Date accessed: April 30, 2003.]

Anderson, C.W., F.A. Rinella, and S.A. Rounds, 1996. Occurrence of selected trace elements and organic compounds and their relation to land use in the Willamette River, Oregon, 1992–1994, USGS Water-Resources Investigations Report 96–4234.

Axness, K.A., J. Potokar, and T. van Drasek, 2002. When the well runs dry: examining the water supply issues in Brown County, Wisconsin, *Water Resources Impact* 4: 4–8.

Beaulac, M.N. and K.H. Reckhow, 1982. An examination of land use–nutrient export relationships, *Water Resources Bulletin* 18: 1013–1023.

Bedford, B. L., 1999. Cumulative effects on wetland landscapes: links to wetland restoration in the U.S. and southern Canada, *Wetlands* 19: 775–788.

Bedient, P.B. and W.C. Huber, 1992. *Hydrology and Floodplain Analysis,* Addison-Wesley Publishing Co., Reading, MA.

Blau, D., 2002. Watershed management plans: bridging from science to policy to operations (San Francisco, California), in *Handbook of Water Sensitive Planning and Design*, R.L. France, Ed., Lewis Publishers, Boca Raton, FL, pp. 459–474.

Booth, D.B., 1990. Stream-channel incision following drainage-basin urbanization, *Water Resources Bulletin* 26: 410–411.

Booth, D.B., D. Hartley, and R. Jackson, 2002. Forest cover, impervious-surface area, and the mitigation of storm-water impacts, *Journal of the American Water Resources Association* 38: 835–845.

Boward, D.P., P. Kazyak, S. Stranko, M. Hurd, and T. Prochaska, 1999. From the Mountains to the Sea: The State of Maryland's Freshwater Streams, EPA 903-R-99-023, Maryland Department of Natural Resources, Annapolis.

Brown, K.B., 2000. Urban Stream Restoration Practices: An Initial Assessment, The Center for Watershed Protection, Ellicott City, MD.

Campbell, C.S. and M. Ogden, 1999. *Constructed Wetlands in the Sustainable Landscape*, John Wiley and Sons, New York.

Canter, L.W. and R.C. Knox, 1985. *Septic Tank System Effects on Ground Water Quality*, Lewis Publisher, Chelsea, MI.

Center for Watershed Protection, 1998. *Better Site Design: A Handbook for Changing Development Rules in Your Community*, Ellicott City, MD.

Center for Watershed Protection, 2003. Impacts of Impervious Cover on Aquatic Systems, Watershed Protection Research Monograph No. 1, Ellicott City, MD.

Chesapeake Bay Program, 2002. The State of the Chesapeake Bay, Chesapeake Bay Program Office, Annapolis, MD.

Clar, M.L., 2002. Applications of low-impact development techniques (Maryland), in *Handbook of Water Sensitive Planning and Design*, R.L. France, Ed., Lewis Publishers, Boca Raton, FL, pp. 297–315.

Council on Environmental Quality, 1996. The 1996 Annual Report of the Council on Environmental Quality, Washington, DC.

DeBusk, T.A. and W.F. DeBusk, 2001. Wetlands for water treatment, in *Applied Wetlands Science and Technology*, 2nd ed., D. M. Kent, Ed., Lewis Publishers, Boca Raton, FL, pp. 241–279.

Doll, B.A., D.E. Wise-Frederick, C.M. Buckner, S.D. Wilkerson, W.A. Harman, R.E. Smith, and J. Spooner, 2002. Hydraulic geometry relationships for urban streams throughout the piedmont of North Carolina, *Journal of the American Water Resources Association* 38: 641–651.

Driscoll, C.T., G.B. Lawrence, A.J. Bulger, T.J. Butler, C.S. Cronan, C. Eagar, K.F. Lambert, G.E. Likens, J.L. Stoddard, and K.C. Weathers, 2001. Acid rain revisited: advances in scientific understanding since the passage of the 1970 and 1990 Clean Air Act Amendments, Hubbard Brook Research Foundation, *Science Links Publication* 1: 1–24.

Dunne, T. and L.B. Leopold, 1978. *Water in Environmental Planning*, W.H. Freeman and Company, San Francisco.

Ehrenfeld, J.G., 2000. Evaluating wetlands within an urban context, *Ecological Engineering* 15: 253–265.

Ehrenfeld, J.G. and J.P. Schneider, 1991. *Chamaecyparis thyoides* wetlands and suburbanization: effects on hydrology, water quality and plant community composition, *Journal of Applied Ecology* 28: 467–490.

Federal Interagency Stream Restoration Working Group, 2001. *Stream Corridor Restoration: Principles, Process, and Practices*, USDA Natural Resources Conservation Service, Washington, DC.

Ferguson, B.K., 2002. Stormwater management and stormwater restoration, in *Handbook of Water Sensitive Planning and Design*, R.L. France, Ed., Lewis Publishers, Boca Raton, FL, pp. 11–28.

Ferguson, B.K. and P.W. Suckling, 1990. Changing rainfall–runoff relationships in the urbanizing Peachtree Creek watershed, Atlanta, Georgia, *Water Resources Bulletin* 26: 313–322.

Finkenbine, J.K., J.W. Atwater, and D.S. Mavinic, 2000. Stream health after urbanization, *Journal of the American Water Resources Association* 36: 1149–1160.

Freeman, A.M., 1990. Water pollution policy, in *Public Policies for Environmental Protection*, P.R. Portney, Ed., Resources for the Future, Washington, DC, pp. 97–149.

Gold, A.J., P.M. Groffman, K. Addy, D.Q. Kellogg, M. Stolt, and A.E. Rosenblatt, 2001. Landscape attributes as controls on groundwater nitrate removal capacity of riparian zones, *Journal of the American Water Resources Association* 37: 1457–1463.

Herson-Jones, L.M., M. Heraty, and B. Jordan, 1995. Riparian Buffer Strategies for Urban Watersheds, Metropolitan Washington Council of Governments Environmental Land Planning Document Series, Washington, DC.

Hewlett, J.D., 1982. *Principles of Forest Hydrology*, University of Georgia Press, Athens.

Holland, C.C., J. Honea, S.E. Gwin, and M.E. Kentula, 1995. Wetland degradation and loss in the rapidly urbanizing area of Portland, Oregon, *Wetlands* 15: 336–345.

Hollis, G.E., 1975. The effects of urbanization on floods of different recurrence intervals, *Water Resources Research* 11: 431–435.

Horner, R.D., D. Booth, A. Azous, and C. May, 1997. Watershed Determinants of Ecosystem Functioning, in Effects of Watershed Development and Management on Aquatic Ecosystems, L.A. Roesner, Ed., Proceedings Engineering Foundation Conference, Snowbird, UT, August 4–9, 1996.

Hunt R.J. and J.J. Steuer, 2001. Evaluating the effects of urbanization and land-use planning using ground water and surface water models, U.S. Geological Survey Fact Sheet FS-102–01.

Jackson, R.B., J.S. Sperry, and T.E. Dawson, 2000. Root water uptake and transport: using physiological processes in global predictions, *Trends in Plant Science Perspectives* 5: 482–488.

Jaskula, J.M. and W.A. Hohn, 2002. Potential impacts of COMM 83 on rural ground water, *Water Resources IMPACT* 4: 10–16.

Jennings, G.D., W.A. Harman, K.L. Tweedy, D. Wise-Frederick, K.A. Hall, and B.A. Doll, 2001. Restoration Case Studies for Incised Rural North Carolina Streams, in Proceedings of the ASCE Wetlands Engineering & River Restoration Conference, D.F. Hayes, Ed., Reno, NV.

Jones-Lee, A. and Lee, G.F. 1993. Groundwater Pollution by Municipal Landfills: Leachate Composition, Detection and Water Quality Significance, Proceedings Sardinia '93 IV International Landfill Symposium, Sardinia, Italy, pp. 1093–1103.

Joyce, L., J. Aber, S. McNulty, V. Dale, A. Hansen, L. Irland, R. Neilson, and K. Skog, 2001. Potential Consequences of Climate Variability and Change for Forests of the U.S., in Climate Change Impacts on the U.S.: Potential Consequences of Climate Change, Report for the U.S. Global Change Research Program, Cambridge University Press, Cambridge, U.K., pp. 489–543.

Kadlec, R.H. and R.L. Knight, 1996. *Treatment Wetlands*, Lewis Publishers, Boca Raton, FL.

Karr, J.R., 1981. Assessment of biotic integrity using fish communities, *Fisheries* 6: 21–27.

Klapproth, J.C. and J.E. Johnson, 2001. Understanding the Science Behind Riparian Forest Buffers: Planning, Establishment, and Maintenance, Virginia Cooperative Extension Publication 420–155.

Klein, R.D., 1979. Urbanization and stream quality impairment, *Water Resources Bulletin* 15: 948–963.

Larson, M.G., D. Booth, and S.A. Morley, 2001. Effectiveness of large woody debris in stream rehabilitation projects in urban basins, *Ecological Engineering* 18: 211–226.

Linker, L.C., R.L. Dennis, and D.L. Alegre, 1996. Impact of the Clean Air Act on Chesapeake Bay Water Quality, International Conference on the Environmental Management of Enclosed Coastal Seas, Baltimore, MD.

Linsley, R.K. and J.B. Franzini, 1972. *Water-Resources Engineering*, 2nd ed., McGraw-Hill, New York.

Magee, T.K., T.L. Ernst, M.E. Kentula, and K.A. Dwire, 1995. Floristic comparison of freshwater wetlands in an urbanizing environment, *Wetlands* 15: 517–534.

Maxted, J.R. and E. Shaver, 1999. The Use of Retention Basins to Mitigate Stormwater Impacts to Aquatic Life, in National Conference on Retrofit Opportunities for Water Resource Protection in Urban Environments, EPA/625/R-99/002, U.S. Environmental Protection Agency, Office of Research and Development, Washington, DC, pp. 6–15.

Meyer, J.L. and C.A. Couch, 1999. Final Report: Influences of Watershed Land Use on Stream Ecosystem Structure and Function, Report to EPA National Center for Environmental Research, Washington, DC.

Mitsch, W.J. and J.G. Gosselink, 2000. The value of wetlands: importance of scale and landscape setting, *Ecological Economics* 35: 25–33.

Morley, S.A., 2000. Effects of Urbanization on the Biological Integrity of Puget Sound Lowland Streams: Restoration with a Biological Focus, M.S. thesis, University of Washington, Seattle.

Munn, M.D. and S.J. Gruber, 1997. The relationship between land use and organochlorine compounds in streambed sediment and fish in the central Columbia River Plateau, Washington and Idaho, USA, *Environmental Toxicology and Chemistry* 16: 1877–1887.

Naiman, R.J. and H. Décamps, 1997. The ecology of interfaces-riparian zones, *Annual Review of Ecology and Systematics* 28: 621–658.

National Academy of Sciences, 1996. *Use of Reclaimed Water and Sludge in Food Crop Production*, National Academy Press, Washington, DC.

National Research Council, Committee on Watershed Management, 1999. *New Strategies for America's Watersheds*, National Academy Press, Washington, DC; Available on-line at http://www.nap.edu/books/0309064171/html. [Date accessed: May 1, 2003.]

National Research Council, Committee on Mitigating Wetland Losses, 2001. *Compensating for Wetland Losses Under the Clean Water Act*, National Academy Press, Washington, DC; Available on-line at http://www.nap.edu/books/0309074320/html. [Date accessed: May 1, 2003.]

Novotny, V. and H. Olem, 1994. *Water Quality: Prevention, Identification, and Management of Diffuse Pollution*, van Nostrand Reinhold, New York.

Paul, M.J. and J.L. Meyer, 2001. Streams in the urban landscape, *Annual Review of Ecology and Systematics* 32: 333–65.

Pielke, R.A. and M.W. Downton, 2000. Precipitation and damaging floods: trends in the U.S., 1932–97, *Journal of Climate* 13: 3625–3637.

Pit, R.,J. Lantrip, R. Harrison, C.L. Henry, and D. Xue, 1999. Infiltration Through Disturbed Urban Soils and Compost-Amended Soil Effects on Runoff Quality and Quantity, EPA/600/R-00/016.

Prince George's County Department of Environmental Resources (PGCDER), 1993. *Design Manual for Use of Bioretention in Stormwater Management*, Division of Environmental Management, Watershed Protection Branch, Landover, MD.

Prince George's County Department of Environmental Resources (PGCDER), 1999. *Low-Impact Development Design Strategies: An Integrated Approach*, Prince George's County, MD.

Richter, K.O. and A.L. Azous, 2001. Amphibian distribution, abundance, and habitat use, in *Wetlands and Urbanization: Implications for the Future*, A.L. Azous and R.R. Horner, Eds., Lewis Publishers, Boca Raton, FL, pp. 143–165.

Riley, A.L., 1998. *Restoring Streams in Cities*, Island Press, Washington, DC.

Rosgen, D.L., 1994. A classification of natural rivers, *Catena* 22: 169–199.

Rushton, B.T. and R. Hastings, 2001. Florida Aquarium Parking Lot: A Treatment Train Approach to Stormwater Management, Final Report, Southwest Florida Water Management District, Brooksville, FL.

Ryan, D.F. and S. Glasser, 2000. Drinking Water from Forests and Grasslands: A Synthesis of the Scientific Literature, G. E. Dissmeyer, Ed., General Technical Report SRS-39, U.S. Department of Agriculture, Forest Service, Southern Research Station.

Sahin, V. and M.J. Hall, 1996. The effects of afforestation and deforestation on water yields, *Journal of Hydrology* 178: 293–309.

Schoch, D., 2002. Hard times on the lower Klamath, *Los Angeles Times*, May 3, 2002.

Schueler, T.R., 1995. *Site Planning for Urban Stream Protection*, Center for Watershed Protection, Ellicott City, MD.

Schueler, T.R., 2000a. The compaction of urban soils, *Watershed Protection Techniques* 3: 661–665.

Schueler, T.R., 2000b. Homeowner survey reveals lawn management practices in Virginia, *Watershed Protection Techniques* 1: 85–86.

Schueler, T.R. and H.K. Holland, Eds., 2000. *The Practice of Watershed Protection*, Center for Watershed Protection, Ellicott City, MD.

Simmons, D. and R. Reynolds, 1982. Effects of urbanization on base flow of selected south-shore streams, Long Island, New York, *American Water Resources Bulletin* 18: 797–805.

Solley, W.B., R.R. Pierce, and H.A. Perlman, 1998. Estimated Use of Water in the U.S. in 1995, U.S. Geological Survey Circular 1200.

Southwest Florida Water Management District, 1996. Northern Tampa Bay Water Resources Assessment Project, Vol 1, Surface-Water/Ground Water Interrelationships, Southwest Florida Water Management District, Brooksville, FL.

Tchobanoglous, G. and F.L. Burton, 1991. *Wastewater Engineering: Treatment, Disposal and Reuse*, McGraw-Hill, New York.

Thompson, P.A., R. Adler, and J. Landman, 1989. Poison Runoff, National Resources Defense Council, Washington, DC.

Tilley, D.R. and M.T. Brown, 1998. Wetland networks for stormwater management in subtropical urban watersheds, *Ecological Engineering* 10: 131–158.

The Trust for Public Land, 1998. *Protecting the Source: Land Conservation and the Future of America's Drinking Water*, The Trust for Public Land, San Francisco, CA.

U.S. Census Bureau, 2001. Housing Starts Report C20/01-1, Washington, DC.

U.S. Department of Agriculture, Natural Resources Conservation Service, 2000. Summary Report: 1997 National Resources Inventory (revised December 2000), USDA Natural Resources Conservation Service, Washington, DC, and Statistical Laboratory, Iowa State University, Ames, Iowa, http://www.nrcs.usda.gov/technical/NRI/1997. [Date accessed: March 3, 2003.]

U.S. Environmental Protection Agency (U.S. EPA), 1988. Criteria for Municipal Solid Waste Landfills, U.S. EPA, Washington, DC.

U.S. Environmental Protection Agency, 1993. Guidance for Specifying Management Measures for Sources of Nonpoint Source Pollution in Coastal Waters, EPA-840-B-92-002, Washington, DC.

U.S. Environmental Protection Agency, 1995. Water Quality Functions of Riparian Forest Buffer Systems in the Chesapeake Bay Watershed, EPA 903-R-95-004.

U.S. Environmental Protection Agency, 1999. Decentralized Systems Technology Fact Sheet: Septage Treatment/Disposal, EPA 832-F-99-068.

U.S. Environmental Protection Agency, 2000a. Nonpoint Source Pollution: The Nation's Largest Water Quality Problem, EPA841-F-96-004A.

U.S. Environmental Protection Agency, 2000b. Water Quality Conditions in the U.S.: A Profile from the 1998 National Water Quality Inventory Report to Congress, EPA841-F00-006.

U.S. Environmental Protection Agency, 2000c. Progress in Water Quality: An Evaluation of the National Investment in Municipal Wastewater Treatment, EPA-832-R-00-008.

U.S. Environmental Protection Agency, 2000d. Low Impact Development (LID): A Literature Review, EPA-841-B-00-005, Office of Water, EPA, Washington, DC.

U.S. Environmental Protection Agency, 2001. Protecting and Restoring America's Watersheds: Status, Trends, and Initiatives in Watershed Management, EPA-840-R-00-001, Office of Water, EPA, Washington, DC.

U.S. Environmental Protection Agency, 2002. Pesticide Industry Sales and Usage 1998 and 1999 Market Estimates, EPA-733-R-02-001.

U.S. Geological Survey (USGS), 1992. Groundwater and the Rural Homeowner, USGS, Denver, CO.

U.S. Geological Survey, 1999. The Quality of Our Nation's Waters: Nutrients and Pesticides, U.S. Geological Survey Circular 1225.

U.S. Geological Survey, 2001. The National Water Quality Assessment Program, Reston, VA.

Valiela, I., G. Collins, J. Kremer, K. Lajtha, M. Geist, B. Seely, J. Brawley, and C. H. Sham, 1997. Nitrogen loading from coastal watersheds to receiving estuaries: new method and application, *Ecological Applications* 7: 358–380.

Voss F.D., S.S. Embrey, J.C. Ebbert, D.A. Davis, A.M. Frahm, and G.H. Perry, 1999. Pesticides Detected in Urban Streams During Rainstorms and Relations to Retail Sales of Pesticides in King County, Washington, USGS Fact Sheet 097-99.

Wang, D., 2002. Successful stormwater management ponds (Massachusetts), in *Handbook of Water Sensitive Planning and Design*, R.L. France, Ed., Lewis Publishers, Boca Raton, FL, pp. 31–46.

Wang, L., J. Lyons, and P. Kaneh, 2002. Effects of watershed best management practices on habitat and fish in Wisconsin streams, *Journal of the American Water Resources Association* 38: 663–680.

Waschbusch, R.J., W.R. Selbig, and R.T. Bannerman, 1999. Sources of Phosphorus in Stormwater and Street Dirt from Two Urban Residential Basins in Madison, Wisconsin, 1994–1995, USGS Water Resources Investigations Report 99-404.

Watershed Forest Ad Hoc Task Force, 1996. Policy Recommendations for the Watersheds of New York City's Water Supply, Watershed Agricultural Council, Walton, NY.

Weidner, C.H., 1974. *Water for a City*, Rutgers University Press, New Brunswick, NJ.

Welch, D.J., 1996. Riparian Forest Buffers: Function and Design for Protection and Enhancement of Water Resources, NA-PR-07-91, U.S. Department of Agriculture, Forest Service, Radnor, PA.

Wignosta, M.S., S. Burges, and J. Meena, 1994. Modeling and Monitoring to Predict Spatial and Temporal Hydrological Characteristics in Small Catchments, Water Resources Series Technical Report #137, University of Washington Department of Civil Engineering, Seattle.

Winer, R., 2000. *National Pollutant Removal Performance Database for Stormwater Treatment Practices,* 2nd ed., Center for Watershed Protection, Ellicott City, MD.

Zhu, W.X. and J.G. Ehrenfeld, 1999. Nitrogen mineralization and nitrification in suburban and undeveloped Atlantic white cedar wetlands, *Journal of Environmental Quality* 28: 523–529.

Zielinski, J.A., 2000. The benefits of better site design in commercial development, *Watershed Protection Techniques* 3: 647–656.

Zielinski, J.A., 2002. Open spaces and impervious surfaces: model development principles and benefits, in *Handbook of Water Sensitive Planning and Design*, R. L. France, Ed., Lewis Publishers, Boca Raton, FL, pp. 49–64.

Zinke, P.J., 1967. Forest interception study in the U.S., in Forest Hydrology: Proceedings of a National Science Foundation International Symposium on Forest Hydrology, W.E. Sopper and H.W. Lull, Eds., Pergamon, Oxford, pp. 137–161.

13 Managing for Fire in the Interface: Challenges and Opportunities

Alan J. Long
School of Forest Resources and Conservation, University of Florida

Dale D. Wade
USDA Forest Service (retired)

Frank C. Beall
University of California Forest Products Laboratory

CONTENTS

13.1 INTRODUCTION

Fire managers define the wildland–urban interface as all areas where flammable wildland fuels are adjacent to homes and communities. With this definition, the wildland–urban interface may encompass a much broader landscape than traditionally perceived. For example, the Tunnel Fire in the Oakland hills in 1991 included a large area that, for practical purposes, could be considered truly urban — the edges of the fire were not far from either downtown Oakland or Berkeley. At the other end of the spectrum, wildland fires also threaten or destroy rural

1-56670-602-5/05/$0.00+$1.50
© 2005 by CRC Press LLC

homes far from the closest population center. No longer is the evening news coverage of subdivisions overrun by wildfire unique to southern California. Today, this is an all-too-common occurrence throughout the U.S. from Washington to Florida and Maine to Arizona. It is also a significant problem in other countries, as demonstrated in Australia during the last two weeks of 2001 when more than 1.5 million acres and 140 homes and businesses were blackened by wildfire.

A curious question arises as we consider fire management and protection throughout this larger wildland–urban interface zone. Why should fire protection currently receive so much attention when the history of this country was a mosaic of small towns and isolated homesteads and cabins within a vast fire-maintained landscape? We believe the answer lies in the fact that fire was once such a common feature of rural life that it was accepted without question. Over the past century, however, technological advances coupled with the unrealistic expectation that humans can control nature led many to believe that this disruptive force could be eliminated from the landscape without dire consequences. Moreover, as our population shifted from a rural to an industrial lifestyle, people moved to cities where they quickly forgot their ancestral ties to fire. Now, as people are rediscovering the benefits of life in the wildland–urban interface, they have to reconnect with this awesome force, and it is a difficult transition (Figure 13.1). The transition is not back to some point in the past because virtually all ecosystems have been altered by human intervention, remarkably so in many instances. Following nearly a century of attempted fire exclusion, fire behavior is now considerably different than it was in the past. Both fire suppression and prevention activities have undergone continual modification in an effort to keep pace with these changes.

This chapter has two objectives: to provide the reader with an overview of the fire-related challenges facing those who live in the wildland–urban interface and the importance of cooperation between these residents and the agencies/organizations charged with managing fire where people and wildlands meet, and to describe some emerging or current strategies these agencies and organizations are using to manage fire hazards and fire itself in the broadly defined wildland–urban interface. We emphasize the importance of both community and individual landowner responsibilities and actions in preparing for the inevitability of fire in wildland ecosystems. A brief discussion of the ecological role and necessity of fire in fire-adapted ecosystems and the impacts of fire suppression on these functions will provide an important starting point for the rest of this chapter.

13.2 HISTORICAL BACKGROUND

13.2.1 Natural Role of Fire

Forest and grassland ecosystems throughout the U.S. are adapted to periodic fires that were historically ignited by lightning or Native Americans. The intentional use of fire was a necessity for these early Americans. They used it to accomplish a wide range of tasks, from protection to improving their standard of living. Where precipitation or rivers, creeks, swamps, and rocky outcrops provided the only barriers to the spread of fire, new ignitions often spread great distances.

In areas such as the Southeast, and especially in Florida, fire undoubtedly spreads across the same piece of ground several times a decade, sometimes as frequently as every few years (Platt et al. 1991). Two important factors that presumably contributed to the fire history in the

FIGURE 13.1 Typical wildland–urban interface home near Santa Rosa, CA, located on a hillside with nearby grass fuels and with evergreen shrubs around the home and beneath overhangs. (Photo by Alan Long.)

Southeast were a much higher lightning frequency than elsewhere in the country along with a large Native American population. Based on early written accounts and plant adaptations to fire, frequent fire must have also been the norm across much of the West, particularly where ponderosa pine (*Pinus ponderosa* Dougl. Ex Laws., the western equivalent to longleaf pine [*P. palustris* Mill.]) dominated vast landscapes. Under a regime of frequent fire, woodlands were open and typically characterized by large pines and a ground cover composed mainly of grasses and forbs (Figure 13.2). To thrive in such an environment, plants had to possess adaptations to frequent fire. For example, longleaf pine, which is thought to have dominated 70 to 95 million acres in the South, is multinodal (having more than one growth spurt during a growing season). It has thick bark, long needles that are concentrated in tufts at the branch tips where they protect a huge bud (high heat capacity) from radiant heat and keep heat from being trapped at the base of the crown, seed germination in the fall as soon as environmental conditions are favorable, the ability to resprout when very young, and a juvenile grass-like stage that can survive periodic fire while developing a good root system until it puts on a burst of height growth to quickly raise the terminal bud above the flames. The grasses and shrubs that characterize these frequent fire regimes also have flowering, seeding, and sprouting capabilities that provide a quick response and recovery after fire.

Because of their frequency, fires were generally of low intensity whether ignited by lightning or Native Americans. These fires served several important ecological functions and maintained open woodlands by preventing the development of dense brush and hardwoods (in the South) or conifers (in the West) that would soon shade out the species-rich groundcover. This frequent thermal decomposition prevents the accumulation of hazardous fuel levels in the understory as well as on the forest floor. Chronic fire regimes helped prevent the invasion of woody plants into grasslands, prairies, and marshes. The regular resprouting of grasses and shrubs using recycled nutrients provides succulent, more palatable,and nutritious wildlife food than would occur in the absence of fire, and this new growth is within easy reach of browsers and grazers.

At elevations above the ponderosa pine zone in the West, the Upper Piedmont of the South, and across much of the Lake States and Northeast, conifer, hardwood, mixed conifer, and conifer/hardwood stands were subject to periodic although less frequent fire. Under hot, dry, windy conditions, fires in many of these plant communities often turned into intense stand-replacing events. Some coniferous species, such as lodgepole pine (*P. contorta* Dougl. ex Loud.) in the West, jack pine (*P. banksiana* Lamb.) in the Lake States and Northeast, sand pine (*P. clausa* [Chapm. ex Engelm.] Vasey ex Sarg.) in Florida, and table mountain pine (*P. pungens* Lamb.) in the Southern Appalachians are well adapted to such stand replacement fires. In fact, fire ecologists have concluded that virtually every terrestrial ecosystem in North America (mangroves and northeastern beech/sugar maple are two exceptions) is characterized by periodic fires, albeit centuries apart in some cases.

13.2.2 Effects of Fire Suppression on Natural Communities

In the late 1800s and early 1900s, severe fires occurred in every region of the country, often started by steam-driven logging equipment operating in forests that had heavy accumulations of residual slash. This debris from logging

FIGURE 13.2 Old-growth ponderosa pine near Bend, OR, where frequent fires have maintained a low, open understory of grasses, shrubs, and herbs. Forest structure is similar to original longleaf pine stands and other ecosystems around the country where fire was a regular ecosystem process. (Photo by Alan Long.)

and/or land clearing played a major role in most of our nation's worst conflagrations between 1840 and 1940. In response to this threat to our natural resources, the federal government set a high priority on reducing these losses through new fire control programs. A policy of complete fire suppression was initiated on public lands by government agencies and was supported on private lands through federal funding of state programs and cooperatives.

Since the 1920s and 1930s when the majority of these programs were initiated (or became effective because of the bulldozer), the area burned by wildfires steadily declined until near the close of the 20th century. Unfortunately, as we applauded these efforts, the composition and stature of our natural ecosystems were changing in response, ever so gradually, yet predictably. Shade-tolerant shrubs and trees began to appear in the understory, shading out the herbaceous groundcover, which in turn resulted in changes in the local fauna. These successional trends have profound fire management implications; they inevitably result in deepening layers of decaying plant material that is colonized by overstory tree roots that are susceptible to drought season fire. In addition, an aboveground woody understory and midstory develops, which provides a combustible ladder for fire to reach treetops (Figure 13.3). These increased live and dead fuel loads have the potential for much more intense and severe fires. Similar forest structures have also resulted from many of our forest management schemes over the past 50 years. The wildland–urban interface is expanding into these significantly modified landscapes, and it is imperative that landowners understand the basic ecology and fire management ramifications of their natural surroundings and the associated risks of living in these areas.

13.3 THE CHALLENGE OF FIRE IN THE WILDLAND–URBAN INTERFACE

13.3.1 Case Studies and General Principles

Two examples of interface fire will demonstrate some important principles that fire suppression agencies accept and landowners should understand. The first principle is that fire is inevitable. We have no control over lightning, little control over arson, and only moderate control, even after substantial education, over human carelessness.

On October 20, 1991, the Tunnel Fire in the hills next to Oakland, CA, burned only 1500 acres, but consumed over 3000 homes with the loss of 25 lives. The hills were an old residential area for both Oakland and Berkeley. Many homes were at least 30 to 50 years old with wood roofs and siding, and with the only access by narrow winding roads. It was much more urban than the typical wildland–urban interface we often see on the news. In many ways, it looked just like a suburb of Oakland, except for the dense, and often natural, cover of pines, eucalyptus, and shrubs on many lots. The Tunnel Fire resulted from an apparent act of carelessness and a rare weather event in which hot dry winds poured over the hills from the east. One local resident described the morning (before the fire began) as very eerie, perhaps a harbinger of what was to come.

In contrast, northeast Florida was subjected to more than 2000 fires over a 2-month period in 1998, started by both people and lightning. These fires burned 500,000 acres of uninhabited woodlands, pasture, and wildland–urban interface, damaging or destroying 330 homes and businesses in the process. An entire county was evacuated at one point and some towns were evacuated several times. In the end, over 100,000 people were evacuated and 214 people were injured; but remarkably, no lives were lost.

FIGURE 13.3 In the absence of fire, this ponderosa pine forest near Mt. Shasta, California, developed into a much denser mixture of pine and other conifer species, with substantial vertical structure. (Photo by Alan Long.)

Up the Atlantic seaboard to the Northeast, westward through the Lake States to the Pacific Northwest, down the Pacific Rim into the Southwest, and then eastward through the Gulf Coast, numerous other examples of wildland–urban interface fires could be presented. A common feature of almost any of these catastrophic interface fires is a rapid rate of spread due to topography, wind, or both, often but by no means always after a prolonged dry season or drought.

Another important principle, well illustrated by comparing the Oakland and Florida fires, is that fire intensity, acreage, and life/resource loss are not well correlated. Natural resource losses from wildfire can be roughly estimated from acreage and fire intensity, but losses at the wildland–urban interface often bear little relation to actual fire size.

One final principle, which must be understood by landowners, is that protection of all structures in the interface may not be possible. Although fire suppression organizations routinely set their highest priority for action on saving lives and property, in areas with many homes and a fast-moving fire, there may not be enough suppression units to protect every home. Vehicular access may be limited, too. Difficult decisions about what to protect have been, and will continue to be, necessary. In such situations, total loss of some homes is almost inevitable.

13.3.2 Fire Suppression in the Interface

One of the most significant features of fire control in the interface is that suppression crews are often faced with fire in both wildland and structural fuels. Fire behavior in the two systems is very different and crews trained for either wildland or structures are often not adequately prepared for dealing with the other. Structural fires are usually fought at close range, often in confined spaces with high concentrations of noxious gases. These superheated combustible gases can literally explode when forced out a doorway by a suddenly introduced stream of water. Structural firefighters thus wear heavy protective gear and self-contained breathing apparatus, which wildland firefighters find too cumbersome and energy-sapping. Wildland fires are usually unconfined and burn in an unrestrictive atmosphere, thereby making it necessary for suppression forces to move along the fireline, sometimes at a very rapid pace in order to make progress. Water is the principal method for suppressing structural fires, but is often unavailable to wildland firefighters in the quantities needed, especially in rugged terrain.

Given the differences in fire behavior, control methods, and protective equipment, suppression of interface fires requires either crews that are trained for both types of fire or close coordination between crews from different organizations. A common ingredient in all fire control organizations is the priority placed on protecting lives and ensuring firefighter safety.

Wildland fire suppression tactics typically focus on surrounding a fire with control lines and using a variety of methods (machines, hand tools, water, foam, retardants, and burning out) to break the fire triangle by removing fuel, oxygen, and/or heat. These tactics usually focus on minimizing acres burned and resource losses. In contrast, fire suppression tactics in the interface often dictate that available equipment and personnel be positioned at threatened homes to knock down the fire as it reaches the residence and put it out if the home ignites. The overall strategy is to protect homes first, rather than minimizing burned acres. Consequently, interface fires can often reach larger sizes than they would if they were burning strictly in wildland fuels. This generally translates into significantly higher suppression costs and can also result in a higher number of homes being affected.

13.3.3 Landowner Expectations in the Interface

Landowners who move from metropolitan areas to the interface have generally enjoyed a variety of services, from police and fire protection to road repair and municipal water systems. Even though most of these services are now more distant or nonexistent, these new interface residents still expect some local or state agency to provide them. In the case of fire protection, they assume the same quality and response time even though narrow, winding, poorly marked roads may prevent fire trucks from reaching their destination or may limit landowner escape routes.

An extreme example of this expectation was demonstrated at a rural hotel outside Bend, OR, which had the protection of 15 fire engines during a 1990 interface fire that burned 22 homes. When the owners were asked subsequently to clean some of the shrubs from their property and needles from the roof, their response was that they did not need to do that since the fire agencies "would protect them again" in the event of another fire (Tom Andrade, Oregon Department of Forestry, personal communication, August 2001).

As naive as this response may seem, it characterizes the expectations of many landowners. They moved to the interface to enjoy "nature," and they place substantial value on their surroundings and privacy (Figure 13.4). They may even build their homes or decks around trees to accentuate the forest ambience. Their values often do not include significant changes in that ambience, and their thinking tends to fall into three possible responses about fire: they will rely on being protected by fire crews; they trust that "it won't happen to them"; or they assume that they can rebuild with insurance money if something does happen.

13.3.4 The Challenges: A Synopsis

The challenges to managing fire in the interface are diverse: vegetative communities that are prone to intense

FIGURE 13.4 Homeowners in the wildland–urban interface value the ambience and privacy of their surroundings and often do not appreciate their fire risk or are unwilling to correct the potential problem. (Photo by Alan Long.)

fires under severe environmental conditions; suppression tactics that focus on protecting structures rather than restricting the spread of fire; homeowners who are reluctant to modify the natural conditions that attracted them to the interface but still expect to be protected; and the necessity of diverse suppression forces to handle structural and vegetative fires simultaneously.

These challenges are being addressed creatively, energetically, with persistence, and with substantial funding. The rest of this chapter describes some of the approaches being used for preventing and managing wildfires. We emphasize the potential of prescribed fire as a fuel mitigation tool in the interface. We give examples of agency and community activities and programs that address these challenges but stress the fact that landowners must ultimately shoulder substantial responsibilities. Public education will play a crucial role in assisting landowners with their responsibilities.

13.4 MEETING THE CHALLENGE

13.4.1 Hazard Assessment

Central to any approach to fire management in the interface is an evaluation of the differences in hazard and risk across vegetation communities and property boundaries. Such assessments are a basic requirement for establishing priorities for allocating fire suppression resources, planning educational programs and fuel modification treatments, and assisting landowners with their landscaping and home design.

Fire risk is simply a quantitative or qualitative evaluation of the likelihood that a particular home or community will be subjected to a wildland fire in the foreseeable future. Fire hazard describes the fuel complex in terms of type, volume, condition, arrangement, and location; it determines the ease of ignition and difficulty of suppression of a wildland fire. Important components of risk and hazard include:

- Types, patterns, and conditions of vegetation and fuels
- Likelihood of an ignition by lightning, humans, or equipment (basic factor affecting risk)
- Design and construction materials of individual homes
- Infrastructure such as roads, signs, water sources, and utilities
- Topography and related environmental factors
- Resource or asset values that would be affected by a fire
- Frequency of adverse weather or climatic conditions.

Depending on the purpose of a particular assessment, these components are used individually or in some combination. Several examples will demonstrate the use and variability of assessments.

California was one of the first states to develop a statewide assessment of fire hazard, with special interest in those communities and watersheds that could be classified as "very high fire hazard severity zones" (Figure 13.5). Vegetation, topography, and asset value in the above list were the primary criteria, along with the potential for extreme fire weather (dry air and strong winds, component 7). The entire state was mapped by the California Department of Forestry and Fire Protection (CDF) using these criteria in a geographic information system (GIS) (California State Board of Forestry 1996). If they accept the CDF classification, communities in very

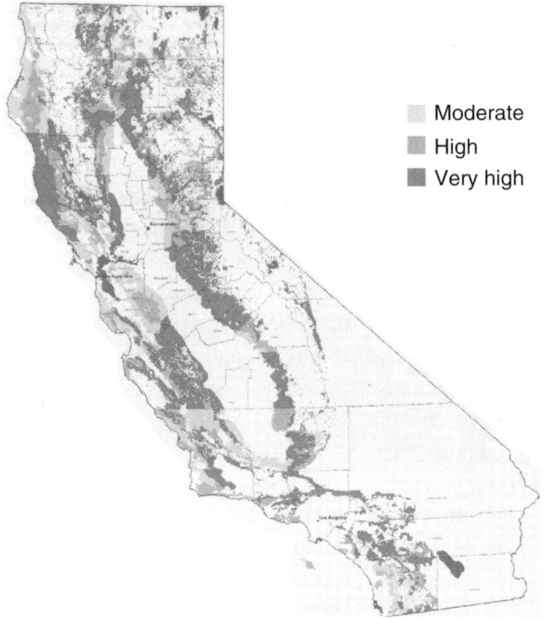

FIGURE 13.5 Statewide hazard assessments such as California's GIS-based system are usually based on vegetation, topography, ignition sources, and population densities. (From California State Board of Forestry 1996.)

high severity zones are required by state law to take appropriate actions to mitigate hazards, reduce risks, and increase fire prevention programs.

Recently, New Mexico prepared a similar GIS-based statewide hazard assessment that utilized vegetation, likelihood of an ignition, and asset value as the main components (Lightfoot et al. 1999). Vegetation communities were classified according to standard fuel models and were assigned low, medium, or high values based on the rate of spread and resistance to control. Ignition probabilities were based on fire statistics for the last 10 years, and asset values were derived from census statistics for population density. The latter two components were also assigned low, medium, and high values, and the actual fire risk was based on these three sets of values, closely following a system used in Virginia (Virginia Department of Forestry 1997) and elsewhere. The state map of fire risk depicts the regions where fire resources will be concentrated for fire suppression and fuel hazard reduction and where fire prevention programs will be increased.

Florida has just completed a similar but much more in-depth statewide assessment that includes most of the components above as well as information about fire history and suppression resources. In addition, the state was divided into 20 weather influence zones and four weather scenarios were used to predict potential fire behavior on each quarter-acre pixel. Then, the proportion of fires over the 20-year database that occurred under each scenario was determined. This information was used to develop a Wildland Fire Susceptibility Index for each quarter-acre pixel statewide. The resulting GIS maps will be used for purposes similar to those of New Mexico as well as to set priorities for mitigation teams that are working on fuel reduction projects around the state.

At a community level, the National Fire Protection Association (NFPA) (1997) publication titled "299 Standard for the Protection of Life and Property from Wildfire" has probably been the most commonly used model or set of guidelines for evaluating wildfire hazards. NFPA 299 recommended a numerical rating system to define the relative contributions to hazard severity of numerous factors, ranging from home construction to weather (Table 13.1).

The NFPA Standard 299 was revised and released as NFPA Standard 1144 in 2002 (NFPA 2002). A similar rating system has been developed by the International Fire Code Institute (IFCI) (2000). Like the NFPA Standards, the IFCI Code provides specifications for water supplies, defensible space, and access in wildland–urban interface areas, and includes a table for rating the severity of the hazard based on vegetation, slope, fire and weather frequency, and fuel models. However, the IFCI Code has not yet been adopted by any jurisdiction because it lacks the flexibility of a standard (Jim Smalley, personal communication, November 2001).

In the last 10 years, a number of states and organizations have developed their own hazard assessment systems, based on the 1991 or 1997 NFPA 299 Standards, the IFCI Code, or their own experience. For example, in 1996, California produced two documents that have been the basis for community and regional programs governing the assessment of wildfire hazards and execution of mitigation projects (California State Board of Forestry 1996; Slaughter 1996). Based on these two broad documents,

TABLE 13.1
NFPA 1144 Hazard Rating System

Possible Points	Contributing Factors
0–25	Roofing materials Siding/deck construction Vegetation types Defensible space Slope Water sources
0–7	Road width and grade Access routes Turnarounds Signs Utility placement
0–10	Topography Weather conditions

Source: National Fire Protection Association (1997).

more recent publications were developed for assessing individual properties: "Structural Fire Prevention Field Guide for Mitigation of Wildland Fires" (California Department of Forestry and Fire Protection 2000a) and "Property Inspection Guide" (California Department of Forestry and Fire Protection 2000b). In Colorado, where interface fires have increased along the Front Range (east slope of the Rocky Mountains), risk assessment procedures have been developed by Boulder County (Boulder County Wildfire Mitigation Group 2001) and the Colorado State Forest Service (1997). Similar programs exist in every other region of the country (e.g., Great Lakes Forest Fire Compact 1996; Virginia Department of Forestry 1997; Florida Division of Forestry 2002).

All of these methodologies have been summarized and compiled into a general assessment procedure that can be used around the country, titled *Wildland/Urban Interface Fire Hazard Assessment Methodology* (National Wildland/Urban Interface Fire Protection Program 1998). Sponsors of this document, which is an important resource for Firewise Communities workshops (described later) and other multiagency planning programs, include the National Association of State Foresters, National Fire Protection Association, U.S. Forest Service, four agencies in the U.S. Department of the Interior and the Federal Emergency Management Agency. The guide is designed to "help users assess the potential of a structure located in a wildland environment to withstand an approaching forest fire without the intervention of fire-fighting personnel and equipment. [It] focuses exclusively on proactive, prefire preventative actions … ." The guide also points to important considerations and actions at the community level.

Many homeowners in the interface may want to see fire risk decreased throughout their community, but they have little control over what is done beyond their property boundary. Thus, their foremost concern should be protection of their own home and property. They can assess their risk, focusing on components 1 and 3 in the list at the beginning of this section (vegetation patterns and building materials) and rating them according to any of the systems listed above. Many states and local organizations or agencies have landowner brochures or publications that do not directly describe risk assessment but list many things a landowner can do in landscaping or home construction to reduce hazardous conditions.

In the wake of the 1998 and 1999 wildfires, University of Florida personnel developed a very simple landscape assessment procedure that landowners can use to determine whether they are at a low, medium, or high level of risk should a wildfire approach their home (Monroe and Long 2000). The assessment only requires a look around their property at the density and continuity of shrubs, grass, young pines, and ladder fuels such as vines, and an evaluation of whether they can see through the vegetation on adjoining properties. For many landowners, such simple procedures may be an important educational method to encourage them to assess their risk and take action to reduce it.

As important as hazard/risk assessment is to all other fire management strategies in the interface, there is an inherent danger in overstressing its utility. The problem is that in any given fire incident, individual properties with a generally low hazard rating may still be at risk in extreme fire situations. Landowners who determine that they are not in a high-risk situation may decide that other landscaping objectives (e.g., backyard wildlife, natural trees and shrubs next to the house, energy conservation) are more important to them. They may, therefore, not carry out some of the simple procedures that would significantly improve protection of their home in case of a fire under unusual circumstances.

Two examples will demonstrate this issue. As the Cerro Grande fire burned through Los Alamos in May 2000, individual homes with wood siding were often ignited by low-intensity surface fires burning through pine needles, dead leaves, or wood piles on the ground next to the structure (Cohen 2000a). Moreover, during the spring 2001 fire season in Florida, the majority of homes that burned were lost in a single afternoon in what is normally considered a low-risk situation: low grass cover with few if any trees or shrubs around the homes. They just happened to be in open areas caught in a series of fast-moving grass fires on a very windy day. In both instances, the majority of these homes could have been saved by an able-bodied person with a garden hose and adequate water or a prefire scratch line that reduced the hazard by removing fuels from the base of the structure. In Australia, fire control agencies offer basic fire suppression and survival training to property owners and encourage them, if they are able-bodied, to stay with their homes when threatened by wildfire rather than evacuate. They have generally found that this is much safer than trying to evacuate everyone and has resulted in significantly fewer homes lost or damaged.

13.4.2 Suppression Strategies and Tactics

Wildland fires, particularly in the interface, usually do not stop at political or fire agency boundaries. Thus, any given fire may cover jurisdictional lines for city, county, state, and volunteer fire departments. In such situations, timely and effective coordination of suppression activities is dependent on prefire planning among agencies and communications systems that allow different agencies to maintain contact.

Prefire planning takes many forms, but most typically includes mutual agreements on who will respond to fires in intermediate areas, compacts to provide mutual support upon request, and lists of equipment and resources available from different agencies. Following the 1998 fires in Florida, the Governor's Wildfire Response and Mitigation Review Committee submitted 90 recommendations that addressed significant issues in wildfire response, recovery, and mitigation/prevention (Governor's Wildfire Response and Mitigation Review Committee 1998). Some of the recommendations relevant to suppression strategies include:

- Prepositioning wildfire suppression resources based on criteria such as drought conditions, wildfire activity, and available resources.
- Adequately equipping rural fire departments for wildland fire.
- Upgrading communications systems for the Division of Forestry and other agencies.
- Increasing wildland fire training for structural and volunteer firefighters.

The emphasis on increasing equipment and training for volunteer departments is especially noteworthy. According to the National Volunteer Fire Council, nearly 23,000 of the 31,000 fire departments in the country have no paid employees and another 5000 use mostly volunteers led by a few paid staff members. Many interface developments and residences are beyond the jurisdiction of municipal fire departments and are dependent on these volunteer suppression crews. Unfortunately, as the number of interface homes and developments is increasing, the general trend for volunteer departments is a decrease in size. This decrease over the past 10 to 20 years is primarily due to a lack of new volunteers to fill the empty spots left by trained volunteers who have been able to find full-time fire/rescue positions. This is a significant nationwide issue.

Some of the reduction in available fire-fighting crews created by the loss of volunteer fire department staff has been compensated for by cross-training structural and wildland fire crews. Increasingly, traditional fire agencies are training their crews in wildland fire behavior and control, and are outfitting them with brush trucks and wildland personal protective equipment. For example, more than 20 percent of the City of Albuquerque, New Mexico, Fire Department staff have been through the suite of basic federal land management courses in wildland fire behavior and suppression. The department has three task forces, all with brush trucks and ATVs for the three major wildland vegetation systems that occur within their jurisdiction (Don Shainin, Battalion Commander and Wildland Coordinator, personal communication, August 2001). The Alachua County, FL, Fire/Rescue Department has similarly cross-trained many of its staff and has equipped them with brush trucks and wildland fire gear (Will May, Alachua County Fire Chief, personal communication, February 2002). This trend will undoubtedly expand in the future, with not only more individuals trained for both structural and wildland fuels but many of those individuals trained at higher levels in the Incident Command System, which is used by many organizations to coordinate emergency response activities. Many other local, state, and federal agencies charged with emergency management in Florida (and elsewhere) are finding advanced-level federal fire management courses valuable for all staff involved with any type of emergency, ranging from hurricanes to citrus canker outbreaks.

Another critical aspect of planning and readiness for interface fires is the prepositioning of suppression and prevention crews before fires break out. As more states complete state- or regionwide risk assessments like those in Virginia, New Mexico, and Florida, they will use the assessments to prioritize the placement of crews and equipment. For example, since the 1998 fires, Florida has been very proactive during the spring fire season, accelerating initial response to most fires and moving prevention crews around the state in response to changes in fire danger. These prevention crews are primarily responsible for reducing the possible sources of human ignitions using techniques such as door-to-door contacts, school programs, displays, and handouts at fairs and malls, but they also have the latitude to undertake other initiatives they deem important for the types of ignitions that might occur in different areas.

13.4.3 Prevention Strategies

The Florida Division of Forestry is convinced that the positioning of fire prevention teams in areas of high fire danger has substantially reduced the number of human-caused fires, but this is only one aspect of fire prevention strategies that range from landowner education to zoning and regulations.

13.4.3.1 Landowner Education

Fire prevention in and around homes has been a traditional message of municipal fire departments and some state forestry organizations for decades. Common venues for these fire prevention programs/displays have been schools, fairs, malls, television and radio public service announcements, and door-to-door contacts. Emphasis has been on preparations for fires started within the home as well as various home exterior and landscape maintenance projects to reduce personal fire risks in case a wildfire threatens. As interface areas expand, these traditional approaches face significant limitations. For instance, municipal agencies often cannot reach new developments outside their jurisdiction where landowners need in-depth information regarding the role of fire and landscaping options that help fireproof their homes while being in harmony with interface ecosystems. Educational programs are needed that explain how to landscape for diverse, and often conflicting, objectives ranging from water and energy conservation to fire preparedness to wildlife habitat improvement in the wildland–urban interface.

In the last two decades, and especially in the last 10 years, fire prevention education has expanded substantially in terms of content, delivery methods, and organizations involved in the programs. State forestry and wildlife organizations have been especially active in developing new programs and materials, often in conjunction with other organizations such as the Cooperative Extension Service, The Nature Conservancy, and various regional cooperatives. Examples of educational materials include:

- *Living with Fire: A Guide for the Homeowner*, a 12-page newsletter that was initially designed by the University of Nevada, Reno, Cooperative Extension Service, which has now been adapted by many western and southern states (Figure 13.6). The major focus of the publication is on making homes defensible in design, construction, and landscaping, but it also includes articles on the role of fire in natural ecosystems, fire behavior in local ecosystems, and what to do when a fire approaches (Smith and Skelly 1999).
- *Creating Wildfire-Defensible Zones* and *Fire Wise Plant Materials*, two landowner publications (eight pages each) prepared by the Cooperative Extension Services in Colorado and New Mexico. Similar brochures in many other states outline details for landowners to follow in home construction, landscaping, and plant selection in interface developments (Dennis 1999a,b).
- *Wildland Fire Education Handbook* (Monroe et al. 2000), a manual with supporting materials (slides, videos, and PowerPoint presentations) created by the University of Florida, Florida Division of Forestry, and The Nature Conservancy for use by extension agents, staff foresters, and fire prevention agencies in programs for landowners. The handbook includes a number of short publications on landscaping, considerations for subdivision development, prescribed burning, and the role of fire.
- *Fire Safe California Community Action Guide* (California Department of Forestry and Fire Protection 1996), produced with the assistance of the Western Insurance Information Service and the California Fire Safe Council. This publication is designed to help interface communities understand why they are at risk and develop recommendations for reducing that risk.

Every one of these educational materials stresses the fact that homeowners must assume responsibility in preparing for the inevitability of a wildland fire and each includes guidelines for carrying out that responsibility through the creation and maintenance of defensible home and landscape conditions. The materials are handed out through agency and extension offices, distributed at landowner/homeowner association meetings and fair booths, and available on many Internet web pages. They are also used as resources for many other newsletters (such as "Creating Fire-resistant Landscapes" in the Santa Fe, New Mexico, Botanical Garden Fall 2000 Newsletter) and media releases. The result is that fire prevention information is much more diverse and more widely distributed today than it was in the past.

Making this information most meaningful for, or adapting it to, particular audiences requires that educators in any organization must understand what landowners already know and what their perceptions and attitudes are toward risk, fire, and prevention options. "Needs assessments" of landowners and the general public were highlighted in the 2001–2002 National Fire Plan as a high priority for additional research and application, and efforts are under way across the country to conduct such assessments. Recent homeowner surveys have already demonstrated additional issues that should be considered in future educational programs. Landowners in Michigan, who experienced several devastating interface fires, including one escaped prescribed burn, were very suspicious of prescribed burning as a legitimate fuel reduction method (Winter and Fried 2000). They also tended to believe that nonresidents were responsible for most of the escaped backyard burns, when in reality most wildland fires result from negligence by permanent residents. These attitudes indicate the importance of educational programs that include not just one-time exposure to fire prevention messages at teachable moments, but repeated descriptions of successful prescribed burns and causes of local wildfires. Fire should be a constant part of the interface landowner's mindset. The lead article in the fall 2001 issue of the Forest Trust Quarterly Report states that it is much more practical for us to learn to live with natural disturbances and become a fire-adapted society than it is to try to change this natural process (Foster 2001).

A survey of 675 people in Florida indicated that a high percentage of both rural and urban homeowners understood that fire is a natural environmental factor in Florida, but they had questions about its effect on wildlife and air quality and were concerned about the effects of air quality and smoke in their immediate vicinity (Monroe et al. 2000). Those issues became focal points for several of the publications and presentation topics in the *Wildland Fire Education Handbook* described previously, and they illustrate the importance of adapting educational programs to audience concerns. A second recent survey in Florida also demonstrated the importance of educational programs for increasing public acceptance of prescribed fire (Loomis et al. 2001).

The increasing role of the Cooperative Extension Service in wildland and interface fire education benefits traditional fire prevention organizations as well as

FIGURE 13.6 "Living with Fire — A Guide for the Homeowner" was originally produced by the University of Nevada Cooperative Extension Service and has since been adapted by a number of other states as a primary educational tool for homeowners. (Adapted from Smith and Skelly 1999.)

landowners. Extension agents and offices are located in practically all counties in the country and they have a well-established rural landowner network. They have a long history of assessing landowner needs and experience with diverse program delivery systems from workshops and demonstrations to one-on-one problem solving. One of the strengths of local extension programs in other disciplines has been backstopping by a variety of state specialists. Although universities may not generally have the same level of expertise in wildland fire as in many agricultural arenas, specialist assistance to extension programs is available through a variety of interagency cooperatives and agreements. One of the things extension does best is provide locally specific education on important public issues. Interface fire management is one such issue, and extension programs nationally and at state levels should continue to play a key role in interface fire education.

13.4.4 Local Policies and Regulations

The ultimate goal of any educational program should be landowner actions that will mitigate risk and enhance firefighting efforts. Integral to these objectives is expanding landowner awareness of the role and effects of fire in natural ecosystems and the ways in which we have affected that role through management and interface development. Such an increased awareness will be invaluable in helping formulate local policies that influence fire management while benefiting landowners at the same time. Local policies may take the form of growth management planning, homeowner association specifications, and/or ordinances that regulate what can and cannot be done on private properties. They counter the only major disadvantage to educational programs, which is that landowner response to those programs is voluntary.

Residential developments frequently have specifications for home design, construction, and landscaping in order to maintain a certain character to the development (Figure 13.7). For example, the Genesee development near Golden, CO, and several communities outside Bend, OR, initially required wood roofs and siding on new homes to fit with the natural surroundings. Each of these communities experienced interface fires directly or nearby in the 1980s or early 1990s. Recognizing that wood roofs may be responsible for a high percentage of

FIGURE 13.7 Home construction near Boulder, CO, with a number of high fire risk features, including wood roof, siding, and overhangs, and both fine fuels and evergreen trees immediately adjacent to the structures. Local ordinances and homeowner association standards now require much more attention to reducing these features. (Photo by Alan Long.)

home ignitions, they have replaced the requirements for wood roofing with fire-resistant roof materials. Other examples can be found around the country where subdivision development guidelines either encourage or discourage fire prevention through green space, landscaping, home construction, or other requirements. If community covenants and restrictions prevent homeowner or association actions that would reduce fire risk, they need to be exposed as a public safety liability. Numerous examples exist by now that demonstrate the fallacy of ignoring the risk and hoping that nothing happens. Homeowner associations, developers, and insurance companies must be involved in any educational programs that focus on interface fire management.

County or other local ordinances are diverse in their detail and intent. After the Tunnel Fire in 1991, the City of Oakland adopted regulations for high fire hazard areas that require fire-resistant materials on new or replacement roofs and siding on new buildings with at least a 1-hour rating for resistance to flammability (Figure 13.8) (Ewell 1995). Along the Front Range in Colorado, building codes have been strengthened to institutionalize fire safety (Johnson and Mullenix 1995), and county governments require a Wildfire Mitigation Plan as part of the Site Plan Review for new development above the 6400-ft elevation contour (called the "Red Zone"). The new revisions for the Growth Management Plan for Alachua County, FL, include specifications for construction, streets and water supplies, and landscaping in high fire risk areas. Similar examples exist across the country for model codes developed at state levels and local ordinances regarding construction, landscaping, and development infrastructure. Less common today, but with considerable future value as prevention tools, are various types of insurance or other incentives (or disincentives) to encourage landowners to take positive actions.

One advantage of the regulatory approach is that it assures fire prevention measures will be taken in areas with an identified fire risk. However, this approach can also be loaded with pitfalls when uniformity is sought across different political jurisdictions. California has probably dealt with this issue longer than any other state, with regulations ranging from the California Building Code Title 24 to many city and county codes and ordinances. Yet, a uniform approach to fire prevention is lacking because of a lack of cooperation between state and local governments (Coleman 1995). Their experience should be an example, and a warning, to other states that fire prevention is a multiorganization effort that will require both educational and regulatory elements working together, as modeled in a number of recent community education and action programs.

13.4.5 Community Action Programs

Interface fires threaten homes, businesses, services, and the many natural resources that are part of ecosystems within the interface. It is, therefore, not only appropriate but imperative that prefire planning involve all stakeholders and interest groups, not just landowners and fire suppression agencies. Planning includes all the elements described in this chapter: risk assessment, prevention strategies and methods, subdivision development and construction standards, landowner and community education, fuel modifications, and fire suppression strategies. We briefly describe three examples of community-based programs below, recognizing that other programs, or adaptations of these three, may be more appropriate in other situations.

FIGURE 13.8 Home construction in the Oakland Hills, CA, following the deadly 1991 fire requires that exterior surfaces be constructed of fire-resistant materials. (Photo by Alan Long.)

Firewise Communities is a program developed by the National Wildland/Urban Interface Fire Program, which is sponsored by a number of state and federal agencies and associations and the National Fire Protection Association (see section on risk assessment). At the heart of the program are regional workshops that ideally include landowners, developers, realtors, and representatives from local government, insurance, banking, emergency management, fire agencies, extension and education, and other interested stakeholders. The goal of the workshops is to encourage the integration of Firewise concepts into community planning at all levels.

The first pilot workshop in the nation was held in Deerfield Beach, FL, in October 1999, under the direction of a national team of instructors and facilitators. The key activity in these workshops involves a set of case studies in which participants evaluate fire hazards, home construction and landscape issues, stakeholder concerns, strategic fuel reduction projects, and educational methods for involving the community in planning. Discussions range from the need for local ordinances to fire suppression coordination. Although the case studies are simulations, complete with GIS maps, the intent is that participants in each workshop should become planners and facilitators for future workshops in their home communities, developing solutions for their real-life situations. As an example, the Florida Division of Forestry has helped conduct workshops throughout the state since the Deerfield Beach workshop. They have been highly successful as judged by the fact that local groups are adopting the suggested methodologies and concepts. Regional workshops have also been held in many other states. The Firewise Communities web site provides information about the workshops, success stories that have followed workshops, and links to many related sources of information (National Wildland/Urban Interface Fire Program 2003).

The second example is the state of California, which has dealt with the specter of interface fires for over 40 years. During that period, fire agencies have used prevention programs, from guidelines for individual properties to large-scale fuelbreaks to building codes, with varying degrees of success. A statewide California Fire Safe Council was formed in 1993 with a mission to preserve California's natural and man-made resources by mobilizing all Californians to make their homes, neighborhoods, and communities fire safe. Through the 50 public and private organizations that make up its membership, it has distributed fire prevention education materials to industry leaders and their constituents, evaluated legislation pertaining to fire safety, and empowered grassroots organizations to spearhead fire safety programs.

The California Fire Plan was initiated in 1995 with the goal of turning fire prevention into a much more proactive and participatory process. The plan outlines a process that defines levels of service for each area of the state, considers values at risk, and includes the public in planning and taking actions before fires occur. Central to the Fire Plan is the formation of local or regional Fire Safe Councils to bring together all interested stakeholders to develop a fire prevention and mitigation strategy for their region or community. To assist the councils, a *Fire Safe California Community Action Guide* (California Department of Forestry and Fire Protection 1996) describes how councils might be formed and function, lists elements of fire-safe communities, neighborhoods, and properties, and presents several case studies. Activities promulgated by Fire Safe Councils include annual clean-up days for communities, fuel reduction in neighborhoods using portable chippers, media events and news releases, educational campaigns, landscaping demonstrations, and working with county officials to incorporate fire safe measures in county general plans (e.g., greenbelts, enforcement of building codes, development of emergency water systems). As of 2001, more than 60 Fire Safe Councils exist in California, with different levels of activity in each. An important element in the continued functioning of any individual council has been people who are willing to take on leadership roles.

The third example is the state of Colorado. The Front Range is home to many interface developments and has experienced a number of disastrous fires in what officials call the "Red Zone." A variety of programs and materials that address fire problems in this zone have been developed in the last 10 to 15 years by counties, federal agencies, and the Colorado State University Cooperative Extension Service. The Community Fire Prevention Partnership combined many of these materials into a FireWise notebook in 1999. The notebook is a primary resource for meetings with landowner associations, builders and developers, city and county officials, and other interested groups.

The above examples illustrate programmatic opportunities to increase awareness of interface fire and encourage actions to mitigate the danger, but an equally important prerequisite for effective fire management planning is the necessity of including all stakeholders, whether they are interested or not! Their ideas and concerns will undoubtedly reveal issues that traditional fire organizations have not considered and their involvement may well provide the support that those same organizations need as they promote the types of mitigation measures described in the following sections.

13.4.6 Landscape-Level Fuel Modifications

Risk assessment, prevention strategies, and community action programs all lead to the next critical aspect of interface fire management — the reduction of fire incidence and intensity. Most fire starts are the result of lightning, negligence and carelessness, or arson, although many other causes (e.g., sparks from equipment or brake shoes, hot mufflers, and spontaneous ignition) contribute to a small percentage of fires. The prevention programs described previously focus primarily on reducing human-caused fire starts, and we will not describe those efforts further here. Rather, we will turn our attention to the many opportunities to protect homes and property, especially by reducing fire intensity or blocking fire spread.

Fuel modification can be divided into two general types; linear and landscape. Linear firebreaks create a relatively narrow break in the path of a fire and are a common feature around many private forests and pastures in the South, where soils are sandy. They are less commonly used elsewhere, except during fire suppression operations. Linear firebreaks are usually created by plows, disks, or bladed tractors that leave mineral soil exposed and all surface fuels pushed to one side or covered by a dirt berm. They will stop many fires, but they lose their effectiveness as fireline intensity increases. Putting in multiple parallel lines will halt most wind-driven fires unless long-distance spotting is a problem. When spotting is a problem, the only effective solution is to rapidly widen the control line by burning out between the line and the approaching fire, before the wildfire reaches the line. Disked or bladed lines provide access to brush trucks and ATVs that are often equipped with drip torches allowing rapid ignition along the upwind side of the prepared break. When used as a preventive measure, linear breaks should be retreated annually, at least in the South, to remove dead fuels that fall on the line as well as to remove anything that has seeded in or resprouted.

In the Southwest, especially the chaparral region of California, wider control lines, ranging from several yards to over 500 ft wide, called fuelbreaks, stretch for miles along ridgetops (Figure 13.9). They are created by removing all the brush along the corridors, and in those cases where trees are present, thinning the overstory to leave widely spaced trees. Interestingly, the extra growth on the residual trees may have compensated for at least part of the cost of creating the fuelbreaks (Grah and Long, 1971). National Forests in California began creating these in the 1950s to provide major barriers to fire spread and to give suppression forces a baseline from which to work. In the following decade, this technique was expanded to remove understory vegetation and thin the overstory along roadways through many of the National Forests in California. In nontimbered areas, fuelbreaks may be created by removing all or most of the shrubs along the corridors, leaving only herbaceous and grass cover. There is much current interest in utilizing this methodology for manipulating fuels around interface developments nationwide, but it is a costly option and should only be contemplated where recent fire history strongly suggests its potential usefulness.

Greenways, greenstrips, or greenbelts are a form of fuelbreak in which the natural groundcover vegetation is replaced with species that will stay green during the fire season, or that are more resistant to fire spread than the natural cover (Davison and Smith 1997). Perhaps, the ultimate greenway is a golf course around an interface community. But even golf courses may not stop a worst-case conflagration because of the plethora of long-distance firebrands. They will, however, break up the head of the fire and serve as an anchor line for suppression forces. Whenever high-intensity fires are forced to the ground in fairly sparse fuels, they lose their ability to produce long-distance firebrands, which is a prerequisite to stopping them as long as unburned fuels are present downwind.

The other general fuel modification technique is landscape-level fuel modification, where large continuous areas of vegetation are substantially altered. Unlike linear

FIGURE 13.9 A ridge-top fireline and fuelbreak in the central California Coastal Range provides protection to wildland–urban interface homes on both sides of the line and a control line for fire suppression crews. (Photo by Alan Long.)

firebreaks, these techniques substantially reduce fire intensity across a large area, and provide an opportunity to correct one of the most important current issues in forest health — the undesirable ecological effects that decades of attempted fire exclusion have wrought. Millions of acres across the country now contain unprecedented forest floor accumulations and overly dense understories and overstories that are prone to insect and disease epidemics. Reducing the stature and/or density of the understory, sometimes in combination with a reduction in overstory density, will not only improve forest health but also remove ladder fuels that allow a fire to reach from surface fuels to overstory tree crowns. Correcting the dense stand conditions may improve both forest health and reduce fire intensities, but to do so requires an immense commitment of time and resources. Priorities for this commitment will undoubtedly favor interface communities in many places because of the high values of property and life in the wildland–urban interface.

13.4.7 Protecting Individual Properties

It is almost always the homes in the interface that result in the huge value losses from wildfire; they are also the asset on which suppression forces generally focus their attention, at the expense of other resources. When faced with an interface fire, suppression crews are often forced into triage because there are more homes threatened than there are suppression forces on the scene. The inevitable result is that some structures are left unprotected. Once homeowners accept this fact, the solution is obvious. The probability of individual homes surviving a wildfire intact is in large part dependent upon what homeowners choose to do to protect their property. It is thus incumbent upon homeowners to plan and carry out protective measures on their property that will improve access, increase defensible space and fire resistance of their home, and maintain those conditions over time. Where lots are less than several acres in size, the probability of a home surviving a wildfire is usually somewhat dependent upon the condition of adjacent properties. It is thus prudent for homeowners to undertake protective activities in concert with their neighbors, either informally or formally through a subdivision or county ordinance. Guidelines for these protective measures are available in brochures (e.g., Federal Emergency Management Agency 1993; Lippi and Kuypers 1998) or on Internet web sites through state, county, and local agencies (e.g., Dennis 1999a; Minnesota Department of Natural Resources 2003; National Wildland/Urban Interface Fire Program 2003). Specific technical help is almost universally available from state and local fire management agencies, and financial help is sometimes available through the types of programs described earlier in this chapter. Key points are summarized below, recognizing that the relative importance of a specific point will vary from region to region.

13.4.7.1 Access for Fire-Fighting Equipment

Provisions for fire-fighting equipment access are critical parameters in risk assessment protocols and community educational programs; if suppression forces cannot quickly reach threatened structures, they cannot adequately protect them. The NFPA 299 Standard (National Fire Protection Association 1997) provides criteria for fire agencies, land-use planners, architects, developers, and local governments for fire-safe development. Local jurisdiction authorities may adjust these criteria (usually more stringently) depending on local conditions. Important features include:

- Road widths that allow two-way traffic
- Vertical clearance above roads and driveways that allows passage of suppression equipment
- Road surfaces and bridge weight limits sufficient for large trucks
- More than one ingress/egress route, with access through unlocked gates
- Road grades, curves, and turnarounds that will accommodate large trucks
- Well-marked roads with clearly visible signs and house numbers
- Water supplies (wet or dry hydrants, reservoirs, storage tanks, swimming pools).

Once suppression equipment gets to an individual property, it is often necessary to move that equipment off the driveway and around structures. Locations of septic tanks, underground water pipes, and other structures such as fences and walls may limit the ability to position equipment where it can be most effective. Where natural gas is available, buried service lines are a significant problem.

13.4.7.2 Home Design and Construction

The second major protective measure, which is dependent upon builders and developers as well as the landowner, is the design and construction of homes and other structures. Many fire-prone interface communities have ordinances that dictate or ban specific building materials, especially roofs and siding. Although wood is often the material of choice for building exteriors in the interface, it is also the most flammable alternative. Wood used for siding, shingles, shakes, and decking can be treated to reduce flammability for a period of time, but brick, stucco, fiber-cement panels, metal, and even logs will be much more resistive to ignition (Slack 1999). However, if heat impinges on metal sheathing long enough, it can cause wood supports behind the siding to ignite. To prevent this, use gypsum sheathing between the siding or roof and the wood supports. New home construction and retrofitting older buildings in many interface developments now requires nonflammable exterior construction. Experience in Florida has shown that fire enters many homes because the fiberglass roof soffits melt (Abt et al. 1987; DeWitt 2000). Slack (1999) gives additional design and construction features to help provide structure protection:

- Noncombustible fine-mesh screens or skirts for subfloor vents, decks, and mobile homes
- Fire-resistive deck construction and roofing materials
- Flat rather than sloped soffits with quarter-inch or smaller wire mesh screening in vent openings
- Thermal (double) paned windows with exterior window covers or shutters
- Simple roof lines and walls or landscape features that will block the direct impingement of radiation or convective gases on the structure and vents
- Spark arresters on chimneys and flues.

In addition to building design and construction materials used, building location on the site and placement of adjacent structures will influence protection from fire. Homes should be set back on slopes so that they are not in the direct path of convection rising from a fire lower on the slope. Fences, outbuildings, woodpiles, and propane tanks should be located well away from the house or, in the case of fences, detached from the house if constructed of wood.

13.4.7.3 Defensible Space

Defensible space is the area between a house and an oncoming wildfire where the vegetation has been modified to reduce fire intensity, thereby providing safer opportunities for firefighters to defend the house. Although there are certain common parameters in most publications about defensible space, precise specifications vary considerably across the country (e.g., Foote et al. 1991; Gresham et al. 1997; Smith and Skelly 1999; Dennis 1999a; Florida Division of Forestry 2000). Landowners can access this information through local agencies and the Internet.

Whether explicitly described or not, defensible space usually includes at least three zones, each designed to serve a slightly different function. The "structure protection zone" (Zone 1, Figure 13.10) is usually the immediate 10 to 15 ft next to the house. Within this zone, any flammable material (dry grass, pine needles, wood mulches, and so on) should be replaced with ground covers (stone to green plants) that will stop ground fires before they reach or go under the house or mobile home. Unfortunately, many interface homes have ignited from burning pine needles or dry grass next to the structure rather than by intense heat or flames burning in other fuels. Other tasks within this zone include removing any branches (dead or alive) from within 10 ft of chimneys, keeping debris such as pine needles off roofs and decks, and removing stacked firewood from porches or from under the roof eaves.

The "defensible space zone" (Zone 2, Figure 13.10) is the next 20 to 100 ft around a building, with the exact dimensions dependent on slope, vegetation types, and prevalent weather. This is the area in which fire-fighting crews will be able to position themselves to knock down

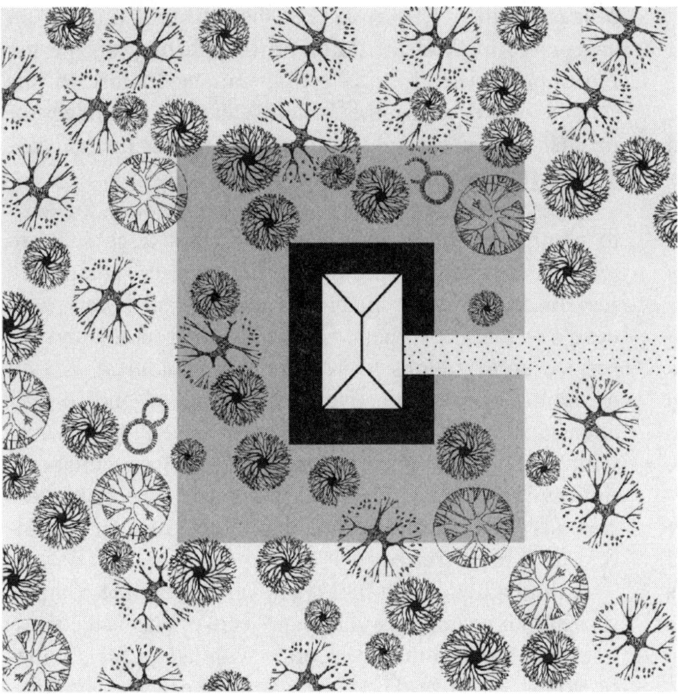

Zone 1
Zone 2
Zone 3

FIGURE 13.10 Defensible space around wildland–urban interface homes is generally designed with at least three zones: Zone 1, immediately adjacent to the structure, contains no flammable fuels; Zone 2 may contain widely spaced hardwood or conifer trees and even a few shrubs in small groups, but it must provide access for fire control crews; Zone 3 allows trees to be more closely spaced, but understory plants are still reduced enough to lower the intensity of approaching fires.

a fire and/or protect the home. This is the most critical zone for protection of structures in the interface. It is one of the main factors that firefighters will consider if they have to make a triage decision as to which structures to protect. The condition of this zone coupled with the exterior home construction materials used often determines whether a home can withstand a fire threat in the absence of fire crews. In addition to providing access, landscaping in this zone should be designed to significantly reduce the amount of radiant heat and convective gases that would impinge on the structure, either of which can lead to ignition of flammable building materials (Cohen 2000b). Common features within this zone include:

- Thinning trees and pruning dead or lower branches to eliminate crown-to-crown contact and foliage near the ground
- Removing ladder fuels (vines, shrubs, and smaller trees that can carry fire into the crowns of the larger trees)
- Removing flammable shrubs (chaparral, eucalyptus, saw palmetto [*Serenoa repens* (Bartr.) Small], wax myrtle [*Myrica cerifera* L.], and gallberry [*Ilex glabra* (L.) A. Gray]), or reducing them to small isolated islands and pruning them to maintain low stature
- Maintaining green lawns or other live ground cover under 6 to 8 in. tall
- Constructing walkways, driveways, fences, or walls with nonflammable materials that provide fuelbreaks of different sizes.

The "forest–woodland zone" (Zone 3, Figure 13.10) extends from the defensible space to property boundaries and beyond. The major purpose of manipulating vegetation within this area is to reduce fuel loading and/or flammability, thereby decreasing potential fireline intensity and production of firebrands. Secondary benefits include biodiversity, wildlife habitat, and soil and water protection. Fuel reduction measures in this zone focus mainly on reducing shrub density and height, thinning dense tree stands, and removing unhealthy and dead trees (although a few dead trees are often left for aesthetic and wildlife purposes). Similar tree and shrub removals should also be considered along power lines, trails, and fire access routes. As in the defensible space zone, slash created by these operations should be chipped, lopped, and scattered, or removed. Leaving large piles of debris simply creates an additional fire hazard, although a few small piles are sometimes left to provide cover for wildlife. Fuel reduction is accomplished with various mechanical tools and equipment, chemicals, prescribed burning, and grazing animals. The benefits and disadvantages of each will be discussed in the next section.

Within each of the three zones, landscaping guidelines often refer to planting or favoring fire-resistant plants. Although there are a few exceptions (such as some succulents), most plant species are potential fuel for fire after prolonged drought. "In fact, where and how you plant may be more important than what you plant" (Dennis 1999b). Many agencies will provide lists of species that have fire-resistant characteristics, but of perhaps more importance are lists of species to avoid. Fire resistance and flammability may be a function of age and size, branching patterns, foliage size and thickness (especially surface area to volume ratios), seasonal changes in foliage, live fuel moisture content, and chemical content. Groups of plants that are more flammable than others include conifers with resinous foliage such as some pines and junipers, shrubs that contain waxes or oils (wax myrtle, gallberry, and saw palmetto) or have aromatic leaves (eucalyptus [*Eucalyptus* sp.] and melaleuca [*Melaleuca quinquenervia* (Cav.) Blake]), shrubs that accumulate large amounts of dead foliage (many chaparral species), plants with high foliar surface-to-volume ratios (sagebrush, bitterbrush [*Purshia tridentata* DC.]) or hairy leaves, and grasses, especially those that grow tall and dense. Less flammable groups include most deciduous hardwoods and succulents, and many annuals and perennials with open branching patterns. For many landowners, however, plant selection is dictated by other objectives such as soil protection, wildlife or human food sources, shade, water conservation, and aesthetics. They can accomplish fire management objectives through the vertical and horizontal distribution of plants (planting pattern) in the landscape rather than by species choice.

It is important that homeowners recognize that the above home protection measures are not one-time projects. Fuels accumulate as vegetation resprouts, continues to grow, or dies. As an example, consider the annual fall of pine needles and other foliage onto roofs and into yards. Without regular maintenance, hazardous conditions are quick to return. Defensible space and homes must be regularly checked and treated to maintain protection from the inevitable fires.

13.4.8 Vegetation Management: Opportunities for Natural Resource Managers

Vegetation management and manipulation in the interface is an exciting, if not daunting, challenge to natural resource managers. As if the many social, cultural, and political issues and conflicts were not enough of a challenge, decisions pertaining to vegetation pattern and ecosystem health result in a whole new arena of resource conflicts and tradeoffs. For example, although fuel reduction is the most crucial step in home protection, as well as necessary to maintain/improve forest health, if not done correctly it may create opportunities for invasive species, increase soil erosion, and degrade water quality. Vegetation management can also be obtrusive to neighbors. Thus, resource managers must determine not only how they need to manipulate ecosystem communities to best reach their objectives, but they must also select an acceptable method for doing so (which may not be the most appropriate from an ecological standpoint). In this section, we describe the benefits and disadvantages of mechanical, chemical, biological (grazing), and prescribed burning methods to modify fuels.

Throughout this chapter, we have referred to the need to manipulate fuels and vegetation on a large scale to restore ecosystem health, reduce fire intensity, and reduce fuel loads. Resource managers should collaborate with developers, home owners, nearby public agencies, and fire service organizations to be sure that ecological as well as anthropogenic objectives are considered and to take advantage of the economies of scale that may be possible with treatments applied to larger areas. These collaborations will also help in hazard or risk assessments that are necessary to determine problematic fuel types and prioritize areas to be treated.

Although problem fuel types vary across the country, one similar characteristic in many ecosystems is a continuous and dense shrub layer, often with ladder fuels (vines or young trees) bridging the gap between surface fuels and the overstory. The gallberry–saw palmetto complex provided much of the fuel in the 1998 wildland–urban interface fires in Florida (Figure 13.11). Both species grow back rapidly after all but the most severe fires. Frequent low-intensity fires historically perpetuated this ecosystem, confining these woody shrubs to the groundcover that was dominated by wiregrass. With attempted fire exclusion, these shrubs have proliferated to the point where even annual dormant season fires have no impact on their density, and high-intensity fires can result within 4 to 5 years. Periodic fire has also maintained the chaparral brush fields of southern California, except that these sites can also experience substantial soil erosion in the brief months before hillsides are revegetated. In the Great Basin, bitterbrush burns intensely but does not resprout as readily after a fire as gallberry and chaparral species. Bitterbrush is a critical browse species for mule deer, and since this plant species recovers slowly after fire, burns that cover large areas could seriously deplete the deer's winter range. These three examples illustrate that although the need for fuel reduction might be a fairly straightforward decision, treatment methods must accommodate other concerns.

Four general methods are potentially available to the resource manager to modify vegetation structure and composition: prescribed burning, machines, herbicides, and livestock grazing. They may be used as single treatments or applied as combination treatments. The applicability of these alternatives is currently the focus of numerous studies nationwide (e.g., Brose and Wade, 2001), but few results have yet emerged.

Prescribed burning is preferred by many landowners because it is the method that historically maintained these

FIGURE 13.11 Dense shrubs and saw palmetto characterize empty lots and adjacent woodlands in many wildland–urban interface communities in Florida. (From Monroe et al. 2000.)

ecosystems. Ideally, forest health will be restored and maintained over the long run by reintroducing fire at frequencies, seasons, and intensities that resemble its natural historical regime. Using prescribed fire, fuels can be significantly reduced over large areas, but this reduction is only temporary. Many examples can be cited where recent burns have dramatically reduced wildfire intensity (Cumming 1964; Helms 1979; Wagle and Eakle 1979; Outcalt and Wade 2000; Thorstenson 2001). In addition to reducing fuel loads, prescribed fire results in numerous other benefits including: wildlife habitat improvement; increased amount, palatability, nutritional quality, and availability of forage; increased fruit and mast production; preservation of endangered plant and animal species; nutrient cycling; disease, insect, and pest control; the ability to time burns to minimize detrimental effects such as air quality or game bird nesting and optimize desired effects such as control of shrub sprouts; slash disposal and site preparation prior to afforestation; improved accessibility; enhanced aesthetics and flowering of groundcover species; reduced suppression costs; reduced firefighter risk; and facilitation of harvesting (Wade et al. 2000). Not only can prescribed fire accomplish multiple benefits, it can generally do so at a cost lower than other vegetation treatments.

Prescribed burning can be defined as the intentional application of fire in accordance with a written prescription under specified environmental conditions in a manner that ensures the fire will be confined to a predetermined area and accomplish planned resource management objectives. Besides describing why, when, where, and how the burn prescription should include a description of current stand and fuel conditions, a smoke management plan, mop-up and control standards, and evaluation criteria (Wade and Lunsford 1989). Tens of thousands of prescription fires are conducted safely to treat over 6 million acres annually across the U.S., including a small percentage in interface areas.

Because prescribed burning is as much an art as it is a science and burns are subject to abrupt unforecast changes in weather, the threat of fire escape or smoke intrusions into smoke-sensitive areas are ever present. Although good planning and prudent execution of the burn plan minimize these threats, a few fires each year experience such problems, sometimes coupled with poor decisions, which result in significant property loss and media attention (such as the 2000 Cerro Grande fire near Los Alamos, NM). These concerns multiply significantly when burning in the wildland–urban interface, which translates into tighter constraints on when fires can be conducted.

A common complaint voiced by prescribed burners is that there are not enough acceptable burn days in a given year. The added constraints at the wildland–urban interface exacerbate this situation, the result being that even if the burn manager effectively uses every acceptable day, only a small portion of the areas that have been identified as needing fuel reduction can be treated in a given year. Resource managers must rely on mechanical and herbicide treatments to treat the rest of the areas in need of fuel reduction.

Perhaps the two biggest drawbacks of prescription fire for experienced southern burners are the potential for unexpected smoke problems and the fact that the fuel reduction achieved is only temporary. Because most woody understory species rapidly resprout after fire, retreatment is necessary every few years. For example, the palmetto/gallberry fuel complex can regain its preburn stature on good sites within 5 years (Brose and Wade 2001). In the wildland–urban interface, where additional resources are generally required to conduct and mop up the burn, costs can be prohibitive unless specific funding for treating wildland–urban interface fuels is available. The only way such funding may materialize is for constituents to apply unified pressure on their county, state, and federal representatives. But first, property owners at risk from wildfire must recognize that they are at risk and that this risk can be mitigated.

Irrespective of any leverage the public might exert, it is incumbent upon organizations charged with fire management in the wildland–urban interface to take a proactive approach to increase landowner knowledge regarding the differences between wildfire and prescription fire and the pros and cons of various fuel reduction treatment options. Public education programs through fire service agencies, extension service offices, and landowner associations will help landowners understand the nature of the ecosystems in which they live and the necessity of periodic fire to sustain them. Public media announcements prior to, during, and after a scheduled interface burn will alert landowners that a burn will occur, provide them with written, verbal, and/or visual information demonstrating that the burn crew can accomplish stated fuel reduction objectives safely and efficiently, and keep the landowners informed of recovery after the fire (Figure 13.12). Small demonstration burns conducted by local and state fire organizations can further serve as an educational program while accomplishing fuel reduction on the demonstration areas. Some agencies have gone so far as to transport interested individuals to a site on the day an operational burn is scheduled, where they are briefed by a member of the burn crew and can see firsthand the complexity involved and the professional manner in which the burn is executed. Because of the inherent complexity of wildland–urban interface burns and the resource values at stake, only experienced burners who have had additional training should conduct such burns.

FIGURE 13.12 Public education is critical for the continued use of prescribed burning as one of the management tools for reducing hazardous fuel levels. This prescribed burn manager is sharing current burn status "live" with a local TV station news crew. (From Monroe et al. 2000.)

Herbicides are attractive to managers because they can be formulated to kill selected species and they are cost-effective, especially since retreatment may not be necessary for much longer periods than the other vegetation management methods. They can also be applied on steep slopes, often with helicopters, where machines cannot operate and where hand-held cutting tools are expensive and dangerous from an operator safety standpoint. However, chemical fuel reduction also has several shortcomings: herbicides can only be applied during the growing season; there is an 18–24-month period after treatment when the dead woody shrubs remain standing and pose an extreme fire hazard; their use is not acceptable to many landowners; and herbicides do nothing to reduce the forest floor. Herbicide treatments provide some protection from soil erosion in steep terrain because the root systems remain in place for several years, and once the dead material falls over, it slows runoff. On flat or gently sloping, ground application methods can include tractor or ATV-mounted equipment, aircraft, or backpack sprayers. Costs are generally reasonable and competitive with mechanical or fire treatments.

Mechanical reduction of fuel loads is frequently chosen in interface ecosystems because a variety of mechanized and hand-held equipment can be used without the weather constraints or potential off-site problems associated with prescription fire or the seasonal limitations of herbicides. Mechanical methods range from hand-held saws, loppers, and other cutting implements to thinning equipment that cuts and removes trees from a site to machines with rotating blades, chains, or rollers that crush and chip everything in their path (Figure 13.13). This mechanized equipment varies from small tractor-mounted implements such as mowers and bush hogs to feller-bunchers and large machines specifically designed for this purpose. As with prescribed burning, there is an immediate reduction in understory stature and potential flame height. Mechanical treatment is more acceptable to landowners who worry about fire or chemicals. Treatments can be applied year-round at a reasonable cost in many situations, at least for the smaller equipment. On the other hand, most equipment is limited to slopes less than 30 to 40 percent, large equipment can be expensive, and safety issues are high with hand-held equipment, especially when it is used on steep slopes. There are also weather constraints on the use of heavy equipment to avoid soil compaction or erosion. To the extent that vegetation is pulled or plowed out

FIGURE 13.13 The Gyrotrac® mowing machine, with teeth mounted on a rotating cylinder, is capable of reducing shrubs and small trees to ground mulch. (Photo courtesy of Larry Korhnak and Florida Division of Forestry.)

of the ground, bare soil can lead to increased soil erosion for a period of time; however, if vegetation is only chopped or cut, it often resprouts, requiring periodic retreatment. Although the fuel is no longer standing, it remains on site as a fire hazard. Disposal of the cut materials can also be a problem unless they are chipped and spread out or burned on site. Fuel reduction projects in interface developments often include an arrangement by which merchantable logs and chips are removed to reduce the cost of treatment.

Livestock grazing is probably the least used method of fuel reduction, although it is also probably the least expensive and most benign in terms of effects on the landscape. There has recently been some renewed interest in this option, and it is currently being field tested in New Hampshire, FL, and elsewhere. Cattle, sheep, and goats can all be used, although their effects will vary depending on the plant species present on the site, slope steepness, and duration of grazing. Goats are notorious for eating almost anything, while cattle and sheep may be more selective, especially if a site is not overgrazed. Plants will resprout after they have been grazed, but livestock can be rotated back onto a site to maintain the vegetation at a low level. Costs associated with livestock include fencing, water sources, supplemental feeds if necessary, and caretakers for the livestock; but many of the costs can be offset when the livestock owner sells animals or when the landowner receives money for grazing lease arrangements with the livestock owner.

13.5 CONCLUSION

Fire is as responsible as any other factor for bringing many issues of the wildland–urban interface into sharp focus. Key issues from a fire management standpoint are vegetation characteristics and fuel loads in interface ecosystems, problems with trying to protect homes, lives, and resources at the same time, landowner values and expectations for services they will receive, and the infrastructure (or lack thereof) to supply those services. A variety of mechanisms is necessary for coping with fire in the interface: risk/hazard assessment, landowner and community education, zoning ordinances and related regulations, cross-training and multiagency cooperatives for fire-fighting organizations, and fuel modifications. Examples of how these mechanisms are administered around the country are many and varied.

No matter what is done to manage fuels, fires are inevitable. There is no silver bullet. Decisions during home construction and landscaping, particularly adopting Firewise recommendations such as defensible space, will help protect individual homes. But the much larger landscapes with continuous fuels in and around interface developments must also be treated. Fuel reduction is possible with various combinations of prescribed fire, mechanical implements, herbicides, and livestock. Each method has a different set of benefits, tradeoffs, and appropriate applications. Prescribed burning is preferred in many situations because it can lead to the reestablishment of natural ecosystem processes, but it presents its own set of challenges to fire management in the interface.

REFERENCES

Abt, R., D. Kelly, and M. Kuypers, 1987. The Florida Palm Coast Fire: an analysis of fire incidence and residence characteristics, *Fire Technology* 23: 230–252.

Boulder County Wildfire Mitigation Group, 2001. The WHIMS (Wildfire Hazard Identification and Mitigation System) Manual, Boulder County, CO.

Brose, P. and D. Wade, 2001. Potential fire behavior in pine flatwood forests following three different fuel reduction techniques, *Forest Ecology and Management* 163: 71–84.

California Department of Forestry and Fire Protection, 1996. Fire Safe California community Action Guide, California Fire Safe Council, Sacramento.

California Department of Forestry and Fire Protection, 2000a. Structural Fire Prevention Field Guide for Mitigation of Wildland Fires, Governor's Office of Emergency Services, Sacramento.

California Department of Forestry and Fire Protection, 2000b. Property Inspection Guide, Governor's Office of Emergency Services, Sacramento.

California State Board of Forestry, 1996. California Fire Plan: A Framework for Minimizing Costs and Losses from Wildland Fires, California Department of Forestry and Fire Protection, Sacramento.

Cohen, J.D., 2000a. Examination of the home destruction in Los Alamos associated with the Cerro Grande Fire, May, 2000, *Wildfire News and Notes* 14: 1, 6–7.

Cohen, J.D., 2000b. Preventing disaster: home ignitability in the wildland–urban interface, *Journal of Forestry* 98: 15–21.

Coleman, R.J., 1995. Structural Wildland Intermix, in *The Biswell Symposium: Fire Issues and Solutions in Urban Interface and Wildland Ecosystems,* D. Weise and R. Martin, Eds., Gen. Technical Report PSW-GTR-158, U.S. Department of Agriculture, Forest Service, Pacific Southwest Research Station, Albany, CA., pp. 141–145.

Colorado State Forest Service, 1997. Wildfire Hazard Mitigation and Response Plan, A Guide, Colorado State University, Fort Collins.

Cumming, J.A., 1964. Effectiveness of prescribed burning in reducing wildfire damage during periods of abnormally high fire danger, *Journal of Forestry* 62: 535–537.

Davison, J. and E. Smith, 1997. Greenstrips: Another Tool to Manage Wildfire, University of Nevada Cooperative Extension Service, Fact Sheet 97–36.

Dennis, F.C., 1999a. Creating Wildfire-defensible Zones, Colorado State University Cooperative Extension Service, Natural Resource Series no. 6.302, www.ext.colostate.edu/pubs/ natres/06302.html. [Date accessed: June 2003.]

Dennis, F.C., 1999b. Firewise Plant Materials, Colorado State University Cooperative Extension Service, Natural Resource Series no. 6.305, www.ext.colostate.edu/pubs/natres/06305.html. [Date accessed: June 2003.]

DeWitt, J.L., 2000. Analysis of the Utility of Wildfire Home Protection Strategies in Central Florida, A final report submitted to the Interagency Fire Science Team, Florida Division of Forestry, Tallahassee.

Ewell, P.L., 1995. The Oakland-Berkeley Hills Fire of 1991, in The Biswell Symposium: Fire Issues and Solutions in Urban Interface and Wildland Ecosystems, D. Weise and R. Martin, Eds., Gen. Technical Report PSW-GTR-158, U.S. Department of Agriculture, Forest Service, Pacific Southwest Research Station, Albany, CA., pp. 7–10.

Federal Emergency Management Agency (FEMA), 1993. Wildfire — Are you prepared? Brochure L-203, Federal Emergency Management Agency, Washington, DC.

Florida Division of Forestry, 2000. Are You Firewise Florida? Florida Department of Agriculture and Consumer Services, Tallahassee.

Florida Division of Forestry, 2002. Wildfire Hazard Assessment Guide for Florida Homeowners, Florida Department of Agriculture and Consumer Services, Tallahassee.

Foote, E.I.D., R.E. Martin, and J. Gilless, 1991. The Defensible Space Factor Study: A Survey Instrument for Post-fire Structure Loss Analysis, in Proceedings 11th Conference on Fire and Forest Meteorology, Missoula, Montana, 1991, Society of American Foresters, pp. 66–73.

Foster, B., 2001. Living with fire: experts discuss fire policy, *Forest Trust Quarterly Report* 25: 1–2, 9–13.

Governor's Wildfire Response and Mitigation Review Committee, 1998. Through the flames ... An Assessment of Florida's Wildfires of 1998, Florida Department of Consumer and Agricultural Services, Tallahassee.

Grah, R. and A. Long, 1971. California fuelbreaks: costs and benefits, *Journal of Forestry* 69: 89–93.

Great Lakes Forest Fire Compact, 1996. Protecting Life and Property from Wildfire: An Introduction to Designing Zoning and Building Standards for Local Officials, Michigan Department of Natural Resources.

Gresham, R., S. MacDonald, and C. Heitz, 1997. The Defensible Space and Healthy Forest Handbook: A Guide to Reducing the Wildfire Threat, Placer County (California) Resource Conservation District.

Helms, J. A., 1979. Positive effects of prescribed burning on wildfire intensities, *Fire Management Notes* 40: 10–13.

International Fire Code Institute, 2000. 2000 Urban–wildland Interface Code, International Fire Code Institute, Whittier, CA.

Johnson, M. and M. Mullenix, 1995. Institutionalizing Fire Safety in Making Land Use and Development Decisions, in The Biswell Symposium: Fire Issues and Solutions in Urban Interface and Wildland Ecosystems, D. Weise and R. Martin, Eds., Gen. Technical Report PSW-GTR-158, U.S. Department of Agriculture, Forest Service, Pacific Southwest Research Station, Albany, CA, p. 33.

Lightfoot, K., M. Martinez, and B. Luna, 1999. Fire in the Wildland Urban Interface Risk Analysis, Forestry Division, New Mexico Energy, Minerals, and Natural Resources Department.

Lippi, C. and M. Kuypers, 1998. Flagler Horticulture: Making Your Landscape More Resistant to Wildfires, Flagler County Extension Publication, University of Florida Cooperative Extension Service.

Loomis, J.B., L.S. Bair, and A. González-Cabán, 2001. Prescribed fire and public support: knowledge gained, attitudes changed in Florida, *Journal of Forestry* 99: 18–22.

Minnesota Department of Natural Resources, 2003. Firewise website, http://www.dnr.state.mn.us/firewise/index.htm. [Date accessed: June 2003.]

Monroe, M. and A. Long, 2000. Landscaping in Florida with Fire in Mind, Circular FOR 71, University of Florida Cooperative Extension Service.

Monroe, M., S. Marynowski, and A. Bowers, 2000. Wildland Fire Education Handbook, Circular 1245, University of Florida Cooperative Extension Service.

National Fire Protection Association, 1997. NFPA 299 Standard for Protection of Life and Property from Wildfire, 1997 Edition, National Fire Protection Association, Quincy, MA.

National Fire Protection Association, 2002. NFPA 1144. Standard for Protection of Life and Property from Wildfire, 2002 Edition, National Fire Protection Association, Quincy, MA.

National Wildland/Urban Interface Fire Program (NWUIFP), 2003. Firewise communities website, http://www.firewise.org. [Date accessed: June 2003.]

National Wildland/Urban Interface Fire Protection Program, 1998. Wildland/urban interface fire hazard assessment methodology, NWUIFP, Quincy, MA.

Outcalt, K.W. and D.D. Wade, 2000. The Value of Fuel Management in Reducing Wildfire Damage, in Proceedings Joint Fire Science Conference and Workshop, L.F. Neuenschwander and K.C. Ryan, Eds., Boise, ID, 1999, University of Idaho, Moscow, ID, pp. 271–275.

Platt, W.J., J.S. Glitzenstein, and D. R. Streng, 1991. Evaluating Pyrogenicity and its Effects on Vegetation in Longleaf Pine Savannas, in Proceedings 17th Tall Timbers Fire Ecology Conference, 1989, Tall Timbers Research Station, Tallahassee, FL, pp. 143–161.

Slack, P., 1999. Firewise Construction Design and Materials, Colorado State Forest Service, www.firewise.org/co/construction.html. [Date accessed: June 2003.]

Slaughter, R., Ed., 1996. California's I-Zone — Urban/wildland Fire Prevention and Mitigation, California Department of Forestry and Fire Protection, and Office of the California State Fire Marshal, Sacramento.

Smith, E. and J. Skelly, 1999. Living With Fire: A Guide for the Homeowner, University of Nevada Cooperative Extension Service, Reno.

Thorstenson, A., 2001. When to burn: Northern Great Plains fire policies, *Distant Thunder* 11: 3,14.

Virginia Department of Forestry, 1997. Wildfire Risk Analysis, Virginia Department of Forestry.

Wade, D.D., B.L. Brock, P.H. Brose, J.B. Grace, G.A. Hoch, and W.A. Patterson, III, 2000. Fire in Eastern Ecosystems, in Wildland Fire in Ecosystems: Effects of Fire on Flora, J.B. Brown and J.K. Smith, Eds., Gen. Technical Report RMRS-GTR-42-Vol. 2, U.S. Department of Agriculture, Forest Service, Rocky Mountain Research Station, Fort Collins, CO, pp. 53–96.

Wade, D.D. and J.D. Lunsford, 1989. A Guide for Prescribed Fire in Southern Forests, Tech. Pub. R8-TP11, U.S. Department of Agriculture, Forest Service, Southern Region, Atlanta.

Wagle, R.F. and T.W. Eakle, 1979. A controlled burn reduces the impact of a subsequent wildfire in a ponderosa pine vegetation type, *Forest Science* 25: 123–129.

Winter, G. and J.S. Fried, 2000. Homeowner perspectives on fire hazard, responsibility and management strategies at the wildland–urban interface, *Society and Natural Resources* 13: 33–49.

14 Forest Health in the Wildland–Urban Interface: Insects and Diseases

Edward L. Barnard
Florida Department of Agriculture and Consumer Services, Division of Forestry

CONTENTS

14.1 INTRODUCTION

Any discussion of forest health can be a complicated exercise, regardless of the particular context: natural forest, tree farm, plantation or commercial production forest, urban "forest," or the wildland–urban interface. Engaging in such an endeavor presumes at least a rudimentary understanding of the subject entities and concepts, or some level of consensus as to the meaning of the various terms. What is a forest? What is forest health? What is the wildland–urban interface, where does it start, and where does it end? These are not, or at least they have not been, short-answer questions. Simple, static, and generally accepted definitions remain elusive due to the very nature of biological systems and the constantly evolving and sometimes conflicting understandings we apply to them.

Helms (2002) recently reviewed the Society of American Foresters' evolving definition and use of the term *forest*. His treatise recounts changes from a utilitarian, timber-production definition to one that reflects a more holistic, ecosystem understanding. He further indicates that functional uses of the term *forest* are often related to "administrative unit (national forest, state forest, reserve forest, industrial forest, or experimental forest), land use…or land cover." My discussion of forest health will center largely on the context of land use; that is, a continuum of varying and often highly altered forests on lands somewhere between the natural forest (wilderness?) and the inner city. These forests in the wildland–urban interface are perceived, influenced, and used by people in a variety of ways, for a variety of purposes, and across a range of intensities (Macie and Hermansen 2002) (Figure 14.1). I hope this chapter takes a holistic approach to forest health, recognizing that there are also operating constraints.

14.2 THE NATURE OF FOREST HEALTH

Like the term *forest*, the term *forest health* has undergone its own evolution from a traditionally narrow, management-focused, utilitarian perspective to a broader more holistic environmental or ecosystem-centered view (Kimmins 1997). Nonetheless, consensus is still elusive as health is often perceived and measured in relationship to one's particular values or objectives. What may be desirable to forest managers emphasizing timber production may well not be to others interested primarily in wildlife habitat or biodiversity. Because forest health is such a value-laden term, defined in part by science, and in

1-56670-602-5/05/$0.00+$1.50
© 2005 by CRC Press LLC

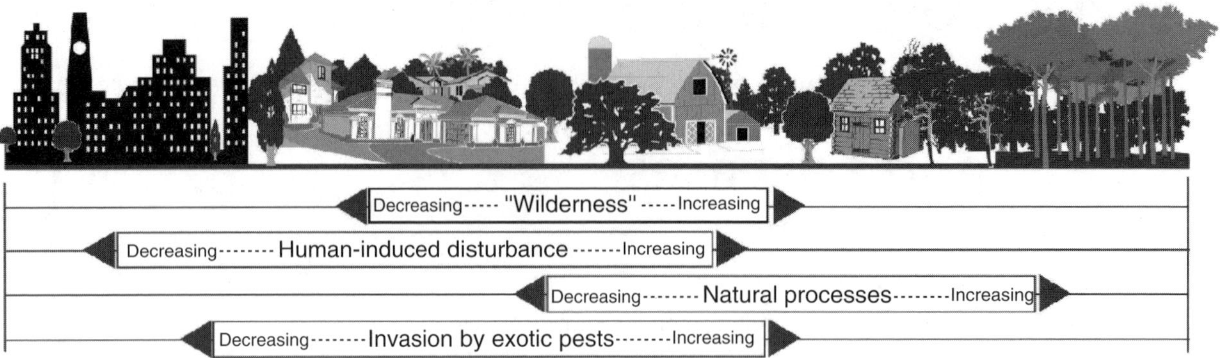

FIGURE 14.1 The sliding scale of the wildland–urban interface. As one moves from wilderness through the wildland–urban interface into the urban "forest," a variety of parameters change (directly or indirectly) in relative proportion to the degree of disturbance imposed on the system(s) by humans, and all have an impact on the health of trees and forest systems.

part by social or personal values and management objectives, it is unlikely that a "one size fits all" definition will be forthcoming anytime soon.

The USDA Forest Service (2003) has defined forest health as

> A condition wherein a forest has the capacity across the landscape for renewal, for recovery from a wide range of disturbances, and for retention of its ecological resiliency, while meeting current and future needs of people for desired levels of values, uses, products, and services.

and has stated that

> A desired state of forest health is a condition where biotic (living) and abiotic (nonliving) influences on the forest (e.g., pests, atmospheric deposition, silvicultural treatments, and harvesting practices) do not threaten resource management objectives now or in the future.

The Society of American Foresters (Helms 1998) has adopted the following definition of forest health.

> … the perceived condition of a forest derived from concerns about such factors as its age, structure, composition, function, vigor, presence of unusual levels of insects or disease, and resilience to disturbance — **note**, perception and interpretation of forest health are influenced by individual and cultural viewpoints, land management objectives, spatial and temporal scales, the relative health of the stands that comprise the forest, and the appearance of the forest at a point of time.

Regardless of the accuracy, completeness, particular emphases, or permanence of forest health definitions, one thing is certain and needs to be emphatically stated. Forest health and tree health are not the same thing. Nonetheless, many are prone to treat them as if they were. In reality, dead, dying, decaying, and insect-infested trees are perfectly natural, and can be components of perfectly healthy forests, depending upon a variety of factors, including the nature of the particular forests and the age, size, numbers, and distribution of dead and dying trees. Some trees are unhealthy; yet, the forests are healthy. In any discussion of forest health, one must be careful to consider the forest as well as the trees. Forest health (good or bad) is a composite result of the natural interactions of the myriad components of forest ecosystems (Figure 14.2).

14.3 THE WILDLAND–URBAN INTERFACE AS A SETTING

Forests in the wildland–urban interface exist in a people-influenced and often a people-dominated landscape. These forests, a product of complex interactions of natural and human factors (Figure 14.2), are varied and constantly changing. By definition, the wildland–urban interface is a human-altered system (Figure 14.1). Forests therein are highly sculpted by the values, perceptions, motivations, and actions of people (Macie and Hermansen 2002; Wear and Greis 2002). Historical and current land uses, site disturbances, system perturbations, the importation and movement of exotic plant species, and applied management practices inevitably have an impact on forest health. Such anthropogenic influences may be harmonious or discordant with organisms and processes in a given place and time, and each or all of these influences may impair, accommodate, or promote the temporal and spatial requirements and dynamics of natural ecosystems.

Forests in the wildland–urban interface are typically highly fragmented, both physically and socially (Figure 14.1). Such fragmentation inevitably affects natural ecological processes, associated management practices (active or passive), and sociopolitically imposed managerial constraints. Forest or woodlot parcels are often small,

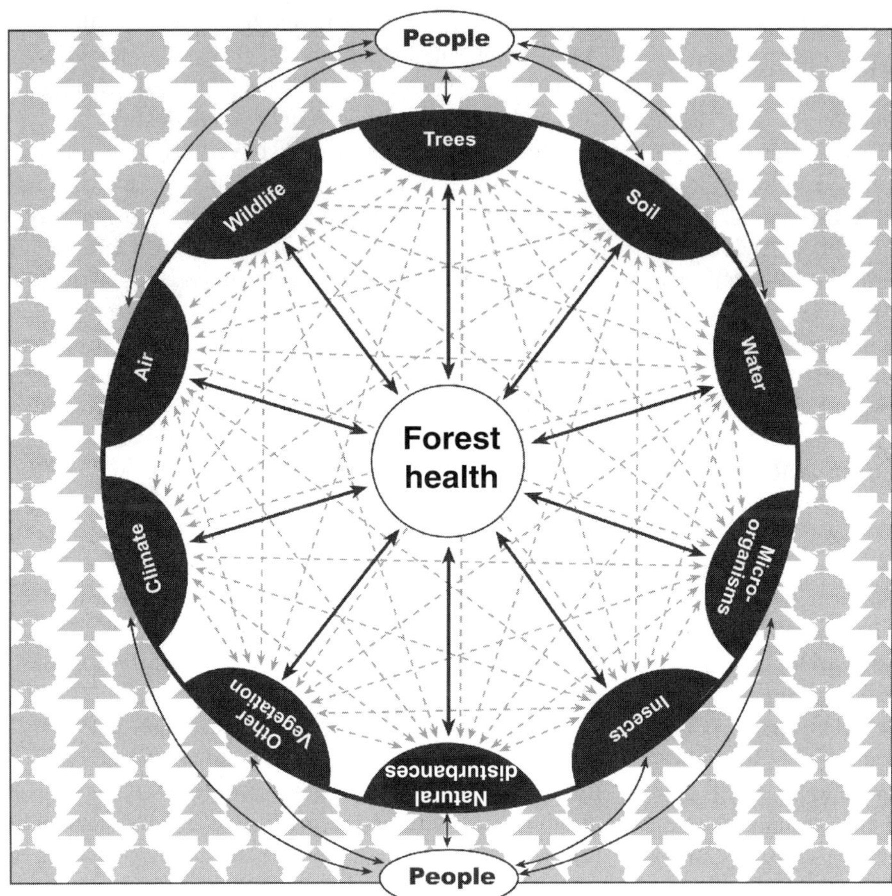

FIGURE 14.2 Interactions among components of forested ecosystems. Each component affects and is affected by every other component, and each affects and is affected by forest health. In the wildland–urban interface, people both affect and are (directly or indirectly) affected by the various components as well as the health of the forests themselves.

creating large perimeters relative to interior land area, consequently affording increased exposure to and opportunities for invasion by damaging indigenous and non-indigenous pest organisms. Edaphic, biotic, and other environmental resources required for healthy trees and forest ecosystems may become tainted or limited by air and water pollution, soil movement and compaction, hydrological alterations, and so forth. Social perceptions and heightened concerns for safety and liability often result in policies or regulations that detract from long-term forest health. Limits on or curtailment of ecosystem-level natural processes (e.g., wildfires, periodic flooding, natural catastrophic stand replacements, and forest succession) and system-mimicking silvicultural practices (prescription burning, intermediate or stand replacement harvests, etc.) often result in unnatural or undesirable changes in species composition and age structure or excessive competition for limited resources, and threaten the health of trees and forests. Limited product and labor markets in the wildland–urban interface may also result in insufficient or ineffective silviculture, thus decreasing forest health. Products of differing natural and human-influenced histories and values, forests in the wildland–urban interface are often unhealthy, and each has different needs and poses different opportunities for management to maintain or promote forest health.

14.4 INSECTS AND DISEASES THAT AFFECT TREES: WHY?

Given the preceding as a contextual framework, I now focus my remarks on tree health as related to insects and diseases in the wildland–urban interface, not forest health *per se*. While tree health is not equivalent to forest health, tree health both influences and is influenced by forest health (Figure 14.2) and, properly interpreted, is often a reasonable indicator of forest health. Tradition has long focused on insects and diseases of trees as causes of poor forest health. Where appropriate, I will stress insects and diseases as reflections or effects of poor forest health. My discussion will be limited to relationships between trees and their respective "pest" organisms, as opposed to a comprehensive treatise of the intricate interactions occurring in forest systems. Other chapters in this book address key components (fire, water, invasive plants, etc.) of the interface forest, all of which are integral parts of a holistic forest health paradigm (Figure 14.2).

14.5 INSECTS AND PATHOGENS THAT AFFECT TREES — MODUS OPERANDI

Both indigenous (native) and nonindigenous (introduced or exotic) insects and pathogens (disease-causing organisms such as bacteria, fungi, or viruses) can pose threats to the health of individual trees and forest systems in the wildland–urban interface. How do they do so, and how do we cope? To begin with, it is imperative to note an important difference in modus operandi between those pest organisms that are indigenous to a particular geographic location or forest system and those that are not.

Indigenous organisms have coexisted with their hosts in their particular ecosystems for thousands of years or more. During this coexistence, host tree species and forest ecosystems have developed mechanisms to defend themselves from or tolerate attack by their "aggressor(s)" (coevolution of host and pest). In fact, many such aggressors are generally normal components of healthy forest ecosystems and contribute positively to ecological functions and natural forest succession (Franklin et al. 1987; Muller-Dombois 1987; Haack and Byler 1993). Furthermore, indigenous pest organisms have also coevolved with their own sets of natural enemies (predators, parasitoids, pathogens, etc.), which tend to keep their populations in check. In these pest- or patho-systems, the pest organisms are usually of little negative consequence to forest health (or the long-term health of their host species). Indeed, many so-called pest organisms in such systems often become pests by responding to the condition or health of their hosts, typically as influenced by their environmental surroundings (Olkowski et al. 1991). Many such pathogens and insects respond to the old age and associated senescence of their host trees and/or the natural environmental or physical stresses to which these potential hosts are subjected. Or, they take advantage of the physiological condition of their hosts or altered populations of natural enemies as influenced by human-imposed cultural practices. In the wildland–urban interface, altered environmental surroundings (e.g., changed drainage and compacted soils), mechanical injury, and other physiological stresses are common. Hence, the wildland–urban interface often fosters or exacerbates insect or disease problems that are typically of little negative consequence in undisturbed forests.

Nonindigenous pest organisms present an entirely different picture. When introduced into forest ecosystems previously unchallenged by their presence, these pests are faced with susceptible species and freedom from the natural enemies that hold them in check in their native ecosystems. Neither the susceptible tree species nor the ecosystems in which they grow have the innate capacity to deal with these invasions; hence, a well-known and growing litany of past and current disasters have taken or are taking place in the forests of North America (including Dutch elm disease, chestnut blight, dogwood anthracnose, Asian longhorned beetle, hemlock wooly adelgid, gypsy moth, white pine blister rust, balsam wooly adelgid, emerald ash borer, and possibly sudden oak death).

How we manage these two very distinct groups of pest organisms in the wildland–urban interface is a matter of substantial importance. Historically, our control or management success when dealing with nonindigenous exotic species has been less than encouraging. While limited local successes have been realized (e.g., retarding the spread of such organisms and their associated impacts), on a grand scale we have lost and will continue to lose the wars once exotics become established. Prevention of pest introductions through carefully crafted, strictly respected and enforced regulatory controls (inspections, certifications, quarantines), and rapid and total eradication of nonindigenous pests at first detection (Childs 2003) are essentially the only utilitarian management/control options at our disposal. Researchers continue to focus on potentially useful biological controls, but such efforts are time consuming, expensive, and seldom 100 percent effective.

Management of indigenous insects and pathogens is quite another story, and one in which lie grounds for optimism. Minimizing the negative impacts of pests in this group begins with and is predicated upon understanding the basic principles of biology and ecology of trees and forests and biology and epidemiology of the insect pests and pathogens of concern. The more we bolster our understanding of such things and apply our understanding to accommodate the biological and ecological needs of trees and forests, the more we promote their health and minimize the negative effects of damaging insects and pathogens.

14.6 PROCESSES AND INTERACTIONS IN THE INTERFACE

Compiling or reading a comprehensive catalog of insects and diseases affecting tree health in the wildland–urban interface would be an arduous task. Extensive lists of insects and pathogens causing damage to trees, which vary geographically with changes in forest ecosystems and their component susceptible species, are widely available (Sinclair et al. 1987; Johnson and Lyon 1991). Rather, I shall focus simply on several indigenous pests or groups of pests (insects and pathogens) that illustrate how insect and disease relations are influenced and what they can portend in the wildland–urban interface.

I have already emphasized that interface forests are altered by human activity. Trees in interface forests are typically subjected to anthropogenic influences to which their counterparts in more rural, less disturbed, or natural forests are not. These influences (often termed tree stressors) include air pollution, mechanical injury, soil-level changes, erosion and/or compaction, reduced tree diversity or otherwise altered species composition or tree age distribution, restricted or increased availability of water and nutrients, and increased vegetative competition.

Site disturbances inherent in the wildland–urban interface (edaphic, mechanical, hydrological via water diversion, impoundment, or draw-down, etc.) and their concomitant stresses on trees often initiate progressive declines in preexisting trees and forests. These declines are of variable and complex etiology, depending on local environmental circumstances, tree species and ages, associated biotic agents, and so on (Manion 1981; Olkowski et al. 1991). For example, while a plethora of basidiomycetous root decay fungi (e.g., *Armillaria*, *Ganoderma*, *Heterobasidion*, *Phaeolus*, *Phellinus*, *Rigidoporus* spp.) are widespread and active in many old or senescent forest ecosystems, these organisms are often effective opportunists when provided with injured or stressed young trees. In fact, these organisms are frequent natural colonizers and destroyers of stumps left behind in tree removals, partial harvests, and site-clearing operations so prevalent in the wildland–urban interface. With the passing of time, more often than not years, stump decay fungi sometimes cause root disease in adjacent, residual (and desirable) trees via root contacts and root grafts (Manion 1981; Shaw and Kile 1991; Slaughter and Rizzo 1999).

Other root disease organisms are often unnoticed by or unfamiliar to most because of their subterranean, microscopic, and therefore invisible habit. Nonetheless, they can play significant roles in the demise of trees in disturbed wildland–urban interface settings. Among these are insect-associated *Leptographium* spp. in coniferous trees and *Phytophthora* spp. in both conifers and hardwoods. Indeed, depending upon the age, species, and surrounding environs of certain tree species, other root-infecting fungi (e.g., *Fusarium*, *Endothia*, and *Phythium* spp.) as well as nematodes and root-feeding and root-colonizing insects may contribute to root dysfunction and poor tree and forest health.

Not only are disturbance-related root diseases an important factor in the wildland–urban interface tree and forest health paradigm, so are disruptions to naturally occurring and mutually beneficial fungus–tree root associations (mycorrhizae; Amaranthus 2003). These highly specialized symbioses typically have very specific and sensitive ecological requirements that may be negatively affected in substantially altered forest systems. With a breakdown in mycorrhizal function, tree nutrition as well as resistance to certain root pathogens may be impaired. It is believed that mycorrhizal deficiencies and/or dysfunctions resulting from extreme soil pH changes (acidic to alkaline) in excessively irrigated Florida landscapes are factors in the state's pine chlorosis and decline problem (Barnard 1995).

When trees in the forest or the wildland–urban interface are physiologically stressed because of environmental or human-caused factors, or they become senescent (Figure 14.3), they typically become hosts for a variety of age- or stress-related pathogens and insects. *Botryosphaeria* spp.; *Biscogniauxia atro-punctata* (Schwein.:Fr.) Pouzar [formerly *Hypoxylon atropunctatum* (Schwein.:Fr.) Cooke] and many *Hypoxylon* spp., for

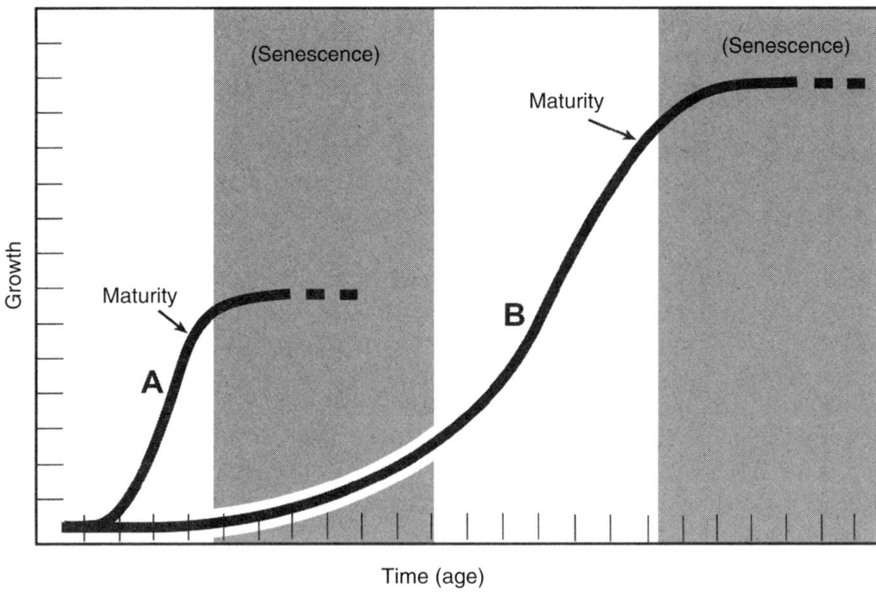

FIGURE 14.3 Comparative growth and aging of a fast-growing, short-lived tree species (A), and a slow-growing, long-lived tree species (B). As each species becomes senescent (respective shaded areas), vulnerability/susceptibility to a variety of old-age-related "pests" such as root decay, heart rots, bark beetles, and wood-boring insects increases.

example, are well-publicized tree canker fungi that are more expressions of water stress than they are aggressive disease-causing organisms on many hardwood species. Other indigenous tree canker fungi simply take advantage of and colonize tree wounds, pruning scars, or broken branch stubs, which generally occur at higher frequencies in people-altered, interface settings. And, as alluded to previously, many decay fungi (causing heart rots, butt rots, and root rots) are not so much pathogenic threats to forests and trees as they are the ecological realities of aging and senescence (Figure 14.3).

Similarly, on the insect side of things, many species of *Dendroctonus* and *Ips* bark beetles, as well as wood-boring insects belonging to the Buprestidae and Cerambycidae, respond to injured or stressed trees. While associated with the demise of large numbers of trees, these insects typically start their population buildups in stressed, dead, or dying trees. Rarely, however, do these pests (or any pests or pathogens for that matter) operate alone. For example, recent widespread mortality of oaks associated with the red oak borer [*Enaphalodes rufulus* (Haldeman)] in Arkansas and Missouri is closely linked to "old age, disease (*Armillaria*) and drought" (Lawrence et al. 2002; Stephen 2002). In Florida, recent outbreaks of the southern pine beetle (*Dendroctonus frontalis* Zimmermann) and *Ips* engraver beetles in wildland–urban interface settings (Figure 15.4) have been intimately associated with overmature, overcrowded, drought stressed, off-site, and sometimes root-diseased pines.

To be sure, not all indigenous insect- and disease-related damage to trees is the result of stress-facilitated parasitism, dependent upon or a function of injured, aging, or stressed trees. Many insects (e.g., caterpillars, sawflies, leaf miners, and aphids) inhabit or consume the foliage of perfectly healthy trees. Similarly, many foliage pathogens (e.g., leaf spot fungi, powdery mildews, leaf rusts, and needlecast fungi) attack healthy trees and cause infections. Normally, however, the severity, timing, and cyclic nature of outbreaks of insects and pathogens of this genre are such that the impact of the organisms on tree and forest health is inconsequential. Healthy trees, especially deciduous trees, are quite able to lose a bit of foliage to pests without any debilitating consequences, unless defoliation is severe and sustained over several years. In fact, some foliage insects and many foliage pathogens operate at such times of the year that the foliage they damage is already senescent and of little photosynthetic or physiological value. As a result, many defoliating insects and foliage pathogens are more high-profile nuisances than they are threats to tree or forest health.

Certain obligately parasitic pathogens such as stem rust fungi (e.g., *Cronartium* spp. on pines and *Gymnosporangium* spp. on junipers) also routinely infect healthy trees. These stem rusts are normally low-frequency and therefore low-impact pests in natural forest settings. However, they can occur at problematic levels and cause considerable damage when their host trees are grown under certain forest management paradigms or situations of reduced species diversity. Fortunately, extensive research has provided effective strategies for minimizing the impacts of these types of diseases in, for example, forest tree nurseries, plantation forests, and Christmas tree plantings. The need for, routine utility of, and cost-effectiveness of such control strategies in other interface settings are subject to question.

Vascular wilts are another group of tree diseases for which physiological stress may not be required, although infection and disease development are often facilitated by injury and environmental factors, respectively (Sinclair et al. 1987). Fortunately, most indigenous vascular wilts, such as those caused by species of *Fusarium*, *Verticillium*, *Acremonium*/*Cephalosporium*, and *Ceratocystis* do not occur at epiphytotic levels in North America. Similarly, although bacterial leaf scorch of certain hardwood species caused by *Xylella fastidiosa* Wells et al. is receiving increased attention as a threat in urbanizing environs (Lashomb et al. 2002), the pathogen is widespread in host tree species (Barnard et al. 1998), and water stress is a major component of disease expression (McElrone et al. 2001). Consequently, like so many tree diseases, bacterial leaf scorch may well be linked to environmental or other predispositional stressors, at least in North America. Like all pathogen–host relationships, however, these situations are subject to change as the dynamics of the wildland–urban interface (e.g., site disturbances, changes in species composition, and movement of plant and tree material) continue their ebb and flow.

Yet another factor that affects insect and disease relationships with trees in the wildland–urban interface is the tendency of people to lavish landscape or ornamental trees with TLC (tender loving care). Trees in the interface are often the recipients of "benefits" such as irrigation, fertilizer (purposeful or inadvertent), and sometimes repeated treatments with pesticides. While such cultural inputs may be well intended and beneficial in certain situations, excessive or continuous application or exposure can facilitate or exacerbate damaging pest scenarios. For example, while certainly not a simple equation, tree fertilization can alter tree chemistry, thereby reducing natural resistance to certain foliage-feeding insects and promoting insect activity (Schowalter and Lowman 1999; Oswald 2001). Excessive or imbalanced fertilization often exacerbates pitch canker infections of pines caused by *Fusarium circinatum* Nirenburg and O'Donnell in commercial production plantations and intensively managed landscapes (Fisher et al. 1981; Anderson and Blakeslee 1984; Myers 1989; Barnard, personal observation). Excessive irrigation of pines can also lead to pH-induced micronutrient deficiencies (Fe, Mn, others), decline, and infestation by secondary insects such as bark beetles, wood borers, and so on (Barnard 1995).

Along a similar vein, excessive or unnecessary use of insecticides in wildland–urban interface settings can result in unwanted and serious insect pest problems (Coppel and Mertins 1977; Barbosa and Wagner 1989). It is believed that excessive insecticide applications may have played a key role in recent problematic outbreaks of Kermes scales (*Allokermes* spp.) on live oaks (*Quercus virginiana* Mill.) in southern interface landscapes (Jim Meeker, personal communication 2003).

14.7 MANAGEMENT COMPLICATIONS IN THE INTERFACE

When tree-threatening or nuisance outbreaks of insects and diseases occur in wildland forests, they are easy enough to ignore or to observe and appreciate as natural ecosystem processes. Salvage harvesting of affected trees in commercial forests or plantations to minimize economic losses is a fairly straightforward process. Response to outbreaks in the wildland–urban interface, however, is not quite so simple. In the interface, safety, aesthetics, access and human perceptions become substantive issues. Salvage or removal of damaged trees growing among buildings, power lines, swimming pools, or septic systems (Figure 14.4) can create logistical nightmares. The costs of tree removal in such settings are often exorbitant, and recouping all or part of them is often unachievable due to limited or nonexistent markets for small volumes of wood in hard-to-get-at places. Further, perceptions of some interface dwellers, such as the belief that harvesting or cutting trees for salvage or pest management is somehow intrinsically evil or environmentally unacceptable, at times preclude effective silvicultural response (Kundell et al. 2002; Tarrant et al. 2002).

Other human-related issues further complicate dealing with insect and disease outbreaks in the wildland–urban interface. For example, well-meaning, yet frequently ill-informed policies to preserve large trees as "ancient" or

FIGURE 14.4 Consequences of an aging, overcrowded, even-aged, drought-stressed, single species pine canopy in the wildland–urban interface in Florida. An outbreak of southern pine beetles leaves hundreds of dead and dying trees scattered through the interface (A; white dots), creating safety hazards, threats to neighboring trees and properties, and in its aftermath, costly cleanup for affected homeowners (B, C). Managing tree stand densities and promoting diversity of tree species and ages are key factors in minimizing risks and impacts associated with such events. (Photos by J.R. Meeker.)

"heritage" specimens may be counterproductive. In many cases, large trees are neither ancient nor heritage. They are simply large trees, perhaps entering senescence and nearing the end of their useful life span (McPherson 2003). As such, they may be particularly vulnerable to insects and diseases (Figure 14.3), a habitat for insect population development or a source of pathogenic inoculum, and a potential aggravation of a larger problem. Further, leaving such trees to die natural deaths may result in the spillover of insect or disease problems to nearby healthy trees, or to trees of adjacent landowners, whose possible legal ramifications and/or liabilities are still uncertain. Mistretta (2002) has noted that the spillover of southern pine beetle infestations from public lands (often managed for preservation purposes) to neighboring private forestlands will be an ongoing problem in the South's increasingly fragmented forests. Such spillover events in Florida's wildland–urban interface have been no small issue in the state's recent southern pine beetle epidemics. Insects and diseases do not recognize or respect property lines or differences in management objectives.

14.8 MANAGEMENT NEEDS AND OPPORTUNITIES

Perhaps the greatest need for promoting and managing healthy trees and forests in the wildland–urban interface is the need for education. Too many people, be they interface dwellers or otherwise, simply do not understand pertinent and critical concepts of tree and forest biology and ecology, let alone specifics of associated insect and disease interactions. Dying trees are too easily blamed on "beetles, which should be sprayed," with little to no consideration of the natural ecological realities that lead to tree mortality. A basic understanding of the essential ecological, environmental, and biotic interactions too frequently labeled bug or disease problems provides the foundation for intelligent resource and risk management and the maintenance of healthy forests.

Educational efforts must focus on a variety of audiences, including planners, policy makers, landowners, developers and builders, resource managers, and homeowners, all of whom have vested interests in the wildland–urban interface. The simple concept so aptly stated by Sir Isaac Newton, "for every action, there is an equal and opposite reaction" perhaps has some application here. If we have learned anything through years (even decades) of research dealing with tree and forest pests, it is that most damaging insects and pathogens are part of natural, healthy forest ecosystems. These organisms are not innately evil, and they characteristically respond to tree and forest conditions and to management activities. To the extent that we dismiss this truth or act as if there is no connection, we will be playing catch-up, dealing with the aftermath of menacing insect and disease outbreaks well into the future. Alternatively, as we begin to understand some fundamental biological and ecological realities and manage our resources accordingly, we can do much to minimize such outbreaks and enjoy the benefits of healthy trees and forests.

Management of forest resources in the interface increasingly becomes entangled in issues of landowner property rights, responsibilities, and liabilities; community attitudes, safety, and welfare; governmental authority and responsibility; and use or nonuse of public funds. Of necessity, understanding and applying sound biological and ecological principles must be integrated into this complex matrix of sociopolitical realities. The task is not easy, but it can be done.

Effective management will be prevention-based and regionally variable, depending upon the silvicultural characteristics and needs of local landscape trees and forest systems. Preventive management will also take into account the biology of and threats represented by regional, host-specific insects and diseases. With appreciation for various complex interactions and situational subtleties, I cautiously offer a few suggestions for developing workable community-based approaches to healthy forests in the wildland–urban interface.

- Create a sociopolitical climate, perhaps with incentives (tax structures, subsidies, etc.), that encourages landowners to practice silviculturally and ecologically sound pest-preventive forestry. Resources may be privately owned, but the benefits of healthy forests are public.
- Promote (via education, incentives, etc.) tree age and species diversity. A diverse forest can help minimize damages in the event of an insect or disease outbreak.
- Develop public policies and regulations that are not counterproductive (e.g., ancient or heritage tree ordinances that unnecessarily fuel insect or disease outbreaks). Allow for and encourage removal and replacement of old, senescent, and pest-vulnerable trees that have exceeded their "useful life-span" (McPherson 2003).
- Seek to develop local markets and technologies that utilize small-dimension, low-quality woody materials, which are produced by risk-mitigating or pest control silvicultural operations, and which are currently "valueless" or "waste" (McPherson 2003).
- Invest in the development and deployment of technology and methods to facilitate the removal of hard-to-access, pest-vulnerable, and pest-infested or damaged trees in interface and urban settings.
- Develop guidelines and encourage practices that enhance tree and forest health before and during development of the interface. Prevention

of tree and forest health problems is more effective than treatment *ex post facto*.

- Support and implement continuous monitoring efforts such as those deployed in the USDA Forest Service's Forest Health Monitoring Program (USDA Forest Service 2003). Systematic monitoring is a good way to assess changing forest conditions, anticipate developing problems, and detect pest outbreaks before they become serious or unmanageable.

In the wildland–urban interface, where preservation, multiple use, production, and urban forest resources and objectives are so commonly interspersed, maintenance of healthy forests requires informed, coordinated, and proactive areawide, community-based management.

ACKNOWLEDGMENTS

I would like to express my sincere appreciation to Ms. Andrea van Loan and Drs. Albert "Bud" Mayfield and Jim Meeker for technical advice and assistance. I also thank Mrs. Angela C. Carter for typing the manuscript, and Ms. Katrina Vitkus for the production of artwork and graphics.

REFERENCES

Amaranthus, M., 2003. Forest primeval and the urban landscape: bridging the gap with mycorrhizae, *Arbor Age* 23: 8–11.

Anderson, R.L. and G.M. Blakeslee, 1984. Pitch Canker Incidence in Fertilized and Nonfertilized Slash Pine Plantations, Forest Pest Management Report no. 84, U.S. Department of Agriculture, Forest Service, Southeastern Area, State and Private Forestry, Asheville, NC.

Barbosa, P. and M.R. Wagner, 1989. *Introduction to Forest and Shade Tree Insects*, Academic Press, New York.

Barnard, E.L., E.C. Ash, D.L. Hopkins, and R.J. McGovern, 1998. Distribution of *Xylella fastidiosa* in oaks in Florida and its association with growth decline in *Quercus laevis, Plant Disease* 82: 569–572.

Barnard, E.L., 1995. Pine chlorosis and decline in Florida landscapes, Forest and Shade Tree Pest Leaflet no. 12, Florida Department of Agriculture and Consumer Services, Division of Forestry, Tallahassee.

Childs, G., 2003. First line of defense: minimizing the impact of non-native invasive insects, *Arbor Age* 23: 10–12, 14.

Coppel, H.C. and J.W. Mertins, 1977. *Biological Insect Pest Suppression*, Springer-Verlag, New York.

Fisher, R.F., W.S. Garbett and E.M. Underhill, 1981. Effects of fertilization on healthy and pitch canker-infected pines, *Southern Journal of Applied Forestry* 5: 77–79.

Franklin, J.F., H.H. Shugart, and M.E. Harmon, 1987. Tree death as an ecological process, *Bioscience* 37: 550–556.

Haack, R.A. and J.W. Byler, 1993. Insects and pathogens: regulators of forest ecosystems, *Journal of Forestry* 91: 32–37.

Helms, J.A., 2002. What do these terms mean? Forest, forestry, forester, *Journal of Forestry* 100: 15–19.

Helms, J.A., 1998. *The Dictionary of Forestry*, Society of American Foresters, Bethesda, MD.

Johnson, W.T. and H.H. Lyon, 1991. *Insects That Feed on Trees and Shrubs*, 2nd ed., Comstock Publishing (Cornell University Press), Ithaca, NY.

Kimmins, J.P., 1997. *Forest Ecology: A Foundation for Sustainable Management*, 2nd ed., Prentice-Hall, Upper Saddle River, NJ.

Kundell, J.E., M. Myszewski and T.A. DeMeo, 2002. Land Use Planning and Policy Issues, in Human Influences on Forest Ecosystems: The Southern Wildland–Urban Interface Assessment, E.A. Macie and L.A. Hermansen, Eds., General Technical Report SRS-55, U.S. Department of Agriculture, Forest Service, Southern Research Station, Asheville, NC, pp. 53–69.

Lashomb, J., A. Iskra, A.B. Gould and G. Hamilton, 2002. Bacterial Leaf Scorch of Amenity Trees: A Widespread Problem of Economic Significance to the Urban Forest, NA-TP-01-03, U.S. Department of Agriculture, Forest Service, Northeastern Area, State and Private Forestry, Forest Health Protection, Morgantown, WV.

Lawrence, R., B. Moltran and K. Moser, 2002. Oak decline and the future of Missouri's forests, *Missouri Conservationist* 63 (7): 11–18.

Macie, E.A., and L.A. Hermansen, Eds., 2002. Human Influences on Forest Ecosystems: The Southern Wildland–Urban Interface Assessment, General Technical Report SRS-55, U.S. Department of Agriculture, Forest Service, Southern Research Station, Asheville, NC.

Manion, P.D., 1981. *Tree Disease Concepts*, Prentice-Hall, Englewood Cliffs, NJ.

McElrone, A.J., J.L. Sherald and I.N. Forseth, 2001. Effects of water stress on symptomatology and growth of *Parthenscissus quinquefolia* infected by *Xylella fastidiosa, Plant Disease* 85: 1160–1164.

McPherson, E.G., 2003. Urban forestry: the final frontier? *Journal of Forestry* 101: 20–25.

Mistretta, P.A., 2002. Managing for forest health, *Journal of Forestry* 100: 24–27.

Mueller-Dombois, D., 1987. Natural dieback in forests, *Bioscience* 37: 575–583.

Myers, T.R., 1989. Influences of Fertility Levels, Host and Pathogen Variability and Inoculation Technique on Pitch Canker Disease in Slash Pine, Ph.D. dissertation, University of Florida, Gainesville.

Olkowski, W., S. Daar and H. Olkowski, 1991. *Common-Sense Pest Control: Least Toxic Solutions for Your Home, Garden, Pets and Community*, Taunton Press, Newtown, CT.

Oswald, M., 2001. Fertile ground: Dr. Dan Herms discusses the relationships amongst tree fertilization, tree growth, defensive compounds and pest attack, *Arbor Age* 21: 24, 27–28.

Schowalter, T.D. and M.D. Lowman, 1999. Forest herbivory: insects, in *Ecosystems of Disturbed Ground,* Vol. 16 of *Ecosystems of the World*, L.R. Walker, Ed., Elsevier, New York, pp. 253–269.

Shaw, C.G., III and G.A. Kile, 1991. Armillaria Root Disease, Agric. Handbook 691, U.S. Department of Agriculture, Forest Service, Washington, DC.

Sinclair, W.A., H.H. Lyon and W.T. Johnson, 1987. *Diseases of Trees and Shrubs*, Comstock Publishing (Cornell University Press), Ithaca, NY.

Slaughter, G.W. and D.M. Rizzo, 1999. Past forest management promoted root disease in Yosemite Valley, *California Agriculture* 53: 17–24.

Stephen, F., 2002. Ozarks under siege by red oak borers, *Society of American Foresters' Forestry Source* 7: 19.

Tarrant, M.A., R. Porter and H.K. Cordell, 2002. Sociodemographics, values, and attitudes, in *Southern Forest Resource Assessment*, D.N. Wear and J.G. Greis, Eds., General Technical Report SRS-53, U.S. Department of Agriculture, Forest Service, Southern Research Station, Asheville, NC, pp. 175–187.

U.S. Department of Agriculture, Forest Service, 2003. Forest Health Protection, www.fs.fed.us/foresthealth. [Date accessed: October 2, 2003.]

Wear, D.N. and J.G. Greis, Eds., 2002. Southern Forest Resource Assessment, General Technical Report SRS-53, U.S. Department of Agriculture, Forest Service, Southern Research Station, Asheville, NC.

15 Invasive Plants in the Wildland–Urban Interface

Sarah Reichard
Center for Urban Horticulture, University of Washington

CONTENTS

15.1 INTRODUCTION

Biological invasions are now recognized as one of the most serious forms of environmental degradation on Earth. As humans have increased their movements around the world and have increased global trade, so have they increased the intentional and unintentional distribution of harmful organisms. While most introduced organisms are beneficial for food, fiber, aesthetics, and other uses, some have the potential to cause great harm. The impact of the harmful species has been demonstrated to be an important component of global change (Mooney and Hobbs 2000).

While it has been known for centuries that competition by weeds for resources needed in agricultural settings is a primary obstacle to successful cultivation of plants, the realization that such competition was also harming wildland ecosystems is much more recent. Although Charles Elton (1958) sounded the alarm in *The Ecology of Invasions by Animals and Plants*, there was little response until the early 1980s when the Scientific Committee of Problems of the Environment (SCOPE) of the International Council of Scientific Unions established an international program to investigate the ecology of biological invasions. This resulted in conferences held around the world, and several books were published (e.g., MacDonald et al. 1986; Mooney and Drake 1986) that were notable in the numbers of questions raised about biological invasions and the paucity of answers. The field of invasion biology was born.

Straightforward competition for resources has been demonstrated to be a harmful impact of plant invasions (e.g., Huenneke and Thomson 1994); but it is by no means the only one. We now know that invasive plants are capable of changing ecosystems by such methods as altering nutrient cycling (Vitousek et al. 1987), increasing fire frequency and intensity (D'Antonio and Vitousek 1992), increasing sedimentation (Blackburn et al. 1982), and hybridizing with native species (Daehler and Strong 1997). These latter impacts are perhaps the most serious because in many cases the effects are not easily reversed. There are also considerable economic impacts. One estimate put the economic cost of invasive plants in natural areas, agriculture, and gardens at $35 billion per year due to loss of productivity, costs of herbicides, and other measures (Pimentel et al. 2000).

For the purposes of this chapter, invasive plants are defined as those species that have spread or are likely to spread into native flora or managed plant systems, develop self-sustaining populations, and become domi-

1-56670-602-5/05/$0.00+$1.50
© 2005 by CRC Press LLC

nant or disruptive to those systems (Figure 15.1). Species that are adventive (naturalize but do not form permanent populations) or are persistent after a long-forgotten planting (such as apple trees remaining from early pioneer plantings) are not considered to be invasive.

The conditions at the wildland–urban interface provide many opportunities for invasive species to establish. The landscape is varied and transitional, with many potential habitats of differing vulnerability to invasion. The urban areas are sinks for many intentional and unintentional introductions, and the movement of humans, birds, and other organisms along the gradient from urban to wildland provides dispersal opportunities. Each of these aspects will be explored in this chapter.

15.2 SUSCEPTIBILITY OF BIOTIC COMMUNITIES TO INVASION

Many factors can affect the probability that a given plant may be able to establish in a new area. Physical factors such as high or low temperatures, precipitation, soil, or photoperiod may prevent an invasion from occurring. Predators may exist in the new location that would make survival and reproduction difficult or impossible. There is some evidence that the species that are already in an area may have important implications for the probability that a newcomer will establish. For instance, Mack (1996) documented several instances where plant invasions were prevented or reversed by native insects or pathogens in the invaded region. This may occur most commonly when there is a close native relative to the invader in the invaded area and therefore organisms that have evolved to prey on that phylogenetic group are present. For instance, Mack found that species that did not have a congener in the native flora were more likely to be invasive than those that do, presumably because there were fewer predators. Lockwood et al. (2001) found that natural area invaders in California were primarily from genera that were not yet represented in that region.

One recent study suggested that it might not be *which* species were already present in an invaded community, but that species-rich communities, in general, may actually be invaded more readily than those with fewer species (Stohlgren et al. 1999). After studying several hundred plots in the Colorado Rockies and in western grasslands, the authors concluded that nonnative species invaded primarily areas of high species richness, meaning that the common belief that only areas with "vacant niches," such as those created by disturbances, could be invaded was incorrect. Their findings indicated that it was the availability of resources, such as water and nutrition, in the native plant communities, independent of species richness, that accounted for the invasions. Conversely, Kennedy et al. (2002) found that species diversity in experimental grassland plots enhanced biotic resistance by increasing crowding and species richness. This was true for both measures used in their research: establishment (in number of invaders) and success (proportion of invaders that were large).

Theory suggests that perhaps the most invaded areas of all, however, are those with intermediate levels of disturbance. These areas receive periodic low-level or small-scale disturbance that removes some competing vegetation, but generally have higher levels of water and nutritional resources because the disturbance has not removed the upper layers of soil or litter. Urban areas are noted for their disturbance to biotic communities through reduction in biomass, changes in hydrology, increased temperature, and other perturbations to the environment.

FIGURE 15.1 Kudzu (*Pueraria montana* var. *lobata*), a vine introduced from Asia, covers thousands of acres of forest and other land in the U.S.

These disturbances encompass natural disturbances such as wind or erosion but are generally anthropogenic, including construction and vehicle and pedestrian traffic. Wildland areas tend to experience some disturbance through both anthropogenic and natural causes, but at much lower levels. The area within the interface is generally intermediate, having patches of disturbance within a less disturbed mosaic. These disturbances might be anthropogenic or they might be gaps in the forest canopy due to natural occurrences such as wind or fire. They may vary in size and in the degree of above- and belowground perturbation.

15.3 HOW DO INVASIVE PLANTS ARRIVE AND DISPERSE?

15.3.1 Unintentional Methods of Introduction

One of the common ways that invasive species have entered the U.S. has been through the contamination of crop seed in rural areas. For instance, it is believed that one of the worst invasive species, *Bromus tectorum* L., or cheatgrass, was a contaminant of wheat seed imported from Eurasia in the 1800s (Novak and Mack 2001). Since 1939 and the passage of the Federal Seed Act, the federal government does regulate the purity of seed and routinely inspects imported seeds for diseases, insects, and weed contamination. However, some contaminated seed inevitably arrives. In 1988, shipments of tall fescue grass seed (*Lolium arundinaceaeum* [Schreb.] S.J. Darbyshire), imported from Argentina and sold through retailers such as K-Mart and Wal-Mart, were found to contain *Nassella trichotoma* Hackel ex Arech. (serrated tussock grass), a federally listed noxious weed (U.S. Congress 1993).

In past years, trade in urban areas was a source of many unintentional plant introductions. For instance, straw was often used as a packing material for goods brought from other countries and, since straw was a mixture of grain stalks and assorted pasture species, it may have been the source of some unintentional introductions. In 1896, a publication noted that some newly arrived species in Denver appeared first at the back door of a crockery shop (Dewey 1896). Presumably, the packing material was discarded near the door. Few imported goods are currently packed in straw. Surveys of soil used as ships' ballast and deposited in ports have found that many nonnative plants likely entered the country as seeds found in the soil. The incidence of nonnative species along the Gulf Coast and later the West Coast appears correlated with the increase in the growth of ports in these areas (National Research Council 2002). Just how many species actually established after being introduced by these methods, however, is open to question.

Accidental introductions are, by their nature, random. Which species will arrive, when they will arrive, and where they will establish are unpredictable events that may change through time. For instance, crop seed contaminants would be presumed to be spread in rural agricultural areas, but the previous example of contamination of grass seed from Argentina suggests that imported seed is used in a number of urban settings as well. The assumption was that many invasions would first occur and be detected in port areas. Cargo was unloaded from ships from all over the world, and plant and insect hitchhikers would be released at the ports. Several years ago, it became the standard to pack cargo in large metal containers, which were then unloaded by cranes directly to trains, to be taken to any number of destinations. Thus, Chicago became the first site for detection of the devastating Asian long-horned beetle, a stowaway on pallets found in the containers. In current times, with current shipping technology, any urban area is increasingly likely to be the first site of a new accidental introduction of an invasive species.

15.3.2 Intentional Methods of Introduction

Invasive plants are often intentionally introduced for a number of beneficial purposes, including food, fiber, medicine, and aesthetics. A previous study (Reichard 1997) found that nearly all invasive woody plants in the U.S. were intentionally introduced, with the largest percentage (82 percent) used for landscape horticulture purposes. The figure is likely smaller when both herbaceous and woody plants are considered — herbaceous species are more likely to have been introduced through accidental means such as crop seed contamination. Horticulturists desire a large palette of plants from which to select and have actively pursued plant exploration around the world (Reichard and White 2001). Plants are also increasingly being introduced and farmed for use as medicinal herbs. Because many medicinal plants have strong secondary compounds that repel insects as well as act on the human body, they may be less attractive to predators. This would increase survivorship and reproduction, facilitating establishment. Little is known about the long-term invasive potential of many of these species. At least one, *Hypericum perforatum*, St. John's wort, is a known noxious weed that causes a number of problems in agricultural lands, but is being farmed in Washington State because of its use as an antidepressant.

15.3.3 Movement of Invasive Species along the Wildland–Urban Gradient

Because many future introductions, especially those of woody plants used for aesthetic purposes, are likely to occur in urban areas, we can expect more invasive species to spread from the urban areas into the interface with wildlands. It is difficult to estimate how many introduced species do become invasive because; in most cases, good estimates of how many total species are introduced into an area do not exist. In addition, we have little information about how the number of individual plants used affects the outcome of invasion. It is intuitive to think that multiple introductions of a species with invasive traits might lead to a greater probability of invasion occurring and there is some evidence from a case study in Canberra, Australia (Mulvaney 2001). Prior to 1909, the area around Canberra was used for dryland agriculture and was sparsely settled. In 1909, Canberra was named the capital of Australia, and development of the city began. Because this was a government project, most plants were obtained from a government nursery or a limited number of commercial nurseries, and records of all plants used were preserved, sometimes with multiple copies. Over 2000 ornamental species were purchased and 2,000,000 plants were installed, with a peak in introductions between 1910 and 1930. Mulvaney, reviewing these introductions decades later, found that 289 species had established by 1990 and that 106 of those species could be considered pests. This number may be lower than what would be expected in other areas because the xeric climate in this region may prevent successful invasion of many species. Mulvaney also found a strong correlation between the number of times a species was planted and its establishment outside cultivation, suggesting that commonly used species may be more likely to become successful invaders.

Invasive species move from urban areas to wildlands along a number of corridors, but roads appear to be a

Box 15.1

Spread of Nonnative *Lonicera* Species from Pittsburgh and Philadelphia

Although it is commonly observed that many invasive plants originate in urban areas and move outward into surrounding wildlands, it has rarely been documented. One way that such spread can be documented is by using herbarium specimens, which are pressed and dried plant parts that have been carefully stored to protect them. Herbarium specimens usually include information about where the material was collected and provide documentation that a species was collected from a certain site at a certain time. While, as a historical record, they may be somewhat unreliable because collectors do not obtain material evenly across the landscape, they do give an indication about which species were present in the landscape at a given time.

The staff of the Morris Arboretum in Pennsylvania took advantage of their Pennsylvania Flora Database to track the movement of nonnative invasive *Lonicera* species in their state (Figure 15.2). They found a clear pattern with a few specimens collected prior to 1880, mostly near Philadelphia and Pittsburgh. By 1920, the number of specimens from these areas had increased, and more appeared to be collected in the surrounding areas. At present, specimens have been collected from nearly every county in southern Pennsylvania, with the largest numbers in and surrounding the major urban areas.

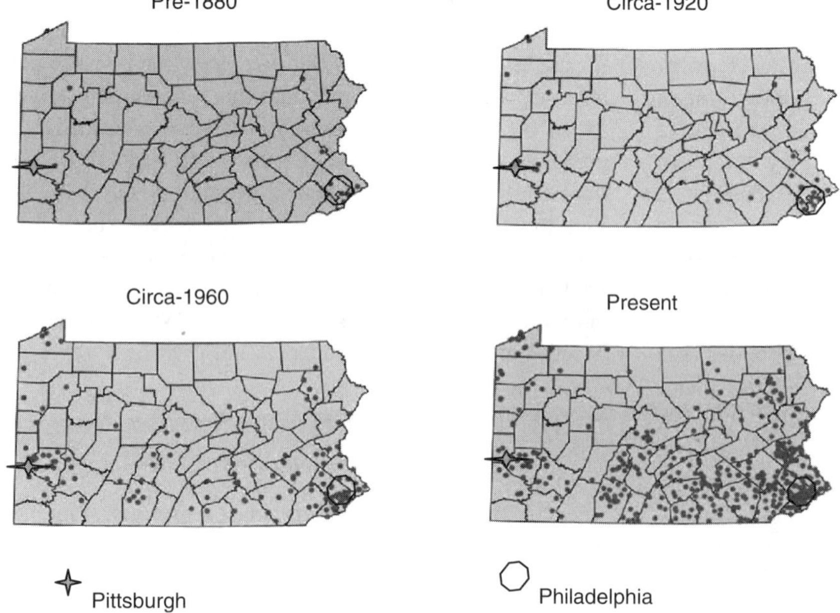

FIGURE 15.2 Spread of nonnative *Lonicera* species in Pennsylvania. (Courtesy of Tim Black and Ann Rhoads, Morris Arboretum.)

major conduit. Roads serve multiple functions in the spread of invasive species. First, the vehicles that use them are often contaminated with the seeds of invasive species. One study found that collecting the mud from one car that was driven more than 15,000 km during one growing season in Germany yielded 3926 plants in 124 species (Schmidt 1989). Mud collected from 75 vehicles in Nigeria contained 714 seeds in 40 species (Clifford 1959). In another study, two years' worth of car wash sludge from an unknown number of cars in Canberra, Australia, produced 18,566 plants in a total of 259 species (Wace 1977). Clearly, automobiles are capable of carrying large numbers of invasive species along the interface, generally within soil on the vehicle. Soil can retain surprisingly large numbers of seeds (Table 15.1).

Roads also provide suitable disturbed habitat for invasive species establishment and contain reservoirs of propagules for future invasions (Parendes and Jones 2000). Railroads are known to provide pathways for the spread of invasive species (Muhlenbach 1979; Forcella and Harvey 1988). In the Pacific Northwest, three pathways of distribution were found by plotting information from the labels of pressed and dried herbarium specimens, particularly collection dates and locations. Junctions in the transportation systems, especially railroads, were major introduction points, and weeds spread along those systems (Forcella and Harvey 1988). Herbarium specimens are not an unbiased source of information since areas near roads and other transportation corridors may be overcollected because of the ease of access; but they do provide one of the few historical records we have of species occurrence (see Box 15.1).

Streams, rivers, and canals also promote the movement of invasive species. In some cases, spread may be attributed to the movement of ships. Twelve years after the invention of the steamboat, botanist William Baldwin found European weeds at every boat landing along the Mississippi River (Foy et al. 1983). In some cases, the riparian zone itself may facilitate migration. A study in Sweden found that flooding extended the potential distribution of many species (Skoglen 1990).

TABLE 15.1
The Number of Seeds per Square Meter Found in Soil in Several Locations

Type of Land	State/Country	No. of Seeds
Arable lands	Minnesota	1000–40,000
	England	28,700–34,100
Freshwater marsh	New Jersey	6405–32,000
Prairie grassland	Kansas	300–800
Annual grassland	California	9000–54,000
Secondary forest	North Carolina	1200–13,200
Coniferous forest	Canada	1000

Source: Silvertown (1987).

Wind disperses a number of invasive plant species, including many trees and most annual herbaceous species (Ridley 1930). Wind-dispersed species tend to have a high production of seeds, a trait considered to be associated with invasive species, because the likelihood of a seed landing in an appropriate site for germination is low. Plants with very small seeds or good adaptations for dispersal such as membranous wings may establish far from the parent plant.

Birds are a significant disperser of many invasive plants, especially shrubs and some trees (Reichard et al. 2000). The invasive species may have fruits that provide greater rewards for the birds and may be preferentially eaten over native species (Sallabanks 1993). The urban environment is often filled with many landscape plants selected for their aesthetic appeal, including abundant brightly colored fruits that make them attractive to many avian dispersers. Landscape plantings with a high proportion of edge and highly visible well-spaced plantings are amenable to birds' visually based food search strategies. Movement through the gut of the bird may scarify hard seed coats and facilitate germination. Moreover, birds, unlike wind, generally deposit the seeds in concentrated doses in sites appropriate for germination, usually below perching sites. Forested communities, in particular, may be threatened by invasive species that are bird dispersed (Reichard et al. 2000).

Finally, the role of hikers within the wildland–urban interface must also be considered. While most accounts are anecdotal, there has been one study published that attempted to quantify the potential of hikers to assist plant invasion (Clifford 1956). Members of an unspecified British expedition were asked to scrape the mud from their boots. Sixty-five plants in 11 species were identified. Most dog owners also recognize that fur is an excellent substrate for many sticky or barbed seeds, and various animals are known to disperse seeds in this way.

15.4 HOW CAN THE INTRODUCTION OF INVASIVES BE PREVENTED?

15.4.1 LEGAL METHODS TO CONTROL INTRODUCTION

There is a widespread misconception that plant species are inspected or tested by the federal government for their invasive potential prior to introduction. In fact, the U.S. does not prohibit the introduction of most plant species unless they are on the federal noxious weed list maintained under the Plant Protection Act. There are currently 109 species listed, although a study done in the early 1990s (U.S. Congress 1993) found at that time 750 species in the U.S. that fit the legal definition of a noxious weed (then defined largely as a plant that is injurious to agricultural interests). With the heightened interest in plant introduction (Reichard and White 2001), many more are likely to be introduced.

Eduardo Rapaport (1991) suggested that if 10 percent of the 260,000 vascular plant species known in the world have the potential to be weeds, there are 26,000 weed species. He estimated that only 10,000 weed species were known and that only 4000 of those had been exchanged between regions of the world (the rest are natives of the areas in which they are considered to be invasive), leaving 22,000 invasive species yet to be introduced. If only 10 percent of those have the potential to become pest problems, there are 2200 potential pests yet to be introduced. With few governmental restrictions to importation and few other resources dedicated to preventing invasive plant introductions, it appears likely that many invasive plants will be introduced to the U.S. and other countries in coming years.

It is possible that the approach the federal government takes in the future may change to be more in line with its approach to the introduction of other organisms, such as biological control insects. This might include having a list of "approved" species that can be imported with few restrictions and others that might be restricted or undergo testing either voluntarily by the importer or by a government agency. In the meantime, the introduction of invasive species can be prevented by importers voluntarily taking a more conservative approach to introduction. Species can be evaluated for invasive potential using a number of methods (Rejmánek and Richardson 1996; Reichard and Hamilton 1997; Pheloung 2001) or can be held and tested for a few years before use in the trade (see Box 15.2).

State noxious weed laws are generally designed to regulate the control rather than importation of invasive and weedy species. State agencies evaluate the invasive ability and potential harm of existing species, usually using a panel of resource experts, to determine if the species should be listed for control measures. These laws vary greatly among the states. In some cases they are strictly advisory, while in others they carry penalties for lack of compliance. In the past, virtually all noxious weed laws, including the federal laws, dealt only with weeds in agricultural settings, while in recent years several states have moved to add weeds of wildlands as well. Some states also have quarantine laws listing species that cannot be sold commercially.

Recent discussions among the federal agencies dealing with invasive plants and state and local governments have revealed some conflicts in jurisdiction that have yet to be resolved. Most detections of new invasions occur on the local level, with county or state agencies responding to control. Invasions are generally regional, however, and transcend political boundaries. What is the role of the federal government in advising surrounding states and coordinating control efforts? For which species is immediate federal intervention needed? These questions will require vigorous discussion before resolution is reached.

Box 15.2

Traits of Invasive Species

Although each invasive plant species has unique combinations of traits that both facilitate its ability to invade and the impact it has on the invaded ecosystem, there are some common attributes that invasive species tend to have and that noninvasive introduced species lack. In general, successful invasive species:

- Have a high reproductive ability, with a short juvenile period, plentiful seeds, efficient dispersal mechanisms such as adaptation for wind dispersal or fruits that appeal to vertebrates, no pregermination needs for the seeds, and high germination rates
- Have the ability to reproduce asexually as well as by seeds, using such methods as rhizomes, stolons, and root suckering
- Have features that make them stress-tolerant, including the ability to fix nitrogen in root nodules, allowing invasion of soils lacking in organic matter, and semievergreen leaves that allow for some leaf loss to relieve stress while still retaining some photosynthetic material
- May be from a wide range of latitude and are likely to be invasive in other parts of the world following introduction.

Noninvasive species may have some of these traits, but it would be rare for them to have many in combination. They are also more likely to be hybrids or have infertile seeds.

15.4.2 Best Practices for the Wildland–Urban Interface

The interface serves as a crucial buffer between wildland and urban areas. By preventing invasions from the urban areas into the interface, the likelihood of introduction into the wildlands is also reduced.

First, planting of native species and species with a long history of cultivation without demonstrated invasive ability should be emphasized in the interface. Species in the latter category include most *Rhododendron* species (excluding *R. ponticum* L.), most other species in the Ericaceae, *Buxus sempervirens* L., and *Camellia* species. One way to determine if a species has a history of being invasive is to check a web page maintained by the government of Western Australia that includes species known to invade anywhere on earth (Department of Agriculture, Western Australia n.d.). When using native species, preference should be given to plants grown from seed collected locally, to reduce the possibility of contaminating local native genotypes. Whole plants should generally only be used if a local salvage operation is occurring nearby because transplanted species tend to have a high mortality rate. Younger plants generally have the highest potential to survive transplant shock while fully mature plants often fail to establish.

Accidental movement of invasive species may also be prevented by sensible hygiene. Vehicles are a major vector of invasive plants (Clifford 1959; Wace 1977; Schmidt 1989), as are hiking boots, mountain bikes, or any type of equipment that carries soil. While it may be difficult to prevent the movement of cars through the interface, several steps can be taken to reduce the movement of seeds. First, place signs in prominent places where visitors to the interface and to wildlands can see them, reminding visitors that they, or their animals or equipment, could be spreading invasive species. Receptacles should be provided for the placement of any seeds or soil removed. Those who work in these areas must also be vigilant that they do not become vectors of invasive plants, especially as they move from infected areas to more pristine ones. Boots should be checked regularly. Mowers, trucks, bulldozers, and any other piece of equipment used should be cleaned before moving to a new site and contractors should be held to that same standard.

Managers may also view the interface as a buffer between more invaded urban areas and the surrounding wildlands. Studies have shown that thick stands of native species are effective in preventing the penetration of wind-dispersed species (Cadenasso and Pickett 2001). Maintaining a dense edge to forested lands can prevent wind-driven invasions into the interior. Bird-dispersed species present a greater challenge, however. Birds can carry a seed in their gut for varying lengths of time, meaning that they can fly some distance before defecating the seed. Seeds are generally dispersed not more than 100 m from the mother plant, but medium to large birds may disperse seeds more than 1 km away (Debussche and Isenmann 1994), and some birds may retain a seed in the digestive tract for more than 100 h (Proctor 1968). Planning to prevent dispersal of seeds by birds is very difficult and it may be more practical to increase monitoring for bird-dispersed species so that invasions are discovered in the earliest stages, rather than trying to prevent them.

If it is not practical to treat the entire interface as a buffer, it should be possible to prioritize key corridors for frequent monitoring. Corridors that are known to facilitate the movement of invasive species include roads, railways, paths, rivers, and streams (Clifford 1959; Wace 1977; Muhlenbach 1979; Schmidt, 1989; Skoglen 1990). Two studies, one in Florida and one in Hawaii, found that in areas where road construction methods led to the introduction of soils different from the native soils, invasives were more likely (Western and Juvik 1983; Greenberg et al. 1997). Such areas should be identified and monitored.

Enlisting the cooperation of landowners in the interface to assist with monitoring the corridors may engage them, as they learn that invasive species may cause management problems as well. Developing early detection methods so that new infestations are quickly located and treated is essential. Models and experience show that species may be eradicated if detected when in only one or few populations, but when populations become numerous and widespread, eradication becomes impossible (Moody and Mack 1988). Because species reproduce at different rates, it is difficult to say how long one has to begin a control program after detection takes place, but in general it is best to act as quickly as possible (see Box 15.3). A reporting system for newly established species should also be developed so that all landowners in the area learn that they have a new species to watch for. This is a role that could be assumed by a local agency, a nongovernmental organization such as one of the many emerging Exotic Pest Plant Councils, or an association of landowners.

Enlisting citizens to help control invasives in urban areas as well as wildlands has proven to be effective in both controlling invasive species and educating the public about the problems associated with invasive species. For instance, the No Ivy League has operated for several years in Portland, OR, to control English ivy (*Hedera helix*). Initiated as a partnership between the city parks and recreation department and a local group wanting to help at a large urban park, the program has broadened to other lands

Box 15.3

Importance of Early Detection and Rapid Response

Invasions generally follow a process beginning with a few small populations that disperse from parent plants. These populations increase in size, and disperse seeds outward to begin even more satellite populations. The more populations that are established, the less likely it becomes that the species will be eradicated, because they will be too widespread for detection and effective control. As Figure 15.3 shows, a large population may disperse seeds outward, but some seeds will likely remain in the parent population.

Although it is important to control large populations, an effective strategy is to identify the smaller populations that may be the leading edge of the invasion and control them first.

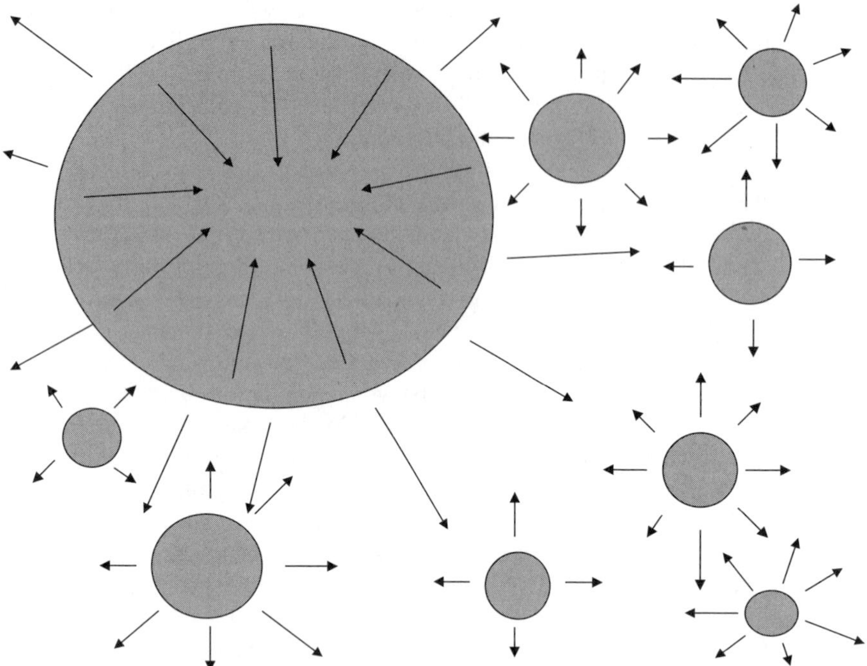

FIGURE 15.3 Satellite populations of invasive plants are generated when seed is dispersed from a population of parent plants. These, if not controlled, create more satellites.

and has as its mission the reduction of the adverse consequences from invasive English ivy in local ecosystems while promoting youth development opportunities, environmental education, and community involvement (Figure 15.4) (No Ivy League n.d.). Many other communities have similar programs to control other invasive species.

Local policies and codes can be drafted to address concern about the movement of invasive species from urban to wildlands. Greenbelt ordinances could incorporate some of the above suggestions, as well as listing regionally appropriate species for planting along the interface. Approval for new projects could include incentives for the removal of invasives and the creation of a buffer along wildland areas.

FIGURE 15.4 A youth crew removes invasive English ivy and restores native habitat at Forest Park in Portland, OR. (Photo by Mary Duryea.)

15.5 CONCLUSION

Invasive species are increasingly recognized as one of the more serious environmental problems for the protection of natural ecosystems. The wildland–urban interface is a front line in the battle to protect wildlands from the invaders that largely originate in urban areas. While efforts to control existing invaders continue to be an important step in protecting wildlands, more resources must be committed to the prevention of the introduction and spread of invasive species. Efforts to improve and apply our understanding of the biology of invasive species, leading to improved methods to analyze the risk of introduced species, may lead to an eventual reduction in the number of invaders that arrive either intentionally or unintentionally. It is unlikely, however, that this will happen in the near future and even less likely that those efforts will completely restrict new invaders. Therefore, a greater effort needs to be made to prevent the movement of species from urban areas out into the surrounding wildlands. Managers and landowners in the interface areas will play an increasingly important role in these efforts, as will local, state, and federal governments.

ACKNOWLEDGMENTS

I thank Susan Vince and Mary Duryea for their extraordinary patience and the faculty and students of the Center for Urban Horticulture for many years of stimulating discussions about urban ecology. I appreciate the suggestions of Alison Fox, Ed Macie, Annie Hermansen, Susan Vince, Mary Duryea, and Kathy Wolf on an earlier version of the manuscript. Finally, I am very grateful to Ann Rhoads and the Morris Arboretum for allowing me to use their information on the spread of *Lonicera* species in Pennsylvania.

REFERENCES

Blackburn, W, R.W. Knight, and J.L. Schuster, 1982. Saltcedar influence of sedimentation in the Brazos River, *Journal of Soil and Water Conservation* 37: 298–230.

Cadenasso, M.L. and S.T.A. Pickett, 2001. Effect of edge structure on the flux of species into forest interiors, *Conservation Biology* 15: 91–97.

Clifford, H.T., 1956. Seed dispersal on footwear, *Proceedings of the Botanical Society of the British Isles* 2: 129–131.

Clifford, H.T., 1959. Seed dispersal by motor vehicles, *Journal of Ecology* 47: 311–315.

Daehler, C.A. and D. R. Strong, 1997. Hybridization between introduced smooth cordgrass (*Spartina alterniflora*; Poaceae) and native California cordgrass (*S. foliosa*) in San Francisco Bay, USA, *American Journal of Botany* 84: 607–611.

D'Antonio, C.M. and P.M. Vitousek, 1992. Biological invasions by invasive grasses, the grass/fire cycle and global change, *Annual Review of Ecology and Systematics* 23: 63–87.

Debussche, M. and P. Isenmann, 1994. Bird-dispersed seed rain and seedling establishment in patchy Mediterranean vegetation, *Oikos* 69: 414–426.

Department of Agriculture, Western Australia, n.d. Permitted and Quarantine Species Lists, http://www.agric.wa.gov.au/progserv/plants/weeds/weeds/weedlist.htm. [Date accessed: October 16, 2003.]

Dewey, L.H., 1896. Legislation Against Weeds, Bulletin no. 17. U.S. Department of Agriculture, Division of Botany, Washington, DC.

Elton, C.S., 1958. *The Ecology of Invasions by Animals and Plants*, Methuen and Co., London, reprinted by University of Chicago Press, Chicago, 2000.

Forcella, F. and S.J. Harvey, 1988. Patterns of Weed Migration in the Northwestern U.S.A., *Weed Science* 36: 194–201.

Foy, C.L., D.R. Forney, and W.E. Cooley, 1983. History of weed introductions, in *Exotic Pest Plant Pests and North American Agriculture*, C.L. Wilson and C.L. Graham, Eds., Academic Press, New York, pp. 65–92.

Greenberg, C.H., S.H. Crownover, and D.R. Gordon, 1997. Roadside soils: a corridor for invasion of xeric scrub by nonindigenous plants, *Natural Areas Journal* 17: 99–109.

Huenneke, L.F. and J.K. Thomson, 1994. Potential interference between a threatened endemic thistle and an invasive nonnative plant, *Conservation Biology* 9: 415–425.

Kennedy, T.A., S. Naeem, K. Howe, J.M.H. Knops, D. Tilman, and P. Reich, 2002. Biodiversity as a barrier to ecological invasion, *Nature* 471: 636–638.

Lockwood, J.L., D. Simberloff, M.L. McKinney, and B. von Holle, 2001. How many, and which, plants will invade natural areas? *Biological Invasions* 3: 1–8.

MacDonald, I.A.W., F.J. Kruger, and A.A. Ferrar, 1986. *The Ecology and Management of Biological Invasions in South Africa*, Oxford University Press, Cape Town.

Mack, R.N., 1996. Biotic Barriers to Plant Naturalization, in *Proceedings of the IX International Symposium on Biological Control of Weeds*, V.C. Moran and J.H. Hoffman, Eds. University of Cape Town, Stellenbosch, South Africa, pp. 19–26.

Moody, M.E. and R.N. Mack, 1988. Controlling the spread of plant invasions: the importance of recent foci, *Journal of Applied Ecology* 25: 1009–1021.

Mooney, H.A. and J.A. Drake, 1986. *Ecology of Biological Invasions of North America and Hawaii*, Springer-Verlag, New York.

Mooney, H.A. and R.J. Hobbs, 2000. *Invasive Species in a Changing World*, Island Press, Washington, DC.

Muhlenbach, V., 1979. Contributions to the synanthropic flora of the railroads in St. Louis, Missouri, U.S.A., *Annals of the Missouri Botanical Garden* 66: 1–108.

Mulvaney, M., 2001. The effect of introduction pressure on the naturalization of ornamental woody plants in south-eastern Australia, in *Weed Risk Assessment*, R.H. Groves, F.D. Panetta, and J.G. Virtue, Eds., CSIRO Publishing, Victoria, Collingwood, Australia, pp. 186–193.

National Research Council, 2002. *The Scientific Basis for Predicting the Invasive Potential of Nonindigenous Plants and Plant Pests*, National Academy Press, Washington, DC.

No Ivy League, n.d. Ivy Removal Project, www.noivyleague.com. [Date accessed: October 16, 2003.]

Novak, S. and R.N. Mack, 2001. Tracing plant introduction and spread: genetic evidence from *Bromus tectorum* (cheatgrass), *BioScience* 51: 114–122.

Parendes, L.A. and J.A. Jones, 2000. Role of light availability and dispersal in exotic plant invasion along roads and streams in the H.J. Andrews Experimental Forest, Oregon, *Conservation Biology* 14: 64–75.

Pheloung, P.C., 2001. Weed risk assessment for plant introductions to Australia, in *Weed Risk Assessment*, R.H. Groves, F.D. Panetta, and J.G. Virtue, Eds., CSIRO Publishing, Victoria, Collingwood, Australia, pp. 83–92.

Pimentel, D., L. Lach, R. Zuniga, and D. Morrison, 2000. Environmental and economic costs of non-indigenous species in the U.S., *BioScience* 50: 53–65.

Proctor, V.W., 1968. Long-distance dispersal of seeds by retention in the digestive tract of birds, *Science* 160: 321–322.

Rapaport, E., 1991. Tropical vs. temperate weeds: a glance into the present and future, in *Ecology of Biological Invasions in the Tropics*, P.S. Ramakrishan, Ed., International Scientific Publications, New Delhi, pp. 215–227.

Reichard, S.H., 1997. Prevention of invasive plant introductions on national and local levels, in *Assessment and Management of Plant Invasions*, J.A. Luken and J.A. Thieret, Eds., Springer-Verlag, New York, pp. 215–227.

Reichard, S.H. and C.W. Hamilton, 1997. Predicting invasions of woody plants introduced into North America, *Conservation Biology* 11: 193–203.

Reichard, S.H. and P.S. White, 2001. Horticulture as a pathway of invasive plant introductions in the U.S., *BioScience* 51: 103–113.

Reichard, S. H., L. Chalker-Scott, and S. Buchanan, 2000. Interactions among non-native plants and birds, in *Avian Ecology and Conservation in an Urbanizing World*, J. Marzluff, R. Bowman, and R. Donnelly, Eds., Kluwer Academic Publishers, Boston, pp. 179–223.

Rejmánek, M. and D.M. Richardson, 1996. What attributes make some plant species more invasive? *Ecology* 77: 655–661.

Ridley, H.N., 1930. *The Dispersal of Plants throughout the World*, L. Reeve and Co., Ashford, Kent, England.

Sallabanks, R., 1993. Fruiting attractiveness to avian seed dispersers: native vs. invasive *Crataegus* in western Oregon, *Madrono* 40: 108–116.

Schmidt, W., 1989. Plant dispersal by motor car, *Vegetatio* 80: 147–152.

Silvertown, J.W., 1987. *Introduction to Plant Population Ecology*, Longman Scientific and Technical, Harlow, Essex, England.

Skoglen, J.S., 1990. Seed dispersing agents in two regularly flooded river sites, *Canadian Journal of Botany* 68: 754–760.

Stohlgren, T.J., D. Binkely, G.W. Chong, M.A. Kalkhan, L.D. Schell, K.A. Bull, Y. Otsuki, G. Newman, M. Bashkin, and Y. Son, 1999. Exotic plant species invade hot spots of native plant diversity, *Ecological Monographs* 69: 25–46.

U.S. Congress, Office of Technology Assessment, 1993. Harmful Non-indigenous Species in the U.S., OTA-F-565, U.S. Government Printing Office, Washington, DC.

Vitousek, P.M., L.R. Walker, L.D. Whiteaker, D. Mueller-Dombois, and P.A. Matson, 1987. Biological invasion by *Myrica faya* alters ecosystem development in Hawaii, *Science* 238: 802–804.

Wace, N., 1977. Assessment of dispersal of plant species: the car borne flora of Canberra, *Proceedings of the Ecological Society of Australia* 10: 167–186.

Western, L. and J.O. Juvik, 1983. Roadside plant communities on Mauna Loa, Hawaii, *Journal of Biogeography* 10: 307–316.

Part V

Conserving and Managing Forests under Different Ownerships

16 Managing Industrial Forestlands at the Interface

Kenneth R. Munson and Sharon G. Haines
International Paper Company

CONTENTS

16.1 INTRODUCTION

Many authors in this book have addressed the wildland–urban interface from perspectives unique to their experience and expertise. The purpose of this chapter is to explore forestland uses at the interface from the perspective of an industrial landowner. This task has some inherent risk in that industrial landowners are not alike, nor do they all share a common motive for owning land. With that acknowledgment, in this chapter we attempt to describe the situation of urban–forest interfaces from the perspective of a large forest products company. We begin with a brief review of land ownership and wood supply in the U.S. to set the context for the role of industrial forests. Then, we present some statistics about land-use changes, followed by a discussion of forest management implications at the interface. The chapter concludes with potential solutions to management of industrial forestland at the wildland–urban interface.

16.2 FORESTLAND OWNERSHIP AND WOOD SUPPLY

Understanding the pattern of forestland ownership and wood supply is useful when considering policies that can encourage or discourage certain forestland uses. There are three general categories of forestland owners in the U.S.: industrial, private nonindustrial, and public. Forest industry owns or controls about 13 percent of the timberlands in the U.S. and supplies 30 percent of the harvested volume (Figure 16.1 and Figure 16.2). Private, nonindustrial owners control 58 percent of the timberlands and supply 59 percent of the total volume harvested. The public, both federal and state, controls 29 percent of the forests while supplying only 11 percent of the wood.

The imbalance between ownership and wood supply is not surprising and reflects differing objectives for land ownership. Industry owns land as part of a business strategy and therefore places priority on the growth of raw materials used for manufacturing paper, packaging, and building materials. The capability of industry to produce wood also reflects technological advances made through private investments in research and development. Private, nonindustrial forest owners have very diverse ownership patterns and attributes. This contributes to a wide range of ownership objectives, with timber production often being a secondary one (see Chapter 17). The contribution of public forestland to wood supply, while potentially significant, is relatively low. This reflects public sentiment regarding both consumptive and nonconsumptive forest uses and resulting policies favoring forest conservation.

It is important to understand this situation as we consider the impacts of continued changing ownerships and land-use patterns on industrial forests. In the southeastern U.S., 26 percent of commercial forestland falls within metropolitan counties with populations exceeding 250,000

1-56670-602-5/05/$0.00+$1.50
© 2005 by CRC Press LLC

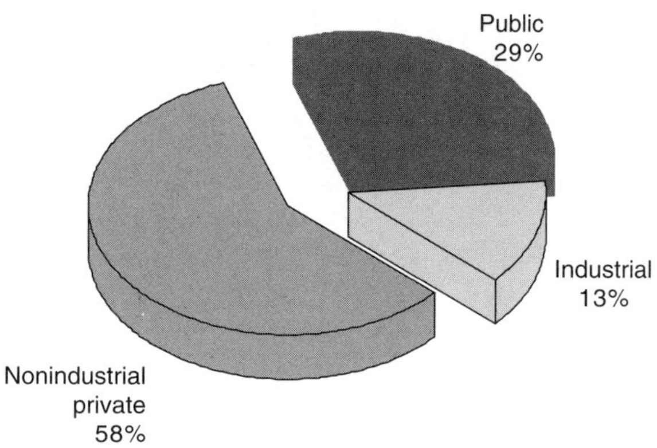

FIGURE 16.1 Timberland ownership in the U.S. Timberland is forestland that is producing or capable of producing crops of industrial wood (more than 20 ft^3/acre/year) and that is not withdrawn from timber utilization. (Adapted from Smith et al. 2001.)

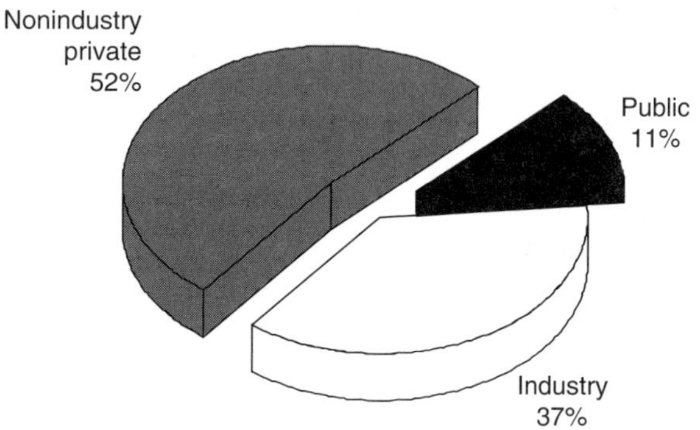

FIGURE 16.2 Source of wood used in industrial manufacturing in the U.S. (Adapted from Smith et al. 2001.)

(DeForest et al. 1991). As much as 43 percent of this acreage may be unavailable for timber production and might be better classified as real estate (Befort et al. 1988). Judging by current trends, industrial forests are clearly vital to the current and future wood supply in the U.S.

16.3 IMPORTANCE OF INDUSTRIAL FORESTS

As discussed in the previous section, industrial forestlands can and do supply a disproportionate share of the raw materials used in the manufacture of paper, packaging, and building materials in the U.S. As various pressures continue to reduce harvesting from public forests in many regions of the world (Braatz 2001), those forests still in commercial production will become increasingly important. Among the reasons for this heightened importance is the higher productivity achieved through active management. This enables greater harvest levels at more frequent intervals. In other words, more wood is obtained from less land.

While industry owns and invests in forests with clear economic objectives, the advent of such programs as the Sustainable Forestry Initiative shows an effort to balance financial goals with environmental and social objectives. Industry has made and continues to make substantial investments in forest resources research, having spent $473.5 million since 1995 (American Forest and Paper Association 2002). This work, much of it in cooperation with scientists at state universities, has resulted in a better understanding of forested systems. This knowledge is the foundation for management regimes that are improving forest productivity and overall forest health. The effects are particularly evident in the southern U.S., where the annual rate of softwood growth on industry lands is nearly 2.5 times that on nonindustrial private forestland (Smith et al. 2001).

Figure 16.3 presents a comparison of loblolly pine growth rates in the southeastern U.S. under three scenarios: unmanaged, low-intensity plantations, and intensive plantations. An unmanaged loblolly pine stand grows at a rate of about 3 t per acre per year and takes 35 years to reach economic maturity. A plantation managed at low intensity grows approximately 5 t per acre per year and matures in 30 years. In contrast, an intensive regime, which might include competition control and fertilization, produces in the range of 9 t per acre per year and matures in 25 years. The economic return on investment, when considering all forms of cost, is better with the more intensive regimes, particularly on high-productivity sites (Sedjo and Botkin 1997). In general, industry lands tend to be managed toward the intensive end of the continuum, public lands toward the unmanaged end, and nonindustrial private lands in the middle, with a leaning toward little management.

The important point is that industrial lands have tremendous timber productivity potential by virtue of the type of land owned and the investment philosophy of industrial forest companies willing to finance more intensive forest management. Developing and encouraging the potential of this land to supply wood for the paper, packaging, and housing needs of our citizens should be a priority. This is not a self-serving statement to inflate the importance of industrial forests. The reality is that wood to meet the demands of a growing world population must come either from plantations managed for this purpose or from other forest types. Industrial forests should be viewed as an environmentally preferable solution in that increasing the output from industrial forests will reduce the pressure to harvest native, old growth, and public forests (Sedjo and Botkin 1997; Braatz 2001).

16.4 LAND-USE CHANGES

The greatest threat to forests in the U.S. is change in land use, especially conversion to developed land where tim-

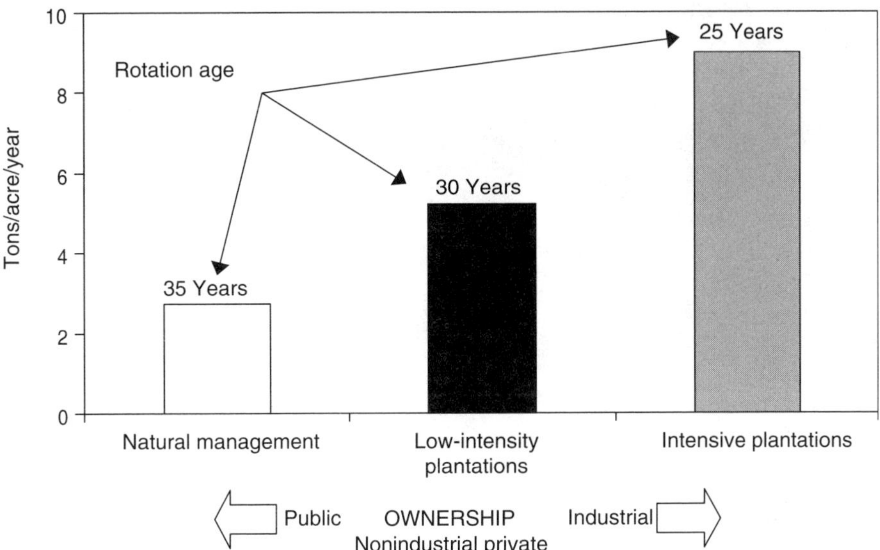

FIGURE 16.3 Loblolly pine growth rates and general rotation ages for three levels of management.

ber productive capability is essentially eliminated or at least significantly diminished (USDA Natural Resources Conservation Service 2000; Wear and Greis 2002). Other authors in this book have provided excellent descriptions of these changes. A few trends are particularly relevant to this chapter and will be highlighted below.

Information from the 1997 Natural Resource Inventory shows that during the 15-year period from 1982 to 1997, about 10 million acres were converted from forests to some form of development (USDA Natural Resources Conservation Service 2000). During the same period, an even greater number of agricultural acres were converted to forests, resulting in a net increase of 3.6 million acres of forest. The question is, will the forest resource continue to increase in the next 15 years, or will there be a net loss of forests as a result of development?

Figure 16.4 shows the change in developed land during the same time period for the four southern states with the most change: Florida, Georgia, North Carolina, and Texas (USDA Natural Resources Conservation Service 2000). The amount of developed land increased during this period by an astonishing 50 percent, apparently as a result of economic and population growth in the southern U.S. This trend is especially graphic in "before and after" aerial photographs. A good example is from an area in South Carolina about 20 mil west of Hilton Head Island. An aerial photograph from 1987 (Figure 16.5A) shows a landscape of very productive forestland with relatively few roads and buildings. In the late 1980s and early 1990s, Hilton Head Island, a popular recreation and retirement area, was almost completely developed. The demand for affordable homes was pushing westward. A 1997 aerial photograph of the same area reveals significant development (Figure 16.5B). What had been a productive

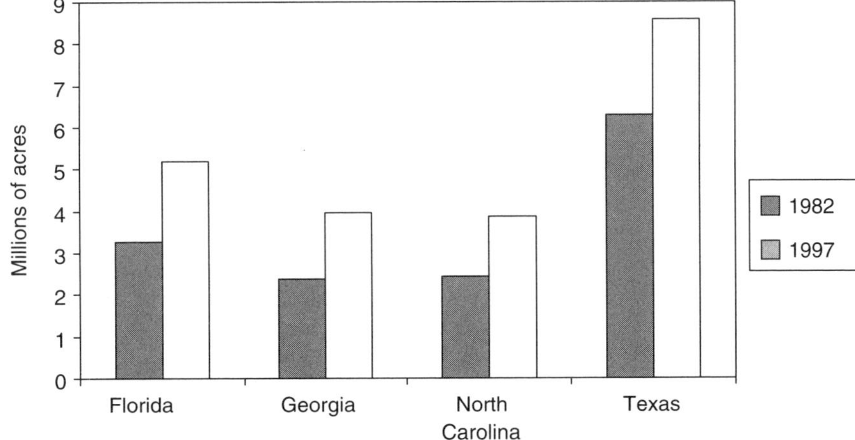

FIGURE 16.4 Change in developed land from 1982 to 1997 for four southern states. (Adapted from USDA, Natural Resources Conservation Service 2000.)

FIGURE 16.5 Aerial photographs of forested land in coastal South Carolina, 1987 (A) and 1997 (B). The same timbered area is outlined in white in both photographs.

forest 10 years earlier is now a planned retirement community. Four-lane highways run north and south, and rooftops are more prevalent than treetops. The only remnants of the forest are trees in yards and along golf course fairways.

Managing forests in the zone of development presents considerable challenges with added costs and risks. These challenges, coupled with rising land values, can cause a company to sell land. Thus, it is not the selling of land that causes development, but the pressure of encroaching development that causes land sales for what, interestingly enough, is called "higher and better use." This is one example, but the pattern is being repeated across the U.S.

16.5 MANAGEMENT IMPLICATIONS

Information in this section came from interviews conducted with International Paper employees who are involved with forest management and forestland sales in the southeastern and northeastern U.S. The interviews were designed to illuminate specific management implications at the interface. The results were variable, caused in part by variation in state and local governance standards and social expectations. This in turn caused variation in on-the-ground cost impacts. Despite the variation, however, some common and insightful trends emerged.

For the purpose of this discussion, cost impacts are divided into five categories: annual administrative costs, timber productivity and associated revenue, taxes, land value, and risks. The cost estimates reflect ranges in the Southeast and offer a reasonable order of magnitude. Clearly, the interface consists of a continuum of impacts from low to high. To facilitate the contrast, forestlands that are distant from the urban interface are referred to as *remote* lands. Those at or near the interface are referred to as *interface* lands.

Annual management costs for remote forest tracts in the South range from $1 to $5 per acre. These costs include forester salaries, office space, vehicles, communications, and other on-going administrative costs, but do not include the cost of specific practices, such as weed control or fertilization. The foresters who were interviewed acknowledged that more administrative time is spent managing lands at the interface. Estimating and allocating administrative time where it is actually spent shows that annual costs at the interface range from $5 to $12 per acre, depending on the circumstances. The sizes of industrial tracts at the interface are usually smaller (Harris and DeForest 1993) with more edges, which means more property line maintenance and neighbor interactions. Neighbor notification is a company requirement for most management operations, such as prescribed burning and application of forest chemicals. It is not uncommon that neighbor or community interactions result in questions, phone calls, or on-site visits, which result in additional administrative time and cost. Occasionally, an event becomes a media story, which consumes several person-days to resolve. More vocal opposition to traditional forest management practices might even take the form of regulations to protect suburban and rural areas from perceived damage, including restrictions on silvicultural practices, requirements for timber harvest permits, and buffer zones (Cubbage and Siegel 1985; Cubbage 1995; Martus et al. 1995).

The second cost category is timber productivity and associated revenues. Management practices, including harvesting, site preparation, weed control, fertilization, and prescribed burning, are used in varying degrees to promote tree growth. In every situation, forests at the interface require more planning and management attention during these activities (Figure 16.6). Some of the incremental costs are captured in the administrative costs described above. Frequently, certain tools that can increase timber productivity are used less often at the interface for fear of criticism from neighbors and the local community. For instance, proximity to homes or commercial development generally precludes the use of prescribed fire. Also, extrawide buffers will often be left along property edges when forest chemicals are applied.

In essence, interface lands tend to be managed under a low-intensity regime, which lowers timber yield (Figure 16.3). In a Virginia study, Wear et al. (1999) found that the transition between rural and urban land use occurred at population densities between 20 and 70 people per square mile and that population effects potentially reduced commercial timber inventories by 30–49 percent. In Mississippi and Alabama, reductions in normal silvicultural harvests close to development and higher population density outweigh the increases in conversion harvests and can lead to short-run timber supply decreases (Barlow et al. 1998). Furthermore, leasing land for hunting and other forms of recreation is often restricted near the interface out of concern for public safety. The net effect of this is to further reduce revenue.

Comparing ordinary revenues from a remote property and an interface property is a useful way to demonstrate the difference in timber productivity. Again, this is highly variable, but holding other factors such as timber volumes and market values constant still offers a reasonable relative comparison. An acre of intensively managed loblolly pine in 25 years will produce a recreation income of $112 and timber revenues of about $4500, using 2001 southern timber prices. Assuming a low-intensity management regime and a loss of half the recreation potential, all other factors being equal, the interface property would yield about $2400 in timber revenue and only $56 from recreation per acre. The difference of $2156 is significant and clearly affects the economics of owning land (Table 16.1).

Another element of cost is property taxes. The calculation and assessment of taxes varies considerably among states and even among counties within states. One constant is that taxes are a function of property valuations and tax rates. Generally, valuation of forestland in rural counties is closely linked to its quality for growing timber. The experts interviewed above offered some ranges from eastern Georgia of $300 to $500 per acre; this is a bare land value, not including timber. The resulting tax on this land can range from a few dollars to $8 an acre (Table 16.2). This range would not include rapidly developing counties and fringe counties in the vicinity of Atlanta. In contrast, forestlands in counties where development is more active may have values of $1500 to $2500 per acre, resulting in taxes of $15 to $35 per acre per year. This is a $10 to $15 annual differential over a rural tract of similar land. To put this in perspective, about 25 percent of the wood grown on an interface acre may go to pay property taxes. Some states, such as Florida, have favorable tax laws that work

TABLE 16.1
Estimated Timber and Recreation Revenue during a 25-year Rotation of Loblolly Pine Planted at Two Locations

	Remote ($)	Interface ($)
Timber harvest	4500	2400
Recreation lease	112	56
Total	4612	2456

TABLE 16.2
Valuation and Annual Taxes per Acre on Forestland (Bare-Land Basis) in Eastern Georgia

	Land	Taxes
Rural county	$300–500	$2–8
Urban county	$1500–2200	$15–20

Note: This does not include highly urban counties in the vicinity of Atlanta.

as an incentive to minimize these effects. Other states, such as Georgia, have laws that benefit private, nonindustrial owners with small acreages, but are not applicable to industrial owners with large holdings. It has been suggested that

> by adopting a tax policy that discriminates specifically against corporate forest landowners while subsidizing thousands who stand ready to buy their land, Georgia is encouraging the decline of the very land uses it says it wants to preserve. Urban sprawl is all but guaranteed (Flick and Newman 1999).

One could argue that tax policy should provide favorable treatment to industries where growing trees is a primary business objective that yields multiple economic and social benefits such as job creation.

The last activity to consider briefly is fire control (see Chapter 13 for a detailed discussion). It comes as no surprise that interface lands are at risk from fire-related threats. First, fires tend to be more frequent when there is greater human contact occurring closer to population centers.

FIGURE 16.6 Weed control in an intensively managed pine plantation results in an open understory (A). In the interface, the proximity of houses may hinder forest management practices such as weed control, leading to dense shrub growth (B). (Photos by Larry Korhnak.)

Second, fires on interface lands tend to be more catastrophic due to higher fuel loads resulting from a lack of low-intensity prescribed fires. Forest managers are often reluctant to use prescribed burning near the interface. Third, once forest fires start, suppression efforts are often directed at saving homes first and forests second. This situation is more problematic with homes scattered on lots as opposed to planned developments with good fuel management standards, road systems, and firefighting capability. Florida and Georgia have had a series of devastating fire seasons recently. An article from the *Savannah Morning News* on May 19, 2001, covering forest fires in South Georgia noted that "firefighters were diverted to protect houses." The point is not that saving homes should not be a priority but that forests at the interface are at a greater risk from losses due to fire than are those in rural areas. This situation further discourages forest ownership and investments in forests at the interface. Of course, the landowner's alternative is to sell the land for further development. This in turn expands the interface and related management issues.

At this point, it is worthwhile to contrast two forms of development at the interface. The first is planned development, such as the retirement community near Hilton Head that we discussed earlier. While this development permanently removed forests from timber production, it has positive attributes, such as a concentrated population, efficient use of space, a well-designed road system, fuel management standards, and fire suppression infrastructure. A contrasting form of development can be labeled "lot schemes." This situation involves relatively unplanned development where homes are built on parcels ranging from less than an acre to several acres. Access roads can be variable in quality and not built to any particular standards. Homes can be located close to property boundaries. Typically, there is minimal fire suppression infrastructure close to these developments. Additionally, it is common for homes to be built among trees and brush with little regard for wildfire safety. Forest development under a "lot scheme" scenario intensifies the problems with management at the interface.

16.6 SOLUTIONS

Education about the benefits provided by industrial ownership of forestlands should be provided for policy makers at all levels of government. The potential for timber productivity of industrial forests is significant, and it can contribute an even greater share of raw materials used for the production of paper, packaging, and building materials for our citizens. Moreover, there is ample evidence for significant environmental benefits of commercial forestland compared to many alternative land uses, particularly in terms of water quality (Ice et al. 1997) and wildlife habitat (Salabanks et al. 2001).

Tax policies should be developed and supported to recognize private investments in forestry and the long-term risks associated with this class of asset. Policies should encourage, not discourage, owners to invest in timber-producing forests. Commercial forests, particularly industrial forests, are actively managed to support domestic manufacturing operations that in turn create jobs and essential products. Industrial forests produce more wood from less land than any other ownership class (Smith et al. 2001).

Closer to the ground, forestland development policy should incorporate principles of smart growth, such as incentives for planned communities and restrictions that discourage "lot-schemes" and related uncontrolled fragmentation of forests.

Finally, policies should encourage the use of conservation easements where development rights are sold and locked up in perpetuity while preserving the option to manage timber and other resources. Conservation easements have become an especially effective mechanism for companies and nongovernmental organizations to achieve complementary objectives in conservation and active forest management (see Box 16.1).

Box 16.1

CONSERVATION EASEMENTS

Conservation easements allow property owners to place legal restrictions on specific activities that may occur on their land by granting the rights to practice those activities to an appropriate party, such as a public agency, a land trust, a historic preservation group, or a conservation organization (Diehl and Barrett 1988). These restrictions may include, for example, the right to subdivide the land, to construct buildings, to harvest timber, or even to access the property. Thus, owners may protect their land from inappropriate development while retaining private ownership.

By 2000, real estate development near Myrtle Beach, SC, was approaching the Davis Farm, a 1200-acre tract of unique forestland in Brunswick County, NC. With floodplains listed among Nationally Significant Natural Heritage Areas, the tract was identified as having an extremely high environmental value. However, its desirability for housing and golf course development also gave it an extremely high commercial value. Using funds from its Clean Water Management Trust Fund, the State of North Carolina purchased a Clean Water Management Easement, assuring a perpetual clean water buffer along Town and Rice Creeks. The North Carolina Division of Forest Resources, through a federal grant, provided funding for a Forest Legacy Easement. Both easements involved a granting of development rights, while allowing International Paper, which owned the land, to apply normal forestry practices in the upland portions of the tract.

16.7 CONCLUSION

Development is the greatest threat to forests in the U.S., especially in the South. Industrial forests, while comprising only 13 percent of the U.S. forests, supply 37 percent of the wood used in our domestic manufacturing operations. Policies should strengthen private ownership and management of forests, and encourage continued investments. It is important to recognize that encouraging and promoting industrial forest management is a solution to reducing the pressure on old growth and nonindustrial forests.

ACKNOWLEDGMENT

The authors gratefully acknowledge the assistance of Dr. James Rakestraw, manager of forest research for International Paper, for his technical contributions to the final version of this chapter.

REFERENCES

American Forest and Paper Association, 2002. The Sustainable Forestry Initiative (SFI) Program 2002 7th Annual Progress Report Summary, American Forest and Paper Association, Washington, DC.

Barlow, S. A., I.A. Munn, D.A. Cleaves, and D.L. Evans, 1998. The effect of urban sprawl on timber, *Journal of Forestry* 96: 10–14.

Befort, W.A., A.E. Luloff, and M. Morrone, 1988. Rural land use and demographic change in a rapidly urbanizing environment, *Landscape and Urban Planning* 16: 345–356.

Braatz, S.M., 2001. State of the World's Forests 2001, Food and Agriculture Organization of the United Nations.

Cubbage, F.W. 1995. Regulation of private forest practices: what rights, which policies? *Journal of Forestry* 93: 14–20.

Cubbage, F.W. and W.C. Siegel, 1985. The law regulating private forest practices, *Journal of Forestry* 83: 538–545.

DeForest, C.E., T.G. Harris, Jr., F.W. Cubbage, and A.C. Nelson, 1991. Timberland Downtown? Southern Forest Resources along The Urban–Rural Continuum, in Ecological Land Classification: Applications to Identify the Productive Potential of Southern Forests, D.L. Mengel and D.T. Tew, Eds., Charlotte, NC, January 7–9, Gen. Technical Report SE-68, U.S. Department of Agriculture, Forest Service, Southeastern Forest Experiment Station, pp. 137–138.

Diehl, J. and T.S. Barrett, 1988. *The Conservation Easement Handbook: Managing Land Conservation and Historic Preservation Easement Programs*, Trust for Public Land, San Francisco, CA, and Land Trust Exchange, Alexandria, VA.

Flick, W.A. and D.H. Newman, 1999. Tax code fuels sprawl, *Atlanta Journal Constitution*, December 11, p. A14.

Harris, T.G., Jr. and C.E. DeForest, 1993. Policy Implications of Timberland Loss, Fragmentation, and Urbanization in Georgia and the Southeast, in Proceedings of the Southern Forest Economics Workshop, Duke University, Durham, NC, April 21–23.

Ice, G.G., G.W. Stuart, J.B. Waide, L.C. Irland, and P.V. Ellefson, 1997. Twenty-five years of the Clean Water Act: how clean are forest practices? *Journal of Forestry* 95: 9–13.

Martus, C.E., H.L. Haney, Jr., and W.C. Siegel, 1995. Local forest regulatory ordinances: trends in the Eastern U.S., *Journal of Forestry* 93: 27–31.

Salabanks, R.,E. Arnett, T.B. Wigley, and L. Irwin, 2001. Accommodating Birds in Forests of North America: A Review of Bird–forestry Relationships, National Council for Air and Stream Improvement Tech. Bull. 822.

Sedjo, R.A. and D. Botkin, 1997. Using forest plantations to spare national forests, *Environment* 39: 15–20, 30.

Smith, W.B., J.L. Vissage, D.R. Darr, and R. Sheffield, 2001. Forest Resources of the U.S., Gen. Technical Report NC-219, U.S. Department of Agriculture, Forest Service, North Central Forest Experiment Station, St. Paul, MN.

U.S. Department of Agriculture (USDA), Natural Resources Conservation Service, 2000. Summary Report: 1997 National Resources Inventory (revised December 2000), USDA Natural Resources Conservation Service, Washington, DC, and Statistical Laboratory, Iowa State University, Ames, Iowa, http://www.nrcs.usda.gov/technical/NRI/1997. [Date accessed: March 3, 2003.]

Wear, D.N. and J.G. Greis, Eds., 2002. Southern Forest Resource Assessment, Gen. Technical Report SRS-53, U.S. Department of Agriculture, Forest Service, Southern Research Station, Asheville, NC.

Wear, D.N., R. Liu, J.M. Foreman, and R.M. Sheffield, 1999. The effects of population growth on timber management and inventories in Virginia, *Forest Ecology and Management* 118: 107–115.

17 Managing Private Nonindustrial Forestlands at the Interface

William G. Hubbard
Cooperative Extension Service Southern Region, University of Georgia

David A. Hoge
Cooperative Forestry Unit, USDA Forest Service Southern Region

CONTENTS

17.1 INTRODUCTION

Recent USDA Forest Service studies estimate that more than 10 million individuals own close to 50 percent of the forestland in the U.S. (Butler and Leatherberry 2003). They are by far the dominant ownership class in the country (Figure 17.1). These individuals have been called "nonindustrial private forestland owners," or NIPF owners, because they are not directly involved in the processing of forest products, nor do they own public lands such as the National Forests or state/municipal forests. Discussions about managing industrial and public forestlands can be found in related Chapters 16 and 18. Understanding and managing the forests of the nonindustrial private landowner at the wildland–urban interface is a complex undertaking that involves creative thinking and innovation.

It is important to note that NIPF ownership varies greatly geographically. Eighty-one percent of the NIPF acreage and 90 percent of the NIPF owners can be found in the North (Lake States and the Northeast) and the South, while the West supports the remaining 19 percent of the land and 10 percent of the owners (Figure 17.2). Fewer of the forests in the West are found in counties classified as nonrural (44.2 percent) compared to the East (79.2 percent) (Smith et al. 2001). Any discussion of the issues and opportunities involving managing NIPFs in the

1-56670-602-5/05/$0.00+$1.50
© 2005 by CRC Press LLC

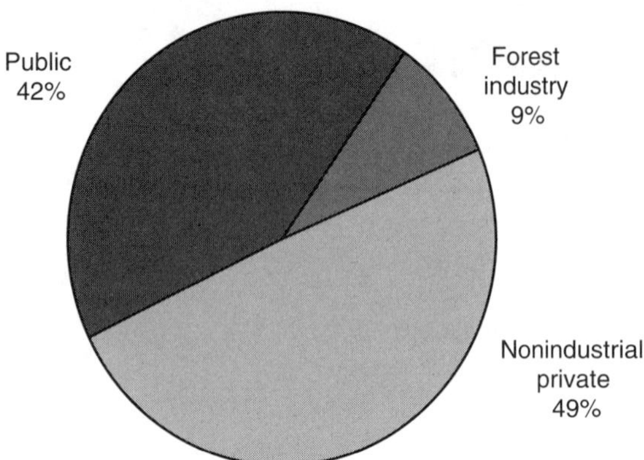

FIGURE 17.1 Forestland ownership in the U. S. (Adapted from Smith et al. 2001.)

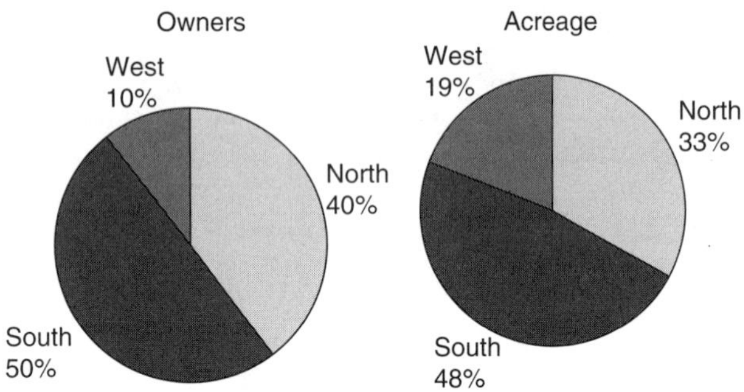

FIGURE 17.2 Regional variation in NIPFs by number of owners and acreage. (Adapted from Birch 1996.)

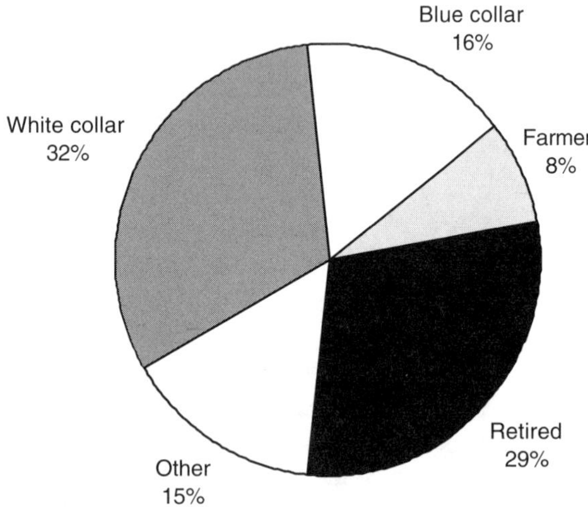

FIGURE 17.3 NIPF ownership by occupational category. (Adapted from Birch 1996.)

interface needs to take these regional variations into account.

Actions or inactions on the part of the owners of these forests will have ramifications not only in the present but also for many years to come. Policies, markets, technologies, and other trends that have an effect on NIPF owners can likewise alter the future landscape. In this chapter, we will review the demographics and characteristics of the NIPF owner in general and discuss how these may be different for the subset of NIPF owners in the interface. We will also summarize some of the barriers associated with forest management on these lands and focus on solutions and opportunities that are either being implemented or discussed within the forestry community.

17.2 WHO IS THE NIPF OWNER?

Owners of NIPFs are as diverse as they are numerous. They are a cross section of America that represents all walks of life (Figure 17.3). Roughly 1 in 28 Americans owns forestland, and the parcel may be as little as 1 acre or as large as 10,000 acres or more (Table 17.1). NIPF owners may live on their forestland or halfway around the world and never set foot on the property. They may own the land for recreational purposes or for a financial return through harvesting timber (Table 17.2). They may have taken advantage of financial, technical, informational, or educational programs offered by service providers, or they may be among the millions who for one reason or another have not utilized these services.

The NIPF owner and resource have been studied for years. Academicians, federal and state forestry agencies, forest industry, nongovernmental organizations (NGOs), and others have worked to understand the extent of this resource, the motivation behind better forest management on these lands, and other significant characteristics and trends. Among the major findings of these studies is a tendency for NIPF lands to change hands fairly regularly and in the process be subdivided and sold in smaller pieces. Birch (1996) found a 28 percent increase in the number of NIPF owners between surveys in 1978 and 1994. During

TABLE 17.1
Number of NIPF Owners and Acres by Acreage Size Class (1994)

	Owners		Acres	
Size (acres)	Number (million)	Percentage	Number (million)	Percentage
1–9	5.8	59	16.6	4
10–49	2.8	28	60.4	16
50–99	0.72	7	47.2	12
100–499	0.56	6	91.6	23
500–999	0.04	0.4	24.5	6
1000+	0.03	0.3	153	39
Total	9.9*	100*	393.4*	100*

Source: Birch (1996).
*Columns may not add exactly to the total due to small rounding errors.

TABLE 17.2
Primary Reasons for Owning Forestland (by NIPF Owner and Acreage)

	Owners		Acres	
Reason for Owning Timberland	Number (million)	Percentage	Number (million)	Percentage
Part of residence	2.6	26.7	32.6	8.3
Enjoyment of owning	1.4	14.1	28.7	7.3
Part of farm	1.2	12.0	38.6	9.8
Estate	0.99	10.0	(data unavail—able)	(data unavail—able)
Land investment	0.92	9.3	39.3	10.0
Recreation	0.88	8.8	37.9	9.6
Farm and domestic use	0.82	8.2	35.8	9.1
Other	0.45	4.5	61	15.5
No answer	0.35	3.6	6.3	1.6
Timber production	0.27	2.7	113.2	28.8
Total	9.9	100	393.4	100

Source: Birch (1996).

TABLE 17.3
Intention of NIPF Owners to Harvest (by Acreage Class, Percentage)

	Time until Intended Harvest			
Size (acres)	1–10 Years	Indefinite	Never	No Answer
1–9	24	24	44	8
10–49	40	32	25	3
50–99	42	37	17	3
100–499	50	36	11	2
500–999	59	30	9	2
1000+	70	20	7	2
Percentage of all NIPF owners	31	28	34	6

Source: Birch (1996).

that same period, the number of tracts in the 100- to 499-acre category and the number of tracts in the 500- to 999-acre category declined by 11 and 9 percent, respectively. This led to substantial increases in the smaller holdings a (43 percent increase in the 50- to 99-acre class and a 115 percent increase in the 10- to 49-acre class). In addition, millions of acres were developed and lost from the NIPF base. Interestingly, the overall average acreage per landowner did not decline substantially (from 42.9 acres in 1978 to 39.7 acres in 1994); however, this was largely due to the inclusion of 60 million acres of southwest and Indian tribal forests in the 1994 study. If these numbers were not included, the decline in average acreage per owner would be more noticeable. Understanding these trends has helped the forestry community begin to position itself for newer owners with smaller tracts of land.

What are these new owners like? Based on historic trends, it is estimated that 35 percent of the forestland currently owned by individuals will change ownership every 10 years (Sampson and DeCoster 1997). Thirty-four percent of the landowners, owning 11 percent of the NIPF land, do not plan to harvest (Birch 1996) (Table 17.3). An additional 28 percent of the NIPF owners, controlling 23 percent of the resource, say that their plans to cut timber are currently "indefinite." As the tract size decreases, the percentage of landowners who plan to harvest in the next 10 years also declines; for example, only 24 percent of the owners of 1- to 9-acre parcels plan a harvest in the next 10 years compared to 40 percent of the owners of 10- to 49-acre tracts. Surveys of landowner objectives are increasingly finding that preserving and appreciating the natural aspects of the land, viewing wildlife, stewardship for future generations, recreation, and investment, among other objectives, rank higher than timber management and harvesting (Table 17.2).

The characteristics reviewed here relate to the general NIPF owner in the country. Are landowners in the interface typically different from those in rural areas? Can we assume that they are younger? Better educated? Do they have more discretionary income? Are they less tied to the land? Do they know less about forest and land management than their parents' generation? Do they have a different set of values and management objectives? Do they desire different types of assistance? Few studies have investigated these characteristics of the interface NIPF owner or compared the differences between an interface NIPF owner and a noninterface NIPF owner. The general consensus in the forestry community is that the average landowner in the interface does indeed satisfy many of the above-mentioned characteristics and has different technical, fiscal, educational, and information needs (Hull et al. in press). It should be noted, however, that there remain a sizable number of traditional NIPFs and owners in the interface and that the various issues they face also need to be addressed.

One study, conducted in North Carolina, investigated some of the differences between urban and rural landowners. Megalos and Cubbage (2000) surveyed NIPF owners in urban and rural counties to determine likeliness to participate in various assistance programs. In general, landowners in urban counties were significantly less likely to embrace educational and technical assistance programs. Rural landowners, in contrast, significantly favored most educational programs. Landowners with holdings in the urban counties were less likely to participate in industry technical assistance, timber price information, and affordable hazard insurance. In summary, more research on the demographics and characteristics of the interface NIPF owners will be needed to understand and more effectively meet their needs.

17.3 WHAT ARE THE ISSUES AND BARRIERS TO FOREST MANAGEMENT AT THE INTERFACE?

Understanding landowner characteristics and demographics is often a first step in addressing the challenges related to NIPF management in the interface. An understanding of the various economic, social, political, and environmental factors that affect landowners and the land is often next. What are the issues in a given local situation? Are there many large, contiguous tracts of forestland in a given area? Are there ordinances or regulations that limit the right to conduct certain forest management activities on the land? Are there local landowner associations or cooperatives that can provide information, technical assistance, and even financial support for small-scale or niche market products? Is there a social stigma attached to forest management practices such as controlled burning, timber harvesting, or aerial fertilizing? Managing private nonindustrial forests at the interface requires a fairly comprehensive understanding of these current and potential socioeconomic and environmental issues. The following is a summary of some of the major issues that are affecting NIPF management in the interface.

17.3.1 FRAGMENTATION AND PARCELIZATION

Land-use *fragmentation* occurs when contiguous blocks of forestland are split into small, isolated tracts intermixed with various land uses such as subdivisions, shopping complexes, farms, or ranches (Figure 17.4). Road building often facilitates forest fragmentation as it opens up previously unbroken tracts of forestland to change. Land-use *parcelization* by comparison is often thought of as a precursor to fragmentation because in this phase, the land is subdivided and sold to owners who may have varying objectives, but the land use has not yet "fragmented" from contiguous forest. Fragmentation and parcelization are not necessarily undesirable in and of themselves, but the potential economic and environmental impacts could be detrimental as forested tracts become smaller and smaller or if they become physically detached from the larger contiguous forest. This is because many aspects of forest and wildlife management are directly related to tract size and connectivity to adjacent forested tracts (see Chapter 11 for a detailed discussion). Larger tracts of contiguous forestland generally produce higher-valued ecological and economic goods or services at lower costs. Examples include adequate habitat needs for desirable wildlife (food, shelter, water), economies of scale for timber management (tree planting, timber stand improvement practices, harvesting), and higher-valued recreational experiences (trails, campgrounds, ecotourism activities). Conversely, the smaller the tract size, the more difficult it becomes to achieve many activities associated with profitable timber and wildlife management (e.g., retaining marketable timber tracts or the ability to perform silvicultural activities such as burning or timber stand improvement). In many parts of the country, these activities are not viable if the tract size is less than 40 to 50 acres (Cubbage 1982, 1983). Management and harvest of high value and specialty trees and alternative forest products can oftentimes be undertaken on substantially less acreage, but equipment needs, potential site disturbance, and sustainability can become concerns.

FIGURE 17.4 Forest fragmentation affects the ecological and social fabric of the landscape.

From a wildlife management perspective, many forest-dwelling animals like wild turkey and quail often require food, water, and shelter that cannot be supported by the typical one- to nine-acre tract held by thousands of interface NIPF owners. From a wildlife nuisance standpoint, forest fragmentation can have undesirable effects as well. Some species such as white-tailed deer have adapted well to the fragmented landscape and have actually become a nuisance due to their high populations (Warren 1997).

From a societal perspective, impacts of fragmentation that are noticeable in the interface include potential degradation of viewsheds and watersheds. Many citizens prefer to view unbroken tracts of land such as a large expanse of forest, pasture, or cropland (Figure 17.5). They relocate their homes or businesses to areas that are visually appealing. They may also choose routes to and from home, work, and other locations that provide a sense of relaxation. As tracts become fragmented, the likelihood of a mixture of forest and nonforest use increases along with the likelihood that the mixed use will contain visually incompatible sites.

17.3.2 Limited Markets for Traditional Forest Goods and Services

Landowners in the interface who are interested in growing and marketing traditional forest products like pulpwood, chips, and sawtimber often face limited markets for these goods. Also affected are those who might be interested in leasing their lands for hunting and recreational purposes. Timber companies, wildlife hunting clubs, recreationists, and other owners are finding it exceedingly difficult to operate with favorable economic returns in the wildland–urban interface (see Chapter 16). Those industries that could create demand for many interface NIPF goods and services face too many economic, legal, or social constraints. Wear et al. (1999) surveyed experts in commercial forestry and found that development in the interface (an area defined as housing between 20 and 70 residents per square mile) decreases the available timber supply between 30 and 49 percent. For the forest industry, larger tracts in the interface that are economical for timber growing are finding more economic value in the real estate arena. Liability, small tract size, proximity to urban centers, and low populations of game also limit participation by hunting clubs in these areas, and many campers and hikers prefer expansive, uninterrupted forested areas and trails that are increasingly found farther and farther from urban centers.

17.3.3 Regulatory and Tax Issues

Local conservation codes and growth management regulations and policies, forest practices acts, and tree preservation ordinances are among the challenges being faced by landowners in the interface. Timber harvesting and controlled burning regulations and ordinances, for example, are appearing in communities and counties surrounding urban centers at rates faster than ever before due to residents' concerns over issues such as smoke and clearcutting (Figure 17.6) (Cubbage 1995). Urban tree preservation ordinances are having impacts on traditional silviculture as well because of expanding jurisdictions and sprawling metropolitan areas that sometimes stretch miles beyond the city limits. Often, in these cases, municipal

FIGURE 17.5 Landowners and society place high value on attractive forested views. (Photo by William Hubbard.)

FIGURE 17.6 Because of residents' concerns over issues such as smoke and clearcutting, timber harvesting in the interface is often subject to more scrutiny and regulation than in traditional rural areas. (Photo by Larry Korhnak.)

planners cannot distinguish where forestry stops and development begins, which causes confusion and increased regulation (Jackson et al. 2003).

Watershed degradation in the wildland–urban interface is a major concern in the U.S. today. Development, mining, agriculture, and to a limited extent silviculture all have impacts on water quality and quantity. Federal, state, and local policies and regulations have been implemented to curb the detrimental effects on water from these land-use practices. State forestry agencies, for example, have overseen the development and implementation of forestry best management practices (BMPs). The agriculture, development, and mining industries have followed suit by self-regulating and following federal, state, and local ordinances. These policies and regulations play an especially important role in the interface due to public scrutiny and the higher intensity of land use and change.

Another issue, brought on by the need for services and amenities associated with new developments in the interface, is increased pressure to raise property taxes. In many states, forest, agriculture, and other conservation lands have been assessed on the lands' inherent ability to produce crops, trees, or conservation values rather than its highest and best-use values. This often leads to lower taxes for farm, forest, and conservation landowners who wish to continue such traditional uses. However, in urbanizing areas, mounting pressures to increase these assessments have resulted in increased property taxes, and these and estate taxes are often cited as reasons why landowners do not retain ownership. In New York, for example, lands are often taxed at the highest and best-use rate, which makes forest management difficult. This often leads to further fragmentation and decreased forest management, as newer owners often do not understand the available forestry opportunities.

17.3.4 Lack of Information and Assistance

NIPF owners in the interface are often unsure or unaware of the information and technical assistance they need on their property. They may also prefer information and assistance that are beyond the realm of traditional public and private service providers such as forestry consultants, extension agents, and state forestry agency personnel. Sampson and DeCoster (1997) describe these owners as "Homesite Managers" or "Forest Attraction Managers," who have different needs such as access to small-scale vegetation management services and general information about tree care, basic forestry concepts, and residence-related topics. Others have interests in ecosystem-based planning opportunities for forest quality, scenic beauty, wildlife habitat, watershed management, clean air, and so on. Access to information and educational, technical, and financial assistance related to these issues or opportunities are critical to these types of owners in the interface.

17.3.5 Social Issues

More individuals and ownership units in a given finite area often lead to greater instances of human–human and human–natural resource conflict. Illegal dumping, private property rights and responsibilities, trespass and timber theft, nuisance wildlife, and aesthetic concerns are just a few of the types of occurrences that must be dealt with in the interface. Conflict resolution, mediation, lawsuits, and other forms of dealing with these issues are costly to the landowner and society and result in inefficiencies that have a substantial impact on the owner, the land, and countless others.

17.4 OPPORTUNITIES

The myriad challenges and issues involved in retaining and managing NIPFs in the interface are daunting and can seem overwhelming to the natural resource professional, environmental specialist, or policy maker. Forests are experiencing increasing pressure in areas where sprawl is prevalent. Development opportunities combined with forest management constraints and a high turnover rate of ownership put many NIPF lands in a state of transition. Regulatory (e.g., ordinances, taxes) and nonregulatory (e.g., market, education/information) options must be considered with the ultimate goal being one of sustainability: balancing the needs of today's citizens and forest landowners with the needs of tomorrow's society. Alternatives need to be investigated at both the individual and community level and include both the producer (landowner) and consumer (citizen) of the forest products and services. These forestlands in the interface yield clean air and water, aesthetics, wildlife habitat, carbon sequestration, biodiversity, and a host of other nonmarket goods and services. Tax incentives and smart growth options, for example, that take into account the inherent value of these amenities to society should be encouraged along with niche markets and alternative forest management schemes for the NIPF owner. The following are some examples of creative and innovative approaches to managing NIPFs in the interface.

17.4.1 Small-Scale and Alternative Forest Management Options

More than 59 percent of the NIPF owners (5.8 million landowners) own less than 10 acres and another 28 percent own between 10 and 49 acres (2.8 million landowners) (Birch 1996). Collectively, these are 87 percent of the landowners but just 20 percent of the forestland. These small forests may be part of someone's backyard, farm, or recreational property. They often exist in close proximity to other small, forested tracts, leading to landscape-scale opportunities for management. Solutions to individual and landscape-scale forest management issues should be investigated. Wilhoit and Rummer (1999) evaluated small-scale, mechanized timber harvesting systems and found some to be quite efficient and effective for owners of small tracts of land. Backyard conservation programs, small-scale and low-impact logging options including use of horses or mules for logging, alternative and niche markets such as pine straw, mushrooms, and forest plants, property tax relief for small ownerships, and other opportunities also need to be considered. These are reviewed in more detail below.

17.4.1.1 Backyard Forest, Wildlife, and Conservation Programs

Owners of small tracts of forestland can benefit from professionals and programs providing on-site advice and assistance directly to them. Many owners are interested in preparing maps of their property, identifying the trees and other vegetation, and understanding the natural processes that occur on their land along with the wildlife habitat, recreation, and aesthetic implications. Washington State's Backyard Stewardship Program (Washington State Department of Natural Resources n.d.), for example, is designed to provide information on small-scale forest and conservation management to the homeowner in a forested setting (Figure 17.7). The program's goals include reducing wildfire risk and damage and promoting forest health and wildlife habitat. Owners can receive "backyard steward status" and a property sign giving them recognition as good stewards of their land. Another recent development in this arena is the start of the international journal, *Small-scale Forest Economics, Management and Policy*, which describes small-scale forestry as a relatively new area of scientific exploration. The journal is the result of a working group of the International Union of Forest Researchers and Organizations (IUFRO) with interests in sharing information about policy and economic issues relating to small-scale forest management (Hyttinen 2002).

17.4.1.2 New and Emerging Markets and Alternative Forest Products

Landowners and natural resource professionals in the interface should also be aware of new and emerging markets. Opportunities such as carbon credits, watershed rights, viewsheds, and ecotourism are becoming marketable amenities and public and private entities are beginning to show an interest (Figure 17.8). The residents of New York City, for example, obtain much of their drinking water from reservoirs in the Catskill Mountains. To ensure the quality of these waters, the city has purchased and manages thousands of acres of watershed forest. In other cases, the city has arranged conservation easements with NIPF owners, allowing the landowners to use their forestlands and obtain tax benefits while giving up some development rights. City officials and environmental groups in New York also work closely with NIPF owners in the region to ensure that they manage their forests in a sustainable manner to keep the reservoirs clean. These programs are beginning to pay off for NIPFs and their owners (City of New York, Department of Environmental Protection 1997).

While the market for carbon credits in the U.S. is still largely undeveloped, the global interest in meeting the Kyoto Agreement relating to carbon pollution reduction continues to increase. Owners of forests and other carbon "sinks" may one day have the opportunity to sell their

FIGURE 17.7 Millions of homeowners across the country own and manage backyard forests for wildlife and conservation purposes. (Photo by William Hubbard.)

FIGURE 17.8 Forest recreation and ecotourism offer landowners and citizens opportunities to interact with nature and to profit economically. (Photo by William Hubbard.)

standing carbon to power companies and others who discharge excess carbon. These credits can basically be bought and sold on the open market, ultimately creating a market-driven solution to a complex global problem. An example of the economic benefit that might accrue to owners of longleaf pine forestlands is given in Chapter 5. It is important to reiterate, however, that these markets currently do not exist, and much work will have to be done before a viable market arises in North America.

In addition to carbon, many alternative forest products can be managed on small- to medium-size tracts of land. A recent book titled *Nontimber Forest Products in the U.S.* (Jones et al. 2002) explores nontimber forest products (NTFP) opportunities that are applicable for NIPF owners in interface areas. An excerpt from the book provides an example of some of the opportunities: "Once found primarily in specialty shops, medicinal herbs and edible wild mushrooms now figure among the offerings on the shelves of mainstream drug and grocery stores. Simultaneously, subsistence and recreational demands for products derived from plant, fungi, and lichen species, gathered in forest, woodland, or savannah ecosystems have also expanded. Examples of NTFPs include ferns, mosses, wild mushrooms, cones, boughs, maple syrup, bark, and hundreds of medicinal plants, such as ginseng and goldenseal" (Jones et al. 2002, xvii).

17.4.2 Land Conservation/Preservation Tools

Conservation easements, transfer or purchase of development rights, and land acquisition are among the tools used by individuals, agencies, and NGOs to conserve and preserve private forestlands. Many of these tools are designed to maintain the natural integrity of the land while allowing specified forest management activities to occur. The advantages and disadvantages of these options are summarized in Table 17.4.

A conservation easement is a legal agreement between a landowner and a nonprofit land trust or government entity that limits the uses of the land for a specified time period in order to protect specified conservation values such as wildlife habitat, ecological diversity, recreational access, and aesthetics (Lind 2001). Various public and private programs, such as the Forest Legacy Program of the USDA Forest Service, work in partnership with states and private entities to protect environmentally sensitive forestland from conversion to nonforest uses through acquisition and conservation easements (Beauvais 2000). A landowner who receives no monetary compensation for a conservation easement may qualify for property tax and income tax deductions. Purchase and transfer of development rights are similar in nature to conservation easements except that the owner actually trades or sells the rights for added monetary or other benefits. More discussion about these tools is provided in Chapter 4 through Chapter 6.

Another preservation/conservation tool that is used is land acquisition, which occurs when a willing landowner donates or sells land to a government agency or nonprofit group. Land acquisition is rapidly increasing as national organizations such as The Nature Conservancy and Trust for Public Lands combine with many state and local entities to purchase lands for societal benefits.

TABLE 17.4
Land Protection and Conservation Options

Land Protection Option	Pro	Con
Donated Conservation Easements are voluntary legal agreements between a landowner and a land trust or local government agency that allow landowners to limit or prohibit development on their property. Conservation easements run with the title so that all future owners of the land are bound by the original agreement.	• Permanently protects land from development pressures • Landowners may receive income, estate,and property tax benefits • No or low cost to local unit of government • Land remains in private ownership and on the tax rolls	• Tax incentives may not provide enough compensation for many landowners • Little local government control over which areas are protected
Purchases of Development Rights are voluntary legal agreements that allow owners of land meeting certain criteria to sell the right to develop their property to local government agencies, state government, or to a nonprofit organization. A conservation easement is then placed on the land. This agreement is recorded on the title to permanently limit the future use of the land to agriculture, forestry, or other open space uses.	• Permanently protects land from development pressures • Landowners are paid to protect their land • Landowners may receive estate and property tax benefits • Local government can target locations effectively • Land remains in private ownership and on the tax rolls	• Can be costly for local unit of government
Transfers of Development Rights are enabled by local ordinances that create sending areas, or preservation areas, and receiving areas, where communities encourage additional growth and development. Landowners in the sending area receive development right credits, which they can sell in exchange for not developing their land. Real estate developers, speculators, or the local unit of government can then purchase the development right credits and use them to increase existing or planned densities in receiving areas.	• Permanently protects land from development pressures • Landowners are paid to protect their land • Landowners may receive estate and property tax benefits • Local government can target locations effectively • Low cost to local unit of government • Utilizes free market mechanisms • Land remains in private ownership and on the tax rolls	• Can be complex to manage • Receiving area must be willing to accept higher densities • Most successful programs typically require a strong real estate market

Source: Thousand Friends of Minnesota (2003).

17.4.3 The Forest Bank™ and Green IRAs

These programs are designed to utilize market investment approaches to conservation and land management and can offer opportunities for forest landowners in the interface. The Nature Conservancy created the Forest Bank in the late 1990s as a unique way to offer landowners the opportunity to deposit or transfer their right to grow, manage, and harvest trees in exchange for a dividend payment and the right to withdraw the value of their timber in cash whenever they would like. The Forest Bank is an idea developed for and marketed primarily to private, nonindustrial landowners. These landowners can make a deposit — or a transfer of the perpetual right to grow, manage, and harvest trees — to the Forest Bank, while retaining fee simple ownership of the underlying land. In structure, the deposit is designed to resemble the familiar CD used by savings banks (Gilges 2000).

The concept of a Green IRA has been promoted as another market investment opportunity for landowners. Although this concept has never been adopted as part of the tax code, the Green IRA would be similar to a regular IRA, with contributions to this account being spent on specific approved forest maintenance and management activities and taxes deferred until the harvest of the forest. For example, a taxpayer could put up to 2000 pretax dollars from nonforest income into a Green IRA account and then use funds from that account for activities that would qualify under the law. These activities could be anything related to maintaining and enhancing the forest (employing foresters or biologists, reforestation, payment of property taxes, precommercial thinnings). The Green IRA would result in tax savings for the forest landowner and more dollars into the local forest economy (DeCoster 1995).

17.4.4 Public and Private Assistance Programs

For years, the debate within the natural resources community has been whether to serve owners or acres. While it has been a philosophical debate at best, the winner has clearly been the owner with the "viable tract of forestland." Taking traditional timber management as an example, forest industry, through landowner assistance programs, has usually served owners with larger tract sizes (greater than 100 acres). These services have included information and technical assistance, subsidized seedlings and tree planting, and a host of other services. State forestry and Extension foresters have served the small to medium tract sizes (25 to 100 acres) with information and technical and financial assistance, and the consulting forester has served the continuum depending on the particular services desired and expected returns. Owners of the micro (1- to 9-acre), and to some extent small (10- to 49-acre), tracts have not traditionally been served. By providing professional development opportunities for service providers who assist these small "boutique forestry owners," state forestry agencies, Extension foresters, and private sector consultants remain relevant and can have an impact in a highly visible and profitable area (Hull et al. 2004). As more than one public forestry official has declared, "the 1 acre landowner has the same number of votes as the 1,000 acre landowner when it comes to electing politicians and steering public forestry programs."

For the larger tracts in the interface, federal programs such as the Forest Stewardship Program, the Forestland Enhancement Program, the Forest Legacy Program, Partners in Flight, and others are designed to assist landowners with their land management and conservation needs. Many of these programs also provide financial support such as cost sharing, technical assistance, educational programs, or purchase of development rights. In some situations, the landowner enters into a formal agreement to manage the land for long-term societal benefits. State and private programs provide similar assistance. Lands within the interface are eligible just as are those in rural areas. Certain minimum and maximum size limits may restrict participation in the programs. The Extension Service in each state also has informational and technology resources that are useful for individuals, community leaders, policy makers, and others interested in retaining and maintaining productive NIPF lands in the interface. More information on these programs can be found by contacting state Extension or forestry agency personnel at the county or state level.

Professional assistance is also needed because lands in the interface are subject to extreme stress and risk each year. Besides the development activity itself, forests in the interface face wildfire and its devastating effects and invasion by exotic insects and plants. Natural resource professionals advocate healthy forests as the first defense against these damaging forces (see Chapter 14 for detailed discussion). Healthy forests are a matter of perspective in many cases, but landowners should take note of the features of a healthy forest and follow professional advice on optimal stocking, maintaining diversity in age and stand structure, soil compaction amelioration, fireline preparation, hazardous fuel load reduction, implementation of BMPs, and inclusion of streamside management zones. The advice and use of a professional natural resource manager regardless of the size of the tract are recommended to develop a management plan that provides a guide for landowners to reduce risk and achieve other objectives.

17.4.5 Landowner Associations, Cooperatives, and Cross Boundary Management Initiatives

Grassroots efforts to provide assistance to owners of forests within the interface are proving to be very

FIGURE 17.9 Short courses, field trips, and demonstration forests are a few of the ways that landowners are learning new management techniques in the interface. (Photo by William Hubbard.)

successful (Figure 17.9). The idea of landowners helping landowners is a common concept that takes on added importance in the interface due to the higher turnover rate and the relatively new owner with little experience in forestland management. Formally organized programs such as the Master Tree Farmer Program, Master Woodland Owner Program, Coverts, Woodlands Advisor Program, and Master Gardener train landowners and homeowners to assist their neighbors and friends with forestry and natural resource problems. Many individuals within the interface are very capable of solving their own problems or creating new opportunities if given access to information and technical assistance. Communication among neighboring landowners locally or via technology (e-mail, internet shortcourses, two-way satellite downlinks) offers a mentoring opportunity and a potential for collaboration.

In addition, local landowner associations regularly schedule events (annual meetings, local workshops), publish newsletters and fact sheets, maintain web sites, and promote forest management to the general public and policy makers. Usually for an annual fee, the landowner becomes a member of an organization whose mission is often closely aligned with his or her own objectives. Membership dues vary widely and opportunities exist to join national, regional, state, and local associations.

Another concept worth noting is the forest cooperative. Much like agriculture cooperatives, labor, equipment, and other resources can be shared to form cooperative ventures that can accomplish multiple objectives. In many cases, these ventures provide higher returns on investment or lower costs of doing business. These cooperatives can involve contiguous or noncontiguous tracts of land. Contiguous cooperatives may even lead to landscape-scale forest and natural resource management opportunities that benefit not just the participating landowners through economies of scale for logging but also the wildlife, water, recreation, aesthetics, and other amenities. A few of the well-established cooperatives exist in the U.S. Midwest (see Box 17.1). Southwest Wisconsin, for example, boasts

Box 17.1

Forest Cooperatives

The *Deerfield Alliance* is an association of neighboring landowners in Deerfield Township, Wisconsin. It was formed in 1998 with two goals: (1) to encourage planned, sustainable management of woodlots in the township and (2) to help absentee landowners with their woodlot management. Martin Pionke, a local dairy farmer and chairman of the *Alliance*, said, "We're a small local group that's trying to get people involved in their woodlands so they won't be so quick to subdivide them. We are helping them manage their woods a little better and maybe get a little more return out of it" (Cooperative Development Services—University of Wisconsin 2002). The *Alliance* works by sharing resources, expertise, and costs. Examples include joint herbicide spraying, tree planting, controlled burning, and small timber harvests. The *Alliance* is part of a statewide effort called the Wisconsin Family Forests, which was organized to support private landowners in learning and applying the concepts of sustainable forestry through the formation of local alliances (Wisconsin Family Forests n.d.).

TABLE 17.5
Prominent Forest Certification Systems in Use in the U.S.

	American Forest Foundation's American Tree Farm System® (ATFS)	Forest Stewardship Council (FSC) Forest Certification & Labeling Program	Sustainable Forestry Initiative (SFI) Program
Stated Purpose of the Standard and Program	To promote the growth of renewable forest resources on private lands while protecting environmental benefits and increasing public understanding of all benefits of productive forestry.	To recognize appropriate, socially beneficial, and economically viable forest management.	To broaden the practice of sustainable forestry to meet the needs of the present without compromising the ability of future generations to meet their own needs by practicing a sustainble land stewardship ethic.
Status of Forestry Standard	ATFS was established in 1941 to promote forest management and to recognize landowners for their commitment to practicing good forestry. Mandatory Standards for Tree Farm Certification have been developed. The process provides individual certification or Third Party Group Certification.	Established in 1993, the FSC has ten Guiding Principles for forest management. These set the framework for development of national/regional standards. In the absence of such standards, audits are conducted against Global Principles.	Initiated in 1992, the original sustainable forestry principles, guidelines, and performance measures have evolved into a formal standard for SFI program participants. SFI compliance is a condition of membership in AF & PA. Its application has been extended to nonmembers through a licensing program.
Geographic Application of Standard	The ATFS is designed for application to the operations of private individual tree farms and/or the managers and members of a Tree Farm Group Certification within the United States.	The FSC is international in scope. It is designed for application on a defined forest area.	The SFI is currently being applied in the U.S. and Canada, and as a demonstration program in Honduras and Central America. The SFI is designed for application to the operations of the organization and also applicable on a defined forest area.
Key Forest Mangagement Elements of Standard	**Standards of Sustainability** 1. Ensuring sustainable forests 2. Compliance with the laws 3. Commitment to practicing sustainable forestry 4. Reforestation 5. Air, water, and soil protection 6. Fish, wildlife, and biodiversity 7. Forest aesthetics 8. Protect special sites 9. Wood fiber harvest and other operations	**10 Guiding Principles** 1. Compliance with laws and FSC Principles 2. Tenure and use rights and responsibilities 3. Indigenous peoples' rights 4. Community relations and workers' rights 5 Benefits from the forest 6. Environmental impact 7. Management plan 8. Monitoring and assessment 9. Maintenance of high conservation value forests 10. Plantation mangement criteria	**Sustainable Forestry Objectives** 1. Broadening the practice of sustainable forestry 2. Ensuring long-term forest productivity and conservation of forest resources 3. Protecting water quality 4. Enhancing wildlife habitat and biodiversity 5. Minimizing the visual impact of harvesting 6. Protecting special sites 7. Continuing improvements in wood utilization 8. Cooperating with others in wood procurement 9. Publicly reporting progress 10. Providing apportunities for the public and others to participate 11.Continuous improvement
Indicators of Conformance with the Standard	The American Forest Foundation Standards of Sustainability for Forest Certification on Private Lands include Performance Measures and Field Indicators required for certification. These newly revised Standards became effective in January 2004.	No regional criteria and indicators currently exist in the U.S. Draft regional criteria and indicators under review by FSC. In lieu of regional standards, FSC-accredited auditing firms have their own proprietary criteria and indicators that are not available to public.	Core indicators have been established for all SFI program objectives and performance measures that all program participants must meet to successfully obtain third-party certification. In addition, all participants must commit to protect Special Sites and Forests with Exceptional Conservation Value.
Status of Forest Certifications (as of 2003)	The American Tree Farm System includes 68,000 private landowners in the U.S. and represents approximately 85 million acres, 26 million acres of which are certified.	Six million acres have been certified to the FSC Standard in the U.S. Fifty thousand acres have been certified to the FSC in Canada.	Approximately 77 million acres have committed to or achieved third-party certification in the U.S. In addition to AF & PA member companies, SFI program participants include public lands, academic institutions, secondary manufacturers, and conservation organizations in the U.S. and Canada.

Source: Forest Focus, spring 2003 newsletter of MeadWestvaco Forestry Division. With permission.

a number of cooperatives that specialize in everything from high-end lumber and wood for use by artists to firewood and pulpwood. These particular cooperatives are set up for marketing sustainable forestry and sustainable forest products and for information dissemination among their members and the general public.

Multiowner and landscape-scale management such as ecosystem or watershed-based management also hold promise for restoring large tracts of fragmented and parcelized private forestlands in the interface (Campbell and Kittredge 1996). Contrary to popular belief in the forestry community, forest owners often have an interest in implementing management practices that cross property boundaries and respond to the larger temporal- and spatial-scale issues of forest ecosystem management (Hull et al. 2000). Legal concerns such as landowner rights/responsibilities and the potential for collusion and antitrust breaches, conflict resolution protocol, and other relationships are important when considering cooperatives and multiownership landscape-scale activities.

17.4.6 Forest Certification

A relatively new concept that may be of interest to landowners in the interface is forest certification. It is a process by which the performance of forest and natural resource management practices is assessed against a predetermined set of standards. Forest certification follows the general hypothesis that products and services harvested from sustainably managed forests will command higher prices in the marketplace than those that are not managed in a sustainable manner. While it is very difficult to predict future social and market policies and their impacts, the concept of forest certification has been integrated into the working culture of many members of the forestry community. Many organizations have developed standards and criteria for certification, with some certifying the land, others certifying the professional forester in charge of management activities for the land, and still others experimenting with group certification options. The actual standards and criteria vary widely by organizational goal and objective. Two of the more showcased programs in the 1990s and early 2000s are the Sustainable Forestry Initiative, or SFI®, and the Forest Stewardship Council's FSC©. They differ on a number of standards, such as the percentage of an owner's property that can be considered pine plantation, the use of certain chemicals, the extent to which timber products are processed locally, and other management practices (Table 17.5). The costs associated with certification can also vary widely. Costs can be prohibitive for the small private forest landowner in the interface; however, opportunities for group certification and certification of the professional forest manager are being investigated by many of the certifying organizations.

17.5 SUMMARY AND CONCLUSION

Managing NIPFs in the wildland–urban interface today is challenging for a variety of reasons. Owners and their managers must have a working knowledge of their management objectives, the current state and biological capacity of their property, and the larger environmental and social context in which they operate. They must also be conversant with current and pending laws, regulations, market trends, and sources of assistance. Natural resource professionals, public policy officials, and the community at large need to work together in combination with private forest landowners to find innovative solutions at the property and landscape scale. Some of these ideas and solutions such as niche markets, preferential tax treatments, alternative forest management schemes, conservation easements, and a host of other options have been presented here. Many other opportunities exist from the local to the global level. More research and outreach in all of these areas will provide solutions to the millions of landowners who provide forest-based goods and services to communities throughout the country and world.

REFERENCES

Beauvais, T.W., 2000. The Role of the Forest Legacy Program in Preventing Forest Fragmentation of Forest Ownerships, in Fragmentation 2000: A Conference on Sustaining Private Forests in the 21st Century, September 17–20, 2000, Annapolis, MD, L.A. DeCoster and R.N. Sampson, Eds., Sampson Group, Alexandria, VA., pp. 359–365.

Birch, T.W., 1996. Private Forest-Land Owners of the U.S., 1994, Resource Bulletin NE-134, U.S. Department of Agriculture, Forest Service, Northeastern Forest Experiment Station, Radnor, PA.

Butler, B.J. and E.C. Leatherberry, 2003. National Woodland Owner Survey — Draft Tables, U.S. Department of Agriculture, Forest Service, Forest Inventory and Analysis, www.fs.fed.us/woodlandowners/results.htm. [Date accessed: July 17, 2003.]

Campbell, S.M. and D.B. Kittredge, 1996. Ecosystem-based management on multiple NIPF ownerships, *Journal of Forestry* 94: 24–29.

City of New York, Department of Environmental Protection, 1997. New York City's water Supply System, http://www.nyc.gov/html/dep/html/ruleregs/finalrandr.html. [Date accessed: July 14, 2003.]

Cooperative Development Services, University of Wisconsin, 2002. *Balancing Ecology and Economics: A Start-up Guide for Forest Owner Cooperation*, 2nd ed., Cooperative Development Services and Center for Cooperatives, University of Wisconsin, Madison, and Community Forestry Resource Center, Minneapolis, MN.

Cubbage, F.W., 1982. Economics of Forest Tract Size: Theory and Literature, General Technical Report SO-41, U.S. Department of Agriculture, Forest Service, Southern Forest Experiment Station, New Orleans, LA.

Cubbage, F.W., 1983. Tract size and harvesting costs in southern pine, *Journal of Forestry* 81: 430–433.

Cubbage, F.W., 1995. Regulation of private forestry practices: what rights, what policies? *Journal of Forestry* 93: 14–20.

DeCoster, L.A., 1995. Maintaining the public benefits of private forests through targeted tax options, in *Maintaining the Public Benefits of Private Forests Through Targeted Tax Options,* R. N. Sampson and M.J. Enzer, Eds., American Forests, Washington, DC, pp. 1–34.

Gilges, K., 2000. The Forest Bank: A Market-Based Tool for Protecting Our Working Forestland, in Fragmentation 2000: A Conference on Sustaining Private Forests in the 21st Century, September 17–20, 2000, Annapolis, MD, L.A. DeCoster and R.N. Sampson, (Eds.), Sampson Group, Alexandria, VA, pp. 340–346.

Hull, R.B., J.E. Johnson, and M. Nespeca, 2000. Forest Landowner Attitudes Toward Cross-Boundary Management, in Fragmentation 2000: A Conference on Sustaining Private Forests in the 21st Century, September 17–20, 2000, Annapolis, MD, L.A. DeCoster and R.N. Sampson, Eds., Sampson Group, Alexandria, VA, pp. 145–153.

Hull, R.B., D.P. Robertson, and G.J. Buhyoff, 2004. "Boutique" forestry: new forestry practices in urbanizing landscapes, *Journal of Forestry* 102:14–19.

Hyttinen, P., 2002. Introducing the new journal, *Small Scale Forest Economics, Management and Policy* 1(1): vii.

Jackson, B., W. Hubbard, J. Stringer, and D. Dillaway, 2003. A look at local forestry ordinances, *Forest Landowners Manual* 62: 42–45.

Jones, E.T., R.J. McLain, and J. Weigand, Eds., 2002. *Nontimber Forest Products in the U.S.*, University Press of Kansas, Lawrence.

Lind, B., 2001. *Working Forest Conservation Easements: A Process Guide for Land Trusts, Landowners and Public Agencies*, Land Trust Alliance, Washington, DC.

Megalos, M.A. and F. Cubbage, 2000. Promoting Forest Sustainability with Incentives: How Landowners Rate the Options, in Fragmentation 2000: A Conference on Sustaining Private Forests in the 21st Century, September 17–20, 2000, Annapolis, MD, L.A. DeCoster and R.N. Sampson, Eds., Sampson Group, Alexandria, VA, pp. 160–169.

Sampson, R.N. and L.A. DeCoster, 1997. *Public Programs for Private Forestry: A Reader on Programs and Options*, American Forests, Washington, DC.

Smith. W.B., J.S. Vissage, D.R. Darr, and R.M. Sheffield, 2001. *Forest Resources of the U.S., 1997*, Gen. Tech. Rep. NC-219, U.S. Department of Agriculture, Forest Service, North Central Research Station, St. Paul, Minn.

Thousand Friends of Minnesota, 2003. *The Land Protection Tool Box*, http://www.1000fom.org/lctools2.htm. [Date accessed: June 1, 2003].

Warren, R. (ed.), 1997. Deer overabundance, *Wildlife Society Bulletin Special Issue* 25(2), The Wildlife Society, Bethesda, MD.

Washington State Department of Natural Resources, n.d. Backyard Forest Stewardship Program Homepage, http://www.dnr.was.gov/htdocs/rp/stewardship/bfs/. [Date accessed: September 23, 2003.]

Wear, D.N., R. Liu, M. Foreman, and R.M. Sheffield, 1999. The effect of population growth on timber management and inventories in Virginia, *Forest Ecology and Management* 118: 107–115.

Wilhoit, J. and B. Rummer, 1999. Application of Small-Scale Systems: Evaluation of Alternatives, ASAE/CSAE-SCGR Annual International Meeting, July 18–21, 1999, Toronto, Ontario, ASAE, St. Joseph, MI, www.srs.fs.usda.gov/pubs/ja/ja_wilhoit002.pdf. [Date accessed: October 1, 2003.]

Wisconsin Family Forests, n.d. Homepage, http://www.wisconsinfamilyforests.org. [Date accessed: October 1, 2003.]

18 The Challenges of Managing Public Lands in the Wildland–Urban Interface

John F. Dwyer
USDA Forest Service, North Central Research Station

Deborah J. Chavez
USDA Forest Service, Pacific Southwest Research Station

CONTENTS

18.1 INTRODUCTION

The wildland–urban interface is an environment that is characterized by diversity, interconnectedness, and change along a number of critical dimensions that include people, resources, resource use, landownership, resource managers, management objectives, and management practices. Change is the rule within the interface, with increasing residential, commercial, and industrial development, as well as associated infrastructure among the major forces for change. Public lands are often the areas in the interface that undergo the least change in land use and cover over time. As such, these lands often experience mounting forces for change in

1-56670-602-5/05/$0.00+$1.50
© 2005 by CRC Press LLC

resource management and use, and in many instances become surrounded by increasingly intensive development and associated infrastructure. Public lands often become "islands of green" within a sea of development. As the largest contiguous holdings under a single ownership within the interface, they may present an especially significant opportunity for managing and protecting endangered plant and animal species, and for providing wildland experiences for urban residents. The value of these public lands and the services that they provide increase significantly over time as the surrounding areas develop more intensively. These public natural areas are where many urban residents experience, use, appreciate, and learn about natural resources and their management.

Public lands in the interface are subject to changing influences and interactions as the complex and dynamic matrix of public and private lands and residential, commercial, and industrial development evolves over time. The wildland–urban interface is not static or stationary. The general pattern is for the periphery of the interface to extend outward from an urban area, and for interior areas to become more developed. However, the development pattern is far from uniform over the landscape, with some developments, particularly residential, concentrated near public land to take advantage of the natural environment and view. More isolated intermixes of developed and natural areas may occur across the landscape as well, and there may be small but significant reversals in the development process when developed areas such as industrial sites and military installations are returned to natural areas. Interface areas may also be found within urban areas as corridors of wildland, often associated with rivers (e.g., the Chicago River Corridor (Gobster and Westphal 1998)). Overall, as urban areas expand across the landscape, the perimeter, which helps define the wildland–urban interface, extends over a larger area and may encompass significant private and public lands.

In this chapter, we describe the types of public lands in the interface, the objectives for managing these lands, the challenges faced by managers of interface lands, and the approaches and critical tools for management of interface public lands. We provide three examples of managing public lands in the interface to illustrate the diversity of public land management.

18.2 TYPES AND OBJECTIVES OF PUBLIC LANDS IN THE INTERFACE

The interface comprises a wide range of public lands that may include natural areas on state and federal holdings, as well as county, municipal, and other local government or special district lands.

18.2.1 STATE AND FEDERAL AREAS

State and federal natural areas in the wildland–urban interface, some of them extensive and others small, may have once been rural in character, but are now in close proximity to, or perhaps partly or totally surrounded by residential, commercial, and industrial development and associated infrastructure. For example, some 14 National Forests around the U.S. managed by the USDA Forest Service were designated as Urban National Forests in 1995 and the number is increasing (Table 18.1). These forests are located within 50 mile of a population center of greater than one million people and demonstrate unique management challenges and opportunities.

Natural areas in the interface held by federal and state agencies are often managed for a wide range of goods and services traditionally provided by these agencies in rural areas, such as wood, water, forage, wildlife, recreation, biodiversity, and aesthetics. For some landholdings, however, the priorities placed on these objectives have changed. Less emphasis is on providing for timber harvesting, mining, use of off-road vehicles, and hunting, and increased emphasis is

TABLE 18.1
The Urban National Forests

Original Urban National Forests	State
Tonto National Forest	Arizona
Angeles National Forest	California
Cleveland National Forest	California
Los Padres National Forest	California
San Bernardino National Forest	California
Arapaho and Roosevelt National Forest	Colorado
Pike and San Isabel National Forest	Colorado
Chattahoochee-Oconee National Forest	Georgia
White Mountain National Forest	New Hampshire
Mount Hood National Forest	Oregon
Wasatch-Cache National Forest	Utah
Uinta National Forest	Utah
Gifford Pinchot National Forest	Washington
Mount Baker-Snoqualmie National Forest	Washington
Additional Urban National Forests	
Eldorado National Forest	California
Sequoia National Forest	California
Sierra National Forest	California
White River National Forest	Colorado
Apalachicola National Forest	Florida
Ocala National Forest	Florida
Osceola National Forest	Florida
Midewin National Tallgrass Prairie	Illinois
Willamette National Forest	Oregon
Caribbean National Forest	Puerto Rico

Note: There have not been official additions to the original 1995 list of Urban National Forests, but the additional National Forests and National Tallgrass Prairie listed here are often considered among the Urban National Forests.

often given to aesthetics, water, environmental education, forest health, biodiversity, and outdoor recreation.

18.2.2 County, Municipal, and Other Local Areas

County, municipal, and other local government or special district lands in the interface may have been acquired to provide open space for locally expanding populations, or may have been obtained at an earlier time for a wide range of purposes. Currently, substantial acquisition of open space at the local level is under way to prevent the loss of natural areas to urban sprawl (Gobster et al. 2000). Locally owned natural areas include city and county parks, nature reserves, wildlife areas, municipal watersheds, transportation corridors, and lands that are a part of institutional holdings such as health care and education facilities.

County, municipal, and special district natural areas in the interface provide a wide range of services that include outdoor recreation, scenic beauty, watershed protection, waste disposal, and biodiversity. In some instances, these lands are similar to state and federal holdings, and may be close to them, but they are usually smaller and give less attention to extractive activities such as timber harvesting, mining, and hunting. Management objectives may vary with the size, ownership, and location of individual landholdings and their proximity to other public lands and to residential, commercial, and industrial developments and associated infrastructure. County, municipal, and special district lands may serve as buffers between different land uses. Also, they may be clustered in riparian or transportation corridors, forming important greenways that extend across the interface and provide linkages between inner city and rural areas.

18.3 THREE EXAMPLES OF PUBLIC LANDS IN THE WILDLAND–URBAN INTERFACE: AN OVERVIEW

Three quite different examples are presented in this chapter that reflect the diversity of public lands in the wildland–urban interface and their management and use. The first focuses on the Angeles National Forest, one of the Urban National Forests and a major natural area that spans the wildland–urban interface near Los Angeles. With its wilderness and other backcountry areas, the Angeles provides a wide range of wildland experiences to nearby urban populations. The second example, the Forest Preserve District of Cook County (FPDCC), is a well-established system of natural areas intermingled with residential, commercial, and industrial areas that are increasingly influenced by the growth of a major metropolitan area — in this case, Chicago. But the FPDCC is a special district of Cook County, IL, and manages these areas under a charter established by the State of Illinois. Like the Angeles, the FPDCC is one of the most heavily used natural areas in the U.S. Our third example, the Midewin National Tallgrass Prairie, also is managed by the USDA Forest Service, is an Urban National Forest, and is found in a metropolitan area (Chicago). However, this significant interface holding is a newly established area and presents some very special challenges and opportunities with the inception of restoration projects to return some 15,000 acres of former industrial sites and agricultural lands to the original tallgrass prairie ecosystem. Taken together, these three examples provide illustrations of the variations in lands, ownership, challenges, opportunities, and management found with public lands at the wildland–urban interface.

18.4 CHALLENGES AND OPPORTUNITIES FACED BY MANAGERS OF INTERFACE LANDS

The challenges and opportunities faced by managers of interface lands have their origins in a number of attributes of these lands and their setting. These include strong, diverse, and dynamic pressures for land and resource use and involvement in management, increasing influence of nearby residents, landowners, developments, and infrastructure, and the increasing overall complexity and dynamics of the wildland–urban interface in which management is carried out. With increasing development and expansion of the interface, these challenges intensify.

18.4.1 Changing Pressures for Resource Use and Involvement in Management

Changing public values and demands over time have especially significant implications for the management of public lands in the wildland–urban interface. Interface areas are often a place of rapid social change. Some of these changes have their origins in changing populations who live in and use the interface, while others are a part of overall societal trends in values (Bengston 1994; Bengston et al. 1999). Shifting demographics in nearby areas (population size, age, race, ethnicity, income) may strongly influence the amount and type of participation in resource management, use of natural areas, and expectations for management and services in the interface (see all three examples, but particularly the Angeles National Forest [Figure 18.1]). Recreational use of natural environments is often correlated with these demographic variables; thus demographic change is likely to bring significant shifts in resource use (Dwyer 1994). The changes in participation in and expectations for management that will accompany demographic changes in the interface are less predictable, but may be particularly significant to managers (Chavez 2000a). When urban

FIGURE 18.1 Picnickers in the Angeles National Forest, CA. Numerous challenges face managers of Urban National Forests, including the diversity and number of visitors. (USDA Forest Service file photo.)

residents move out to interface areas, they may bring urban values and approaches to planning, involvement, and management. Numerous challenges are being faced by managers of public lands in the wildland–urban interface (see Box 18.1).

There are increasing pressures for a wide range of uses of public natural areas in the interface, many of them linked to the character of the natural environment and the proximity of significant urban populations (see all three examples). Public interface areas often experience novel recreational activities not yet seen in more rural areas. New technologies for transportation and outdoor recreation are likely to appear in the interface before they are found in backcountry areas. For example, bicycle travel in natural areas has tended to spread from the interface to more rural locations. This can create new opportunities for tourism and economic development in local communities, but can also create conflict with established roadway and trail users (Chavez 1996, 1997; Hoger and Chavez 1998). With these and other intense and diverse user pressures, and the associated demands for resources, resource management becomes challenging for the managers of public lands in the interface. These challenges are likely to bring about substantial increases in expenditures for operations and maintenance.

There tends to be increasing public interest in, scrutiny of, and involvement with the day-to-day management and use of public areas in the interface. More people are present to observe, be influenced by, and at the same time influence management. Some are neighbors of the public lands and have a particularly close association with these lands and their managers and users. The activities of volunteers and a wide range of not-for-profit groups may be an important component of planning and management activities; for example, the growth of community-based ecological restoration activities on public lands (Gobster and Hull 2000). In Illinois, the Volunteer Stewardship Network (VSN) is an important force for recruiting, organizing, and training volunteers. The VSN has played a key role in restoration of areas in the FPDCC and promises to play a strong role in the restoration of the Midewin National Tallgrass Prairie (Box 18.2).

18.4.2 Nearby Lands, Developments, Landowners, and Infrastructure

The development of nearby private and public lands tends to have increasing influences on the management and use of public interface lands. Land-use and cover changes in and near interface areas are often frequent and continuing. These influences are multidimensional and may be reflected by changes in:

- Traffic and noise
- Air and water quality
- Surface and subsurface flows of water
- Impacts on plants, animals, and soil
- Frequency and severity of fire
- Invasive plant and animal species
- Limitations on public and private access
- Encroachment on public lands
- Conflicts between new and established uses and users
- Community growth, development, and sustainability
- Social problems extending to public lands (e.g., crimes, drugs, timber theft, vandalism, and arson).

Box 18.1

Challenges and Opportunities in the Urban National Forests

Urban National Forests are located within 50 mi of populations greater than one million people, and demonstrate unique management challenges and opportunities:

- They are affected by demographic shifts from agrarian communities and middle and upper classes' "escape" from suburbia. Urban and rural values may conflict and increase management complexities.
- Community-generated demands are on the rise.
- Traditional recreation programs are not adequately meeting the needs of the urban populations and the objectives of ecosystem management. There may be intense recreational use, a focus on day-use, and competition for open space.
- These forests often experience the newest recreation fads. Administration is complicated because management guidelines may not yet exist.
- The demands for public access to and through National Forest lands continue to grow, placing additional strain on recreation and land programs.
- Information strategies are complex due to language, cultural, and class diversity.
- The ability to provide for both employee and public safety is being challenged by increased public use and value differences.
- Intermingled ownership patterns in combination with metropolitan needs significantly increase management complexities.

Box 18.2

Volunteer Stewardship Network

The Volunteer Stewardship Network (VSN) is a force of over 5000 people from numerous walks of life who take care of nearly 40,000 acres of rare prairie, oak savanna, wetlands, and woodland ecosystems in Illinois. It has been estimated that the value of the time and skill donated by VSN volunteers was well over $2.5 million for the years 1990–1994. While the focus of VSN is restoration, it is a substantial organization with individuals engaged in a number of other tasks, including education, publicity, and monitoring (Miles 1996).

Development-related changes at or near the interface may be seen differently by various groups of stakeholders. While some may be concerned over the implications for natural resources, others may view increasing development as bringing improved roads and other infrastructure, as well as additional opportunities for employment and economic development. Some new owners may make substantial investments in the area's infrastructure, including its stock of housing, and also work to support improved management of local natural resources (Rickenbach and Gobster 2003).

Builders or developers often choose to locate residential and sometimes commercial developments close to public lands to take advantage of the natural environments provided. Public lands are valued for their amenities, such as aesthetics, solitude, wildlife, outdoor recreation, and biodiversity, as well as the assurance of continued natural conditions in the future. This increases the pressures placed on public lands and their management by those located nearby and creates new challenges for resource management and for providing public access to the public lands. Changing private land ownership in the interface may bring more landowners, more posting to prevent public use, and subsequently more pressure for the use of public lands. Increasing numbers of relatively small private holdings adjacent to public holdings create major challenges for the management of public lands, including managing the multiple interfaces between these many holdings and the public land, meeting the expectations of nearby residents, and building coalitions among landowners to manage landscapes and land uses beyond public land boundaries.

Planners seeking to provide expanded infrastructure for new developments at or near the interface often look to public lands to provide space for these services (e.g., roads, water supplies, waste disposal, sewage, and power lines). A single public landowner is easier to deal with when locating these facilities than numerous owners of small parcels. When utilities are accommodated on public lands, a major challenge is to minimize disruption of natural systems and resource use, while at the same time capitalizing on the corridors that are created. Fragmentation of natural areas may be a problem accompanying infrastructure development on public lands.

18.4.3 Influences on Management of Natural Resources

The proximity of residents and high levels of site use often call for restrictions on management practices such as timber cutting, prescribed burning, and use of herbicides. Activities such as hunting, trapping, and use of off-road vehicles are often limited for these same reasons. Given that the interface resides between urban and rural areas, management practices, resource uses, and user expectations from these two quite different kinds of environments tend to come together at the interface. The result may be some difficult decisions on what approach will prevail in particular circumstances. For example, cutting firewood from dead and downed trees may be a highly desirable practice in rural areas but not in the interface, where high demands could decimate the supply of downed wood and the cutting could disturb or endanger some of the large number of users and neighbors. Downed wood might also be needed to retain functioning ecosystems.

The wildland–urban interface environment can be a costly setting for managing public natural areas because it includes increased costs of acquisition, infrastructure, and management. All of these have their origins in increasing public and private demands for the management and use of these lands, which includes accommodating and dealing with large groups of users and implementing special management approaches. Such increases make it difficult for public agencies with lands in rural areas as well as in the interface to make equitable allocations of scarce resources across their interface and rural holdings.

Public land managers working in interface areas have limited opportunities for land acquisitions due to rising land prices and increasing competition for residential, commercial, and industrial use. This reduces opportunities to diffuse public demands for outdoor recreation and open space over a wider area of public lands and intensifies the pressures on existing holdings. For a discussion of the complex considerations involved in making public land purchases in interface areas, see Ruliffson et al. (2002, 2003).

Identifying the appropriate spatial scale of management can be a major challenge. Some specific sites may be heavily used and may require specialized management; examples include intensively used outdoor recreation facilities. At the same time, the entire landscape in the interface area is critical to the provision of aesthetic quality, wildlife habitat, water and watershed management, outdoor recreation, and many other important goods and services. In addition, efforts to reduce the risk of fire to homes and communities in the interface often involve activities at the homesite, in the community, and in nearby and more distant forestlands. Protection is most effective when these efforts are coordinated across the landscape. Consequently, interface management plans and programs must operate at multiple scales and usually involve a number of jurisdictions.

18.5 APPROACHES AND TOOLS FOR MANAGEMENT OF INTERFACE LANDS

Promising approaches and tools for managing public lands in the interface include practices such as control of access and restoration of ecosystems and approaches such as collaborative management, adaptive management, and improved involvement of others in natural resource planning and management.

18.5.1 Control of Access

Controlling access to public lands in the interface is often critical to effective management given the large numbers of individuals living nearby, sometimes as neighbors who share a common boundary, as well as potential users from urban areas. Controlling the use of motor vehicles and where they are parked is often a key management consideration, as is keeping users out of some areas after dark to provide for public safety and to reduce damage to resources and disruption of nearby residential areas (Wendling et al. 1981; Dwyer 1983). Peaking of use on weekends and during particular seasons and weather conditions can cause significant congestion and associated challenges for managers (Dwyer et al. 1985; Dwyer 1988a,b,c; Chavez 2001). Trails are often high-demand resources in interface areas, and control of trail use and keeping users on trails are major issues, particularly where trails go through sensitive areas. Often, managers must decide how to provide for pedestrian, equestrian, cycle, skateboard, and motorized use of interface trails. Chapter 10 provides a detailed discussion of ways to create management zones that offer a variety of recreational experiences and also maintain environmental quality.

18.5.2 Restoration of Ecosystems

High levels of use and nearby developments and associated infrastructure, as well as prior land use and restrictions on some management practices, can have a significant detrimental impact on public natural resources over time. Lack of fire and other disturbances sometimes lead to extensive proliferation of exotic plants. Significant attention must often be given to restoration of public lands at the interface, which may have been degraded through heavy use or by the expansion of exotic invasive plants and animals. In addition, lands that were formerly developed may be purchased by public agencies for greenspace or recreation purposes, thereby requiring significant restoration of the natural environment. Yet, some restoration techniques, such as use of fire, herbicides, and animal, tree, or plant removal, may generate public concern, particularly when they are carried out close to residential areas or when the purposes or outcomes of these techniques are not well understood by users and residents (Gobster 1997; Gobster and Hull 2000). In some instances, salvage may be impractical or impossible, resulting in a negative impact on forest health (e.g., access to harvest areas attacked by the pine beetle or to reduce fuel loads, limitations on the use of fire in fire-dependent ecosystems). See the FPDCC example for an illustration of the potential controversy associated with ecological restoration and some suggestions for resolving it.

18.5.3 Collaborative Management

The fragmented interface landscape with diverse owners calls for collaborative management among public and private landowners to provide important goods, services, and environments and to make management cost-effective (Maser et al. 1994). Critical to collaboration is the ability to build effective relationships, negotiate and resolve conflicts, and leverage funds and other resources across organizations (Selin and Chavez 1995). Critical partners in these collaborative efforts include communities and not-for-profit groups as well as landowners and users. The Chicago Wilderness organization (Chicago Region Biodiversity Council), a coalition of public and private organizations (Ross 1997), is involved with both the FPDCC and the Midewin National Tallgrass Prairie (see Box 18.3).

Public agency collaboration with or assistance to other government agencies or private landowners to manage their lands can be a key component of resource management in the interface. The interface is a complex environment, in which there may be many approaches to meeting public needs. In some instances, public agencies may find it cost-effective to work with other owners to help them meet important public needs. Some good examples include the efforts of the USDA Forest Service State and Private Forestry Programs and the USDI National Park Service's (NPS) Rivers, Trails, and Conservation Assistance Program. Some examples of areas where NPS lands and programs are serving as key catalysts for efforts in urban areas include the Mississippi National River and Recreation Areas, Minneapolis/St. Paul, Minnesota; Potomac Heritage National Scenic Trail, Washington, DC.; and the Boston Harbor National Recreation Area (Wink Hastings, National Park Service, personal communication, August 2001). The work of Chicago Wilderness to develop and implement a Biodiversity Recovery Plan for northeastern Illinois is also an excellent example of collaboration (see Box 18.3).

Box 18.3

Chicago Wilderness

Chicago Wilderness or the Chicago Region Biodiversity Council is a collaboration of more than 140 public and private organizations dedicated to the protection, restoration, and stewardship of the natural communities of the Chicago Region through fostering their compatibility with human communities whose lives they enrich. Chicago Wilderness is the Council's name to designate the natural resources and lands stretching from southeastern Wisconsin, through northeastern Illinois, to northwestern Indiana that contain globally significant concentrations of tallgrass prairies, woodlands, wetlands, and waters. The forest preserves, state parks, federal lands, and private conservation holdings of the Chicago region total more than 200,000 acres. Some of this landscape is legally protected, but it suffers from the pressures of struggling to survive amidst millions of people.

Chicago Wilderness developed and implemented a Biodiversity Recovery Plan for northeastern Illinois (Chicago Wilderness 1999) whose goals were:

- Foster a sustainable relationship between society and nature in the region.
- Involve the citizens, organizations, and agencies of the region in biodiversity conservation efforts.
- Strengthen the scientific basis of ecological management.
- Protect globally and regionally important natural communities.
- Restore natural communities to ecological health.
- Manage natural communities to sustain native biodiversity.
- Develop citizen awareness and understanding of local biodiversity to ensure support and participation.
- Enrich the quality of lives of the region's citizens.

18.5.4 Adaptive Management

The rapidly changing matrix of people, resources, management agencies, and land uses at the interface calls for adaptive management (Chavez 2002). The Applewhite Picnic Area of the San Bernardino National Forest has been the site of adaptive management in recreation since 1993. The site, built for 250 visitors in the 1940s, came to be utilized by more than 1700 in the 1990s and was redeveloped and redesigned to meet the needs of an urban minority population. Follow-up studies and recommendations were made, followed by additional changes by managers, and then more research was conducted. More recently, the Falls Picnic Area (San Bernardino National Forest) has become another site of adaptive management in an Urban National Forest. This process requires careful monitoring of resource management decisions and their physical, biological, and social outcomes so that management can be modified to produce desired outcomes. Planning processes such as those carried out by the National Forests and Chicago Wilderness (Biodiversity Recovery Plan) encourage adaptive management. A common limitation on adaptive management is lack of information on trends in resources and their use over time, pointing to a strong need for inventory and monitoring (Dwyer et al. 2000b). A critical need is formal monitoring of forest resources in urban and urbanizing areas (Dwyer et al. 2000a,b).

18.5.5 Citizen and Local Government Involvement

Involvement of nearby residents, local government, and other interested groups in planning and actually carrying out management has proven to be a highly effective aid in a number of instances (Ross 1994; Westphal and Childs 1994). This involvement can help reduce conflict over management, get substantial work done at a reduced cost, and provide highly desirable and rewarding experiences for the individuals and groups who volunteer. Grese et al. (2000), Schroeder (1998, 2000), Westphal (1995), and Winter (1998) document the benefits to volunteers who participate in natural resource management projects in and near urban areas.

Diverse populations, varying familiarity with natural resources and their management, different cultures, multiple languages, and varied economic interests and standings complicate communication and education efforts and call for innovative approaches (see Box 18.4 and Chapter 9) (Hodgson et al. 1990). At the same time, these efforts have the potential to reach a large number of people and influence their perceptions of resource management across the urban to wilderness spectrum (see Box 18.5) (Dwyer and Schroeder 1995; Chavez 2000b, 2001). Consequently, environmental education is often a priority with public lands in the interface (Figure 18.2). The urban constituency brings a new level of communication that is further challenging land management. What will be the role of the Internet in communicating resource management issues on public lands?

Box 18.4

Communication Innovations

One example of an innovative technique for communication is the Forest Information Van (FIV). Since research indicated that some forest visitors in southern California, particularly Hispanics, were not utilizing traditional communication tools such as visitor centers, the FIV was developed to take information to the visitors at the sites where they were recreating. The FIV was stocked with information that visitors indicated they desired. These items included information about other sites, things to see and do in the area, and facts concerning the area's flora and fauna. Bilingual volunteers drove the FIV.

Box 18.5

Hawkins Park

In South Central Los Angeles, a park exists today where there was once a pipe graveyard. The park, on 8.5 acres, is located in an area known for gang activity. This did not deter a collaborative effort to construct a natural area on this acreage. On site now is a combined exhibit/education center and ranger residence at the park's main entry. Activities include opportunities for free bus trips to other local parks and forest settings. From creating jobs and after-school programs to changing attitudes toward stewardship of the landscape within and outside the neighborhood, this natural resource project is having clear cultural impacts (Sorvig 2002).

18.6 Examples of Managing Public Lands in the Wildland–Urban Interface

18.6.1 Managing a Large Federal Holding that Spans the Wildland–Urban Interface: The Angeles National Forest

The Angeles National Forest, located in southern California, is within a 1-hour drive of more than 13 million people. California's population has shifted over the years so that it is no longer a majority white/Euro-American state. The residents of the state are culturally and racially diverse, and within the state Los Angeles County is among the most diverse.

This National Forest was not always urban. The population growth in southern California has made it urban, and now it is among the most urban of wildland places in the U.S. Most of Los Angeles County is highly developed, with the Angeles National Forest making up about 72 percent of the open space in the county. Like other urban areas, the population is most dense near the central city, with suburban and semirural areas in the foothills adjacent to the Angeles and in the north part of the county. Even so, there is a great deal of encroachment or development next to the Forest's boundaries.

The San Gabriel Mountains, which make up most of the Angeles National Forest's approximately 694,000 acres, have an extremely rugged terrain. Only 4 percent of the Forest is accessible; most of that is riparian, where wildlife, riparian vegetation, and recreational users compete for access to streams that are important for water supply. Most of the threatened and endangered species in the Angeles National Forest depend on riparian areas. The Forest is characterized by heavy year-round use. It ranked

FIGURE 18.2 Students connect to one another with string to simulate the "web of life" during an educational field trip at the Midewin National Tallgrass Prairie. Environmental education is often a management priority on public lands in the interface. (Courtesy of Marta Witt, Information Officer.)

fifth in recreational use among National Forests in 1995, with 9.8 million recreational visitor days. Anticipated trends include more use as the population increases and more diversity in visitor demographics, reflecting the changing community.

Among the many issues or challenges in managing the Angeles National Forest are landfills on adjacent land, exotic species, threatened and endangered species that depend upon riparian areas frequented by recreational users, and water supply (both quality and quantity). Other issues being faced by this interface forest include the development of nearby lands, limitations on management practices (e.g., controlled burns), air pollution, light pollution (which interferes with observation towers), and visual quality (e.g., telecommunications equipment). The Angeles National Forest also has to consider social issues, including pressures for diverse use (especially in recreation), changing social values, tribal government relations, residential value, special use permits, urban viewsheds (e.g., power transmission lines), quality of life, dumping, vandalism, sustainable and community-supported tourism, and social equity.

Although much time and effort is involved in addressing the myriad of physical and resource challenges, the social problems being faced by managers and planners on the Angeles especially require innovative techniques and tools. The Angeles National Forest has worked to address these challenges in three ways: research, outreach, and partnership. First, Forest leaders were centrally involved in developing a USDA Forest Service research work unit in southern California that houses research social scientists. The Forest Supervisor, along with the supervisors of three other forests in southern California, recognized the changing populations in southern California and the influences of those changes on the National Forest to include visiting populations, land-use patterns, and community needs. The resulting research work unit, developed in 1987, conducts social science research (most related to outdoor recreation), which is then transferred to the Angeles and other Urban National Forests through reports and presentations that provide recommendations for implementation, such as the redevelopment and redesign of the Applewhite Picnic Area mentioned in the Adaptive Management section. The unit provides demographic information, research on group size and activities, frequency of visitation, and effectiveness of communication media. Researchers also evaluate the impacts of social ills on the forest (e.g., crimes) and the likelihood of new and emerging uses of the Forest. Research results from this unit have included valuable management aids such as collaboration guidelines (Selin and Chavez 1995), adaptive management tools (Chavez 2002), ways to improve involvement of the public (Winter 1998), and innovative communication suggestions (Hodgson et al. 1990; Chavez 2001).

The second tool is an urban forester (a USDA Forest Service employee), whose task is to build collaborative relationships with local urban forestry organizations and participate in urban forestry programs. Through these relationships, the Angeles National Forest gains a better understanding of the nearby community, and the community learns more about the services available from the National Forest, as well as how to sustain forests. The urban forestry programs include urban camping programs where inner-city youth are introduced to the outdoors, with training on safe and responsible use of wildlands, Earth Day events in the local community that provide bilingual, multicultural environmental education workshops, and elementary and secondary school programs that focus on resource restoration. ECO-teams, an innovative communication tool, were developed in partnership with the California Environmental Project. Team members are Los Angeles community teens or young adults who are seeking employment opportunities with natural resource agencies. Most of the team members are bilingual, with Spanish as their native language. The team members deliver ecological messages to forest visitors, with emphasis on visitors of Hispanic/Latino background. The messages they deliver vary but usually center on litter control and fire dangers (Absher et al. 1997). Most recently, the messages have focused on switching from barbecues using coals to those using gas because of the fire danger from abandoned hot coals. The messages are delivered in Spanish or English as needed by the visitors and are taken directly to the visitors on-site.

The innovative tools being used by the Angeles National Forest are a great beginning. Now, the Forest needs to focus attention on the relevance of public lands to everyone. It might do this by using messages on health issues, quality-of-life issues, and riparian and water-quality issues.

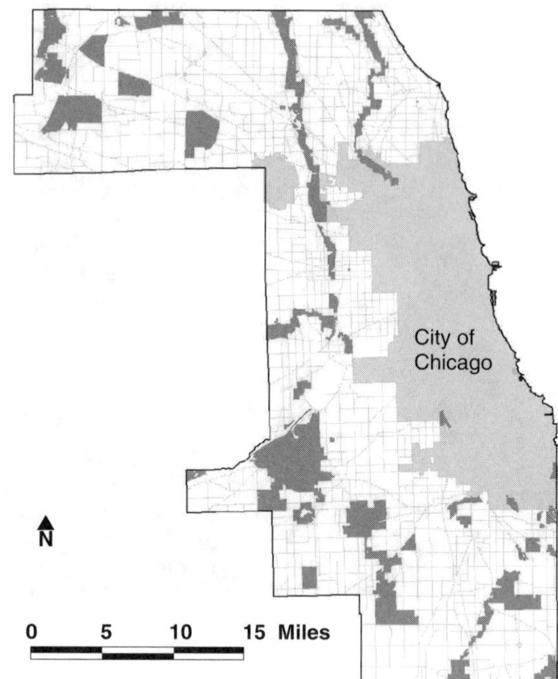

FIGURE 18.3 Forest preserves in Cook County, IL (see darker gray patches). (Courtesy of the CIS Section of the Planning and Development Department of the Forest Preserve District of Cook County.)

18.6.2 Managing County Forestlands in the Interface: The Forest Preserve District of Cook County, IL

The FPDCC is an example of how public natural areas in highly populated interface environments can furnish significant opportunities for recreation and education while being protected, preserved, and enhanced (Wendling et al. 1981). The District manages some 68,000 acres of natural areas around Chicago (Figure 18.3). District holdings make up 12 percent of the land area in a county with 5.5 million residents, and receive an estimated 40 million annual visits. Sites near preserves continue to be popular for residential developments. Residents of those developments are often heavy users of nearby preserves and may carefully monitor resource management and use. Some residents work as volunteers in District management programs. One of the District's major management challenges has been to find ways to accommodate large numbers of visitors while protecting the resource from overuse and abuse.

The District's holdings are not parks but rather forested sanctuaries — large natural reservations with facilities for recreational use on their fringes. Individual tracts range from 40 to 15,000 acres in size, and all interface with or are close to residential, commercial, or industrial developments. Most are along rivers, creeks, and wetlands, thus influencing the character and health of the region's water resources and riparian environments. The forest is the backbone of the preserves, and its preservation, restoration, and care are paramount in all management efforts. Significant amounts of outdoor recreation and environmental education are provided in a manner consistent with the preservation of the forest resource.

In carrying out its mission to preserve and restore the resource in a highly urban context, hunting and trapping are not allowed, and dead trees are left to provide wildlife habitat and complete the nutrient cycle. No timber or firewood is cut from standing or downed trees. Exceptions are made only for picnic and special-use areas and along roads or parkways where dead trees or limbs are removed if they present a hazard to users or pose risks of disease and insect attack. With improvement of habitat, beaver and deer have returned to the preserves in large numbers, and fox and coyote are seen more frequently than in the past (see Box 18.6).

Opportunities for intensive outdoor recreation are limited to the fringe areas, while the densely forested interior portions are left in their natural state. Fringe developments provide for picnicking, relaxing, sightseeing, sunbathing, socializing, golf, softball, sledding, tobogganing, snowmobiling, skiing, boating, fishing, swimming, ice-skating, and bicycling. Overnight camping is limited to designated group youth camps.

Forests are protected through control of access, and by encouraging visitors to choose areas that can withstand heavy use. Large groups of users are encouraged to picnic

Box 18.6

BIODIVERSITY

The Forest Preserve District of Cook County is home to the following:

- 5000 species of insects
- 300 species of prairie plants
- 365 species of birds
- 80 species of fish
- 78 species of trees and shrubs
- 41 species of wild mammals
- 32 species of reptiles
- 18 species of amphibians

in the numerous small open areas, with specimen trees and groves as well as shelters, rather than under the canopies of more heavily forested areas. Permanent meadows adjacent to picnic areas accommodate large groups for team games and other group activities. Permits specifying a particular time and place are required for large groups, thus helping to distribute use and reduce damage. Picnic areas under a forest canopy resulted in heavy damage to the forest and less than optimal experiences for users. Picnic facilities were removed from these areas and the forest environment was restored.

As most users arrive by automobile, limitations of access and parking are effective in visitor management. The establishment of numerous small picnic areas on secondary roads, with parking spaces along the road rather than in large lots, reduces concentrations of recreational users and the associated heavy use. Ditches, plantings, and other permanent barriers along roads limit parking to designated areas. Access to the interiors is exclusively by trail.

Alcoholic beverages are allowed, except within 50 ft of parking areas and roads. This policy tends to reduce user conflicts and disperse users away from parking areas and roads. The preserves are closed at sundown and the district employs its own police force. There has been a transition from traditional ranger patrols to a police organization in response to social conflicts created by a large urban population. The police, who are empowered to make arrests, work in cooperation with law enforcement personnel from nearby areas.

The District program in environmental education is closely allied to outdoor recreation and forest preservation. Year-round programs are offered, and five nature centers located throughout the preserves serve a whole range of urban residents. Environmental education is also provided in the form of school visits, teacher institutes, and training courses for youth group leaders. Adults are reached through lectures to various groups, field trips, and nature bulletins.

Tree planting and resource protection have been hallmarks of management programs since the District's inception, as much of the land had previously been farmed. However, in more recent years, the District has emphasized restoration of its natural and built features (Dwyer and Stewart 1995; Stewart 1995). After more than 80 years of protection from fire and other disturbances, many of the District's holdings were overrun with exotic invasive plants such as European buckthorn, which precluded the regeneration of native trees and other plants. Working in cooperation with volunteer groups, not-for-profits, and other agencies, the District developed a program of restoring some of its forests, woodlands, savannas, and prairies to more natural conditions. This has involved physical removal of some plants, as well as the use of fire, cutting, and herbicides. These efforts are also taking place on other holdings in the Chicago area under the leadership of Chicago Wilderness (see Box 18.3). There has been good public support for the restoration of Forest Preserve areas, as well as some controversy over the use of some vegetation and wildlife management techniques, especially when they are carried out near residential and recreational areas, or if the purpose and extent of these treatments are not clear.

After reviewing the controversy concerning restoration of District lands, Gobster (1997) made the following suggestions for reducing future debate, most of which focus on the human dimensions of resource management:

- Plan restoration in a landscape context that considers adjacent areas and nearby residents
- Design restoration sites with people in mind
- Promote two-way communication involving managers, users, and nearby residents
- Encourage involvement in planning and management by users, nearby residents, and other interested individuals and groups.

For more than 80 years, residents of the Chicago area have given strong support to the District's acquisition and management programs, and have made heavy use of the forest preserves. This helps the District deal with persistent issues concerning efforts to take District lands for other uses such as firehouses, water storage tanks, schools, athletic fields, highways, and utility rights-of-way. Public support has been the key in finding the appropriate management strategies for maintaining natural areas in an urban environment and in getting funds. Faced with limited financial resources, the District has emphasized strategic planning, cooperative partnerships, and grants to help meet future challenges (Stewart 1995). These challenges are likely to focus on protecting and restoring these valuable natural environments that interface with a complex matrix of residential, commercial, and industrial areas.

18.6.3 Converting Industrial Areas to Natural Areas at the Wildland–Urban Interface: The Midewin National Tallgrass Prairie

The Midewin National Tallgrass Prairie was established in 1996 as the first national tallgrass prairie in the country (Figure 18.4). It is administered by the USDA Forest Service, in close cooperation with the Illinois Department of Natural Resources and with support from hundreds of volunteers, partner agencies, businesses, and organizations (USDA Forest Service 2001a,b). As a major portion of the peacetime conversion of the former Joliet Army Ammunition Plant (Figure 18.4), it is an example of restoration of a large area (15,000 acres) of industrial and agricultural land to a native ecosystem. Midewin is the largest prairie restoration ever attempted in the U.S. — a major physical, biological, and social challenge — and its outcome promises to have significant implications for the 8.5 million people who live in the Chicago Metropolitan Area as well as for visitors from afar.

The restoration job is substantial in that less than 3 percent of the 15,000 acres of land transferred to the USDA Forest Service is covered with remnants or patches of native vegetation. Large areas continue to be used for farming, and structures (including large concrete bunkers formerly used to store ammunition), roads, and other surfaces from the former ammunition facility are found throughout the area. However, Midewin provides habitat for a rich assemblage of plants and animals, including 18 species listed as threatened or endangered by the State of Illinois. Nearby areas that were also once part of the ammunition plant are being converted into a national veterans' cemetery, two industrial parks, and a county landfill. These land uses will contribute to the complex wildland–urban interface that characterizes the context in which Midewin is managed and used.

Midewin is the key parcel of the Prairie Parklands, an area of approximately 40,000 acres composed of public, private, and corporate lands that are significant for habitat conservation. The Illinois Department of Natural Resources administers the Des Plaines Conservation Area, Goose Lake Prairie State Natural Area, and Heidecke Lake Fish and Wildlife Area; Will County manages a number of Forest Preserves within the Parklands; and corporate owners of Parkland sites include Commonwealth Edison, General Electric, Exxon-Mobil, BP Amoco, Stephan, and Dow Chemical. In all, 22 sites owned by state, county, and local governments, corporations, and interested private landowners are located within 12 mile of Midewin. The Prairie Parklands provides a unique opportunity to protect, restore, and manage the largest prairie ecosystem east of the Mississippi River.

Given its size, Midewin can provide unique opportunities to the Prairie Parklands that smaller units may not be able to provide. In addition, the proximity of the other Prairie Parklands units can enable integrated recreational planning such as the creation of trail linkages among the parcels. Since many management issues will cross administrative boundaries and will need to be addressed at a landscape scale, cooperation among components of the Prairie Parklands is essential to the management of Midewin.

Midewin is also the largest parcel of protected open space in northeastern Illinois and a key component of the Chicago Wilderness. Located just 40 mile southwest of Chicago, Midewin represents an unprecedented opportunity for urban dwellers to experience the wide-open spaces that characterized the Prairie State 200 years ago.

In the summer of 1840, while traveling through the area near Joliet, IL, Eliza Steele recorded the following passage in her journal, later published as *A Summer Journey in the West* (Steele 1875):

FIGURE 18.4 The USDA Forest Service manages Midewin National Tallgrass Prairie (Wilmington, Illinois) with the goal of restoring former industrial sites and agricultural lands to the original tallgrass prairie ecosystem. An abandoned structure is a vestige of the Joliet Army Ammunitions Plant. (Photos by Dave Klenosky, Purdue University.)

> I started with a surprise and delight. I was in the midst of a prairie! A world of grass and flowers stretched around me, rising and falling in gentle undulations, as if an enchanter had struck the ocean swell, and it was at rest forever .... We passed whole acres of blossoms all bearing one hue, as purple perhaps, or masses of yellow or rose; and then again a carpet of every color intermixed, or narrow bands, as if a rainbow had fallen upon the verdant slopes. When the sun flooded this mosaic floor with light, and the summer breeze stirred among their leaves, the iridescent glow was beautiful and wondrous beyond anything I had ever conceived.

This then is the overall vision for the Midewin National Tallgrass Prairie — to return the Prairie lands, as much as is realistically possible, to the splendor and wonder first recorded by Eliza Steele. The vision, articulated in the legislation establishing Midewin, includes restoring the natural habitats and processes of the Prairie, thereby promoting sustainable ecosystems; conserving populations of fish, wildlife, and plants; providing for scientific, environmental, and land-use education and research; and providing a variety of recreational opportunities that enhance the visitor's appreciation of the prairie ecosystem.

The complexity of the job faced by the USDA Forest Service as it begins to achieve this vision is reflected in the goals that have been proposed as part of the collaborative planning process for Midewin (USDA Forest Service 2001a). These goals are grouped into seven major areas:

1. Ecological sustainability
2. Recreation, interpretation, and scenic integrity
3. Heritage resources
4. Facilities and transportation
5. Lands and special uses
6. Fire management
7. Air quality and smoke management

The restoration projects to achieve this vision, such as the prescribed burn shown in Figure 18.5, will be accessible by the residents of the Chicago Metropolitan Area, many of whom will become involved as individuals or as members of groups. The Midewin restoration is a key component of major efforts to maintain or restore natural landscapes in northeastern Illinois, including the Prairie Parklands and the Chicago Wilderness. Moreover, what is learned at Midewin is likely to help guide restoration of industrial and agricultural lands at the interface in many other parts of the U.S.

18.7 SUMMARY AND CONCLUSIONS

Public lands in the wildland–urban interface are diverse in terms of ownership, management, land use, and land cover. They are managed in the complex and dynamic environment that characterizes the interface. The interrelated physical, biological, and social changes of the interface have strong impacts on public lands and their management and use. These islands of green are among the most highly valued and intensively used natural areas in the country. Experience with these areas shapes how a substantial portion of the U.S. population views natural resources and their management and use.

Many types of landowners at the federal, state, and local levels manage the public lands. Some of these lands represent long-held natural areas that are experiencing increasing urban pressure, such as the Angeles National Forest and the FPDCC, Illinois, while others are areas that were once developed and are now being restored to natural conditions, such as the Midewin National Tallgrass Prairie.

The challenges of managing public lands in the interface focus on balancing protection of the extensive natural areas that often characterize these lands with local needs for diverse and complex uses of these lands as well

FIGURE 18.5 A prescribed burn helps maintain and rejuvenate the Midewin Tallgrass Prairie ecosystem. (Courtesy of Marta Witt, Information Officer.)

as the goods and services that they can provide. This challenge is best met by involving citizens in collaborative and adaptive planning and management; controlling access to the area (particularly for motor vehicles); modifying management practices to meet interface conditions while facilitating management of ecosystem processes; restoring damaged ecosystems, and partnering with local, state, and federal governments. Managing these public lands requires expertise in managing ecosystems receiving substantial use, restoring degraded environments, and working with diverse large populations. The lessons learned in existing interface areas may well provide guidance for the future management of other areas that will experience urban expansion.

REFERENCES

Absher, J.D., P.L. Winter, and K. James, 1997. Delivering environmental education and interpretive messages in urban-proximate field settings: "Lessons" from southern California, *Trends* 34: 30–37.

Bengston, D.N., 1994. Changing forest values and ecosystem management, *Society and Natural Resources* 7: 515–533.

Bengston, D.N., D.F. Fan, and D.N. Celarier, 1999. A new approach to monitoring the social environment for natural resource management and policy: the case of U.S. national forest benefits and values, *Journal of Environmental Management* 56: 181–193.

Chavez, D.J., 1996. Mountain biking: direct, indirect, and bridge building management styles, *Journal of Park and Recreation Administration* 14: 21–35.

Chavez, D.J., 1997. Mountain bike management: resource protection and social conflicts, *Trends* 34: 36–40.

Chavez, D.J., 2000a. Invite, include, and involve! Racial groups, ethnic groups, and leisure, in *Diversity and the Recreation Profession: Organizational Perspectives,* M.T. Allison and I.E. Schneider, Eds., Venture Publishing, State College, PA, pp. 179–191.

Chavez, D.J., 2000b. Wilderness visitors in the 21st century: diversity, day-use, perceptions and preferences, *International Journal of Wilderness* 6: 10–11.

Chavez, D.J., 2001. *Managing Outdoor Recreation in California: Visitor Contact Studies 1989–1998*, General Technical Report PSW-180, U.S. Department of Agriculture, Forest Service, Pacific Southwest Research Station, Albany, CA.

Chavez, D.J., 2002. Adaptive management in outdoor recreation: serving Hispanics in southern California, *Western Journal of Applied Forestry* 17: 129-133.

Chicago Wilderness (Chicago Region Biodiversity Council), 1999. Biodiversity Recovery Plan, http://www.chiwild.org/pubprod/brp/index.cfm. [Date accessed: May 5, 2003.]

Dwyer, J.F., 1983. Management Technologies for Outlying Forests: A Summary and Synthesis, in Proceedings of the Seminar on Management of Outlying Forests for Metropolitan Populations, P. F. Folliott and W. H. Banzhaf (tech. coords.), Temperate Forests Directorate of the U.S. Man and Biosphere Program, Milwaukee, WI, pp. 27–31.

Dwyer, J.F., 1988a. Levels and patterns of urban park use, *Trends* 25: 5–8.

Dwyer, J.F., 1988b. Predicting daily use of urban forest recreation sites, *Landscape and Urban Planning* 15: 127–138.

Dwyer, J.F., 1988c. A site specific model for predicting daily use of urban forest recreation sites, *Journal of Park and Recreation Administration* 6: 38–53.

Dwyer, J.F., 1994. *Customer Diversity and the Future Demand for Outdoor Recreation*, General Technical Report RM-252, U.S. Department of Agriculture, Forest Service, Rocky Mountain Forest and Range Experiment Station, Fort Collins, CO.

Dwyer, J.F., G.M. Childs, and D.J. Nowak, 2000a. Forestry in urban and urbanizing areas of the U.S.: connecting people with forests in the 21st century, in *Forest and Society: The Role of Research.* Sub Plenary Sessions Vol. 1, B. Krishnapillay, E. Soepadmo, N, Arshad, A. Wong, S., Appanah, S., Chik, N. Manokaram, H. Tong, and K. Choon, Eds., XXI IUFRO World Congress, Kuala Lumpur, Malaysia, pp. 629–637.

Dwyer, J.F., D.J. Nowak, M.H. Noble, and S.M. Sisinni, 2000b. Connecting People with Ecosystems in the 21st Century: An Assessment of our Nation's Urban Forest, General Technical. Report PNW-490, U.S. Department of Agriculture, Forest Service, Pacific Northwest Research Station, Portland, OR.

Dwyer, J.F., H.W. Schroeder, and R.L. Buck, 1985. Patterns of Use in an Urban Forest Recreation Area, in Proceedings 1985 National Outdoor Recreation and Tourism Management Conference, Myrtle Beach, SC, February 24–27, 1985, Clemson University, Clemson, SC, pp. 81–89.

Dwyer, J.F., and H.W. Schroeder, 1995. The human dimensions of urban forestry, *Journal of Forestry* 92: 12–15.

Dwyer, J.F. and S.I. Stewart, 1995. Restoring Urban Recreation Opportunities: A Review with Illustrations, in Proceedings of the Fourth International Outdoor Recreation and Tourism Trends Symposium and the 1995 National Recreation Resource Planning Conference, J.L. Thompson, D.W. Lime, B. Gartner, and W.M. Sames, Eds., University of Minnesota, St. Paul, pp. 606–609.

Gobster, P.H., 1997. The Chicago Wilderness and its critics: the other side, *Restoration and Management Notes* 15: 32–37.

Gobster, P.H., R.H. Haight, and D. Shriner, 2000. Landscape change in the Midwest: an integrated research and development program, *Journal of Forestry* 98: 9–14.

Gobster, P.H. and R.B. Hull, 2000. *Restoring Nature: Perspectives from the Social Sciences and Humanities,* Island Press, Washington, DC.

Gobster, P.H. and L.M. Westphal, 1998. People and the River: Perception and Use of Chicago Waterways for Recreation, Chicago Rivers Demonstration Project Report, U.S. Department of Interior, National Park Service, Rivers Trails, and Conservation Assistance Program, Milwaukee, WI.

Grese, R.E., R. Kaplan, R.L. Ryan, and J. Buxton, 2000. Psychological benefits of volunteering in stewardship programs, in *Restoring Nature: Perspectives from the Social Sciences and Humanities,* P.H. Gobster and R.B. Hull, Eds., Island Press, Washington, DC, pp. 119–142.

Hodgson, R.W., R.E. Pfister, and D.E. Simcox, 1990. Communicating with Users of the Angeles National Forest — Report no. 2, unpublished draft supplied by authors.

Hoger, J.L. and D.J. Chavez, 1998. Conflict and management tactics on the trail, *Parks and Recreation* 33(9): 41–56.

Maser, C., B.T. Borman, M.H. Brooks, A.R. Kiester, and J.F. Weigland, 1994. Sustainable forestry through adaptive ecosystem management is an open-ended experiment, in C. Maser, Ed., *Sustainable Forestry: Philosophy, Science, and Economics,* St. Lucie Press, Delray Beach, FL, pp. 304–340.

Miles, I., 1996. Prairie Restoration Volunteers: The Benefits of Participation, Master's thesis, Natural Resources and Environmental Sciences, University of Illinois, Urbana-Champaign.

Rickenbach, M.G. and P. H. Gobster, 2003. Stakeholders' perceptions of parcelization in Wisconsin's Northwoods, *Journal of Forestry* 101: 18–23.

Ross, L.M., 1994. Illinois volunteer corps: a model program with deep roots in the prairie, *Restoration and Management Notes* 12: 57–58.

Ross, L.M., 1997. The Chicago Wilderness: A coalition for urban conservation, *Restoration and Management Notes* 15: 17–24.

Ruliffson, J.A., R.G. Haight, P.H. Gobster, and F.R. Homans, 2003. Metropolitan natural area protection to maximize public access and species representation, *Environmental Science and Policy* 6: 291–299.

Ruliffson, J.A., P.H. Gobster, R.G. Haight, and F.R. Homans, 2002. Niches in the urban forest: organizations and their roles in acquiring metropolitan open space, *Journal of Forestry* 100: 16–24.

Schroeder, H.W., 1998. Why people volunteer, *Restoration and Management Notes* 16: 66–67.

Schroeder, H. W., 2000. The restoration experience: volunteers' motives, values, and concepts of nature, in *Restoring Nature: Perspectives from the Social Sciences and Humanities,* P.H. Gobster and R.B. Hull, Eds., Island Press, Washington, DC, pp. 247–264.

Selin, S. and D.J. Chavez, 1995. Developing a collaborative model for environmental planning and management, *Environmental Management* 19: 189–195.

Sorvig, K., 2002. The wilds of South Central. *Landscape Architecture* 92: 66–75.

Steele, E., 1875. *A Summer Journey in the West*, Arno Press, New York.

Stewart, S.I., 1995. Challenges in Meeting Urban and Near-Urban Needs with Limited Resources: A Summary of the Workshop Discussion, in Proceedings of the Fourth International Outdoor Recreation and Tourism Trends Symposium and the 1995 National Recreation Resource Planning Conference, J.L. Thompson, D.W. Lime, B. Gartner, and W.M. Sames, Eds., University of Minnesota, St. Paul, pp. 603–605.

USDA Forest Service, 2001a. Midewin National Tallgrass Prairie: Proposed land and resource management plan, U.S. Department of Agriculture, Forest Service, Wilmington, IL.

USDA Forest Service, 2001b. Midewin National Tallgrass Prairie: Draft Environmental Impact Statement, U.S. Department of Agriculture, Forest Service, Wilmington, IL.

Wendling, R.C., S.J. Gabriel, J.F. Dwyer, and R.L. Buck, 1981. Forest Preserve District of Cook County, *Journal of Forestry* 79: 602–605.

Westphal, L.M., 1995. Participating in urban forestry projects: How the community benefits, in Inside Urban Ecosystems: Proceedings 7th National Urban Forest Conference, C. Kollin, and M. Barratt, Eds., American Forests, Washington, DC, pp. 101–103.

Westphal, L.M. and G.M. Childs, 1994. Overcoming obstacles: creating volunteer partnerships, *Journal of Forestry* 92: 28–32.

Winter, P.L., 1998. San Gorgonio wilderness volunteer association, unpublished report.

Index

A

B

C

D

M

N

V

W

Z

90 0820240 4

WITHDRAWN
FROM
UNIVERSITY OF PLYMOUTH
LIBRARY SERVICES